Mycobacterium avium subspecies *paratuberculosis* Infection in Ruminants—Agent, Disease, Diagnosis, and Control

Mycobacterium avium subspecies *paratuberculosis* Infection in Ruminants—Agent, Disease, Diagnosis, and Control

Guest Editors

Karsten Donat
Heike Köhler

MDPI

Basel • Beijing • Wuhan • Barcelona • Belgrade • Novi Sad • Cluj • Manchester

Guest Editors

Karsten Donat
Clinic for Reproduction and Neonatology of Animals
Justus-Liebig-University Giessen
Giessen
Germany

Heike Köhler
Institute of Molecular Pathogenesis
Friedrich-Loeffler-Institute
Jena
Germany

Editorial Office

MDPI AG
Grosspeteranlage 5
4052 Basel, Switzerland

This is a reprint of the Special Issue, published open access by the journal *Animals* (ISSN 2076-2615), freely accessible at: https://www.mdpi.com/journal/animals/special_issues/mycobacterium_avium_subspecies_paratuberculosis_ruminants.

For citation purposes, cite each article independently as indicated on the article page online and as indicated below:

Lastname, A.A.; Lastname, B.B. Article Title. *Journal Name* **Year**, *Volume Number*, Page Range.

ISBN 978-3-7258-6826-1 (Hbk)
ISBN 978-3-7258-6827-8 (PDF)
https://doi.org/10.3390/books978-3-7258-6827-8

Cover image courtesy of Markus Pfau

Contents

About the Editors

Karsten Donat

Karsten Donat currently acts as managing director of the Thuringian Animal Disease Fund, an institution by law responsible for compensation in case of infectious disease outbreaks and for operating the regional Animal Health Service. His general interests are infectious disease control and epidemiology, and managing infectious and other production diseases in cattle and pigs. As an adjunct professor at the Justus-Liebig-University Giessen, he teaches production disease management in dairy cows. Besides paratuberculosis, his current projects deal with selective dry cow treatment, metabolic disorders of transit cows, health data management, and on-farm communication skills. He is chair of the further education committee of the Thuringian veterinarians' board and a member of the national committee for vaccination in veterinary medicine.

Heike Köhler

Heike Köhler, DVM, PhD, is a senior research scientist at the Friedrich-Loeffler-Institut Jena, Germany. She is a veterinary immunologist with special interests in the pathogenesis of mycobacterial infections, vaccination, diagnostics, and test validation, and leads the German National Reference Laboratory for Paratuberculosis.

Preface

Paratuberculosis, caused by infections with *Mycobacterium avium* subspecies *paratuberculosis* (MAP), is a serious risk to animal health in ruminants worldwide. The infection occurs in industrial countries as well as in countries with low and middle incomes, affecting the economic performance of the dairy and beef industry. Nonetheless, due to its long incubation period and chronic nature, the disease is often unnoticed. Moreover, there is no specific treatment available at present. Control strategies are the only adequate way to deal with the disease. They consist generally of test-and-cull strategies and improvement of herd management practices, but the chosen focus differs between countries and regions depending on local MAP prevalences; economic impact; and risk assessment of farmers, veterinarians, and animal health authorities.

This is a reprint of the Special Issue '*Mycobacterium avium* subspecies *paratuberculosis* Infection in Ruminants—Agent, Disease, Diagnosis, and Control' published in *Animals* between 2023 and 2024.

We believe that this special collection has made a valuable contribution to the knowledge about paratuberculosis and its causative agent. While the prevalence of paratuberculosis and effective diagnostic and control strategies in dairy and beef cattle have already been extensively studied, relatively little is known about the situation in goats and sheep. It is therefore very pleasing that five studies in this collection focus on paratuberculosis in small ruminants. Two of them investigate the seroprevalence of paratuberculosis on a regional level, in dairy goat farms in northern Italy (Gaffuri et al.) and in small ruminants in Austria (Schrott et al.), and indicate risk factors for seropositivity. A study from Germany provides evidence that the genotypes of MAP isolates from goats are not unique but can also be found in cattle and other animal species in this country (Pickrodt et al.). The occurrence of MAP in different locations in a dairy goat barn and potential transmission sites to young stock were identified (Pickrodt et al.). Stefanova et al. demonstrated that vaccination of goats with an inactivated vaccine against MAP has not only positive effects on the extent of paratuberculosis-related lesions, but also on concomitant, not MAP-related, gastrointestinal and respiratory lesions. Corbiere et al. provided some evidence that vaccination of young calves against MAP has a beneficial effect on fecal shedding.

Three articles in this collection address risk factors for the MAP status of cattle herds (Krieger et al.), for within-herd transmission in dairy farms (Field et al.), and for the persistence of MAP in the environment (Rochard et al.). New approaches to improve the serological diagnosis of paratuberculosis in cattle, either involving alternative lipid antigens in ELISAs or the prediction of future paratuberculosis test results using machine-learning algorithms, are presented by Mason et al. and Imada et al. The importance of host genetics for the control of MAP infection is supported by the results of an in vitro study by Canive et al. Data from Matthews et al. indicate that MAP infection causes temporal changes in the composition of the intestinal microbiome of calves. Last but not least, Jolly et al. reviewed experimental laboratory animal and cell-based models for paratuberculosis research, their advantages and limitations, to offer a vision of their potential use.

We are grateful to all contributing authors for their outstanding research efforts. Without their dedication, this collection would not come into existence. Finally, we want to thank the editorial board members and staff of *Animals* for their support of this Special Issue.

Karsten Donat and Heike Köhler
Guest Editors

animals

MDPI

Article

Paratuberculosis, Animal Welfare and Biosecurity: A Survey in 33 Northern Italy Dairy Goat Farms

Alessandra Gaffuri [1], Filippo Barsi [1,2,*], Edoardo Magni [3], Stefania Bergagna [4], Debora Dellamaria [5], Matteo Ricchi [1,2], Livia De Paolis [4], Giorgio Galletti [1], Norma Arrigoni [1,2], Valentina Lorenzi [1,6], Francesca Fusi [1,6], Alice Prosperi [1] and Chiara Garbarino [1,2]

1 Istituto Zooprofilattico Sperimentale della Lombardia e dell'Emilia Romagna "Bruno Ubertini", 25124 Brescia, Italy; alessandra.gaffuri@izsler.it (A.G.); matteo.ricchi@izsler.it (M.R.); giorgio.galletti@izsler.it (G.G.); norma.arrigoni@izsler.it (N.A.); valentina.lorenzi@izsler.it (V.L.); francesca.fusi@izsler.it (F.F.); alice.prosperi@izsler.it (A.P.); chiaraanna.garbarino@izsler.it (C.G.)

2 Italian National Reference Centre for Paratuberculosis, Istituto Zooprofilattico Sperimentale della Lombardia e dell'Emilia Romagna "Bruno Ubertini", 29027 Piacenza, Italy

3 Official Veterinary Service, ATS Montagna, 23100 Sondrio, Italy; edoardo.magni101@gmail.com

4 Istituto Zooprofilattico Sperimentale del Piemonte Liguria e Valle d'Aosta, 10154 Torino, Italy; stefania.bergagna@izsto.it (S.B.); livia.depaolis@izsto.it (L.D.P.)

5 Istituto Zooprofilattico Sperimentale delle Venezie, 35020 Padova, Italy; ddellamaria@izsvenezie.it

6 Italian National Reference Centre for Animal Welfare, Istituto Zooprofilattico Sperimentale della Lombardia e dell'Emilia Romagna "Bruno Ubertini", 25124 Brescia, Italy

* Correspondence: filippo.barsi@izsler.it; Tel.: +39-0532-730058

Simple Summary: Paratuberculosis is a chronic incurable bacterial infection widespread all over the world in ruminants. The disease impacts animal health and welfare and causes significant economic losses in animal productions. This survey investigated the spread of paratuberculosis in northern Italian dairy goat farming by serological testing. Contextually, a welfare and biosecurity assessment through a standardized protocol was conducted. More than half (19 out of 33, 58%) of the investigated farms were infected, with a mean intra-herd prevalence of 7.4%. Welfare assessment showed quite favorable average results, although in 24% of the farms the welfare level was poor. On the contrary, 58% of the farms showed an unsatisfactory biosecurity level. Our results provide information on the spread of paratuberculosis in dairy goat farms of northern Italy. For this relevant disease, official prevalence data in goat breeding are still scarce. Moreover, the present work highlighted the low level of biosecurity measures implemented by the farmers.

Abstract: Paratuberculosis is a notable infectious disease of ruminants. Goats appear to be particularly susceptible. The survey aimed to investigate the spread of paratuberculosis in Italian goat farming and evaluate whether the presence of the disease could be influenced by welfare and biosecurity deficiencies. A serological survey for paratuberculosis in 33 dairy farms in northern Italy was conducted. Contextually, animal welfare and biosecurity were assessed, using a standardized protocol of 36 welfare indicators and 15 biosecurity indicators which assigns to each farm a welfare and biosecurity score from 0 (any application) to 100% (full application). An overall result of less than 60% was considered insufficient. Nineteen farms (58%) tested positive for paratuberculosis, with a mean intra-herd seroprevalence of 7.4%. Total welfare ranged from 39.56 to 90.7% (mean 68.64%). Biosecurity scores ranged from 10.04 to 90.01% (mean 57.57%). Eight farms (24%) showed poor welfare conditions (welfare score < 60%) and 19 (58%) an unsatisfactory biosecurity condition (biosecurity score < 60%). With respect to the explorative character of the study, an indicative association between seven welfare and biosecurity indicators and paratuberculosis seropositivity was identified. The presence of paratuberculosis in northern Italy dairy goat farms was confirmed. The welfare and biosecurity assessment protocol proved to be an accurate tool, capable of identifying critical points for managing health, welfare and productivity.

Keywords: *Mycobacterium avium* subsp. *paratuberculosis*; MAP; dairy goats; serology; biosecurity; animal welfare assessment

1. Introduction

Paratuberculosis is a chronic incurable enteritis of ruminants caused by *Mycobacterium avium* subsp. *paratuberculosis* (MAP) [1]. The transmission primarily occurs through the ingestion of infected feces.

The disease is important because of its impact on the economy, on the animal welfare and for public health in general [2]. Among domestic ruminants, goats appear to be particularly susceptible [3]. It has been observed that goats are naturally more susceptible to MAP infection than sheep and cattle and may play a more important role than sheep in the transmission and maintenance of the disease [2,4,5]. In goats, the onset of clinical signs is most common between two and three years of age, whereas subclinical infection is most often seen in the early years [2,6]. Indeed, paratuberculosis in this species is insidious and symptoms are usually not clearly evident. As a consequence, it is often diagnosed only at the latest state of disease when it has spread to most animals of the flock. Infected individuals often do not show diarrhea but non-specific signs as weight loss, exercise intolerance and decreased milk production [7]. Sardaro et al. [8] reported that economic losses and consequent profit inefficiency caused by the disease in breeding of small ruminants are due to decreased milk production, diagnostic and disease control costs, culling of affected animals and low carcasses values at slaughter. Surveillance and control of paratuberculosis can be of critical importance in some developing countries where small ruminants play a vital role in the livelihood of poor communities, as well in worldwide disadvantaged areas and in the increasing sector of intensive goat breeding [9,10].

Another important reason to investigate the disease in animals is related to the detection of MAP in humans affected by different chronic diseases, such as Chron's disease. These observations suggested a hypothetical zoonotic role for MAP that thus far, has not been confirmed or denied [11]. In this context, since several studies detected MAP in goat cheeses, often made from raw milk [12–15], from a health-risk point of view, contamination with MAP of foods of animal origin should be prevented.

According to the Regulation (EU) 2018/1882 [16], paratuberculosis is subjected to surveillance in cattle, buffaloes, sheep, goats, camelids and cervids.

Although the Regulation (EU) 2018/1882 clearly reports the obligation to notify the disease, because of scarce knowledge or difficulties in diagnosing subclinical infections [5], the disease is often underreported [2].

All over the world, paratuberculosis has been reported in goats [2]. In Europe, Nielsen and Toft [17] reported an inter-herd prevalence of infected goat herds over 20%. 5. Jiménez-Martín et al. [5] performed a cross-sectional investigation on 83 sheep farms and 70 goat farms in Andalusia (southern Spain) and detected an apparent seroprevalence of 90% in goat flocks and 66.3% in sheep flocks. In the same study, the estimated individual true seroprevalences were 8.4% for sheep and 25.2% for goats.

In Italy, goat farming is still considered marginal despite the fact that the presence of the species is recorded throughout the country (about 1,000,000 heads were reared in 2022, of which 300,000 dairy goats (https://www.vetinfo.it/, accessed on 11 April 2023) and it is expanding.

In Italy, paratuberculosis in goats was reported in Tuscany region (central Italy) [18] and Apulia region (southern Italy) [10]. This last epidemiological study was carried out in 419 semi-extensive dairy goat, sheep and mixed flocks and reported a true seroprevalence at flock level ranging from 63.8 to 92.4% in flocks with different species of small ruminants. Moreover, the same study reported, at individual level, statistically significant higher seroprevalence in goats, confirming the great sensitivity of this species to MAP infection [10]. To the authors' knowledge, no other published studies are available, underlining the scarcity of these data for most of the Italian regions, especially those where the breeding of goats represents an important local industry because of the cheese production and other typical products.

Notably, for its impact on goat health, paratuberculosis is one of the diseases–together with caseous lymphadenitis and caprine arthritis encephalitis–specifically considered for their overall effect on goat welfare [19]. On the other hand, management deficiencies in terms of animal welfare and biosecurity could lead to the introduction and spread of the disease on the farm. In fact, the application of biosecurity measures is directed to prevent the introduction and spreading of MAP infection in the farm, whereas welfare assessment focuses on farm management measures and structural characteristics, both potentially impacting on the spread and progression of the disease.

In the last decade, the assessment of animal welfare at farm level received increasing attention but, for some species, such as goat, there is still no specific legislation both at European and Italian level. Since 2015, the Italian Reference Centre for Animal Welfare, located at Istituto Zooprofilattico Sperimentale della Lombardia e dell'Emilia Romagna (IZSLER-CReNBA), has implemented a specific protocol for the on-farm animal welfare and biosecurity assessment in dairy goats.

The aim of this study was to carry out an exploratory investigation on the occurrence of paratuberculosis in Italian goat farming and to evaluate whether the presence of paratuberculosis could be influenced by herd management in terms of welfare and biosecurity measures. For this purpose, we conducted a survey in 33 dairy goat farms spread across four regions in northern Italy providing welfare and biosecurity assessment data by a specific protocol and collecting data about paratuberculosis status by ELISA test.

2. Materials and Methods

This study was carried out in the frame of a research project funded by the Italian Ministry of Health from 2019 to 2021. Thirty-three dairy goat farms were involved in this study. Farmers were contacted through veterinary practitioners who already had relationships with the laboratories. Farmers showing interest in the project voluntarily joined it and their farms were included in the sampling. Enrolled flocks were located in four different regions of northern Italy: Piedmont (5), Trentino-Alto Adige (3), Emilia-Romagna (5), and Lombardy (20).

Selected farms were visited once during 2019. In the same occasion, blood samples were taken, and animal welfare and biosecurity measures inspections were performed by trained veterinarians using the specific protocol of IZSLER-CReNBA (see Section 2.2).

The milk yield of each farm was recorded for one year starting from the first visit, so the average milk production per head per day was calculated.

2.1. Serological Assay

Samples for serological analysis were collected by the Official Veterinary Services in the frame of mandatory regional brucellosis control plans. Therefore, it was not necessary to collect additional samples for the present investigation. Blood samples were taken by jugular venipuncture into vacutainer tubes without anticoagulant. Samples of the animals over 12–18 months of age were stored at refrigeration temperature (4 °C) and analyzed in a few days.

Serological analysis was performed by an ELISA commercial kit (ID Screen® Paratuberculosis Indirect, ID-vet, Montpellier, France), according to the manufacturer's instructions: inconclusive ($0.6 < S/P$ ratio < 0.7) and positive ($S/P \geq 0.7$) samples from the screening test were submitted to confirmatory testing by an additional ELISA commercial kit (ID Confirmation® Paratuberculosis Indirect, ID-vet, Montpellier, France). Samples with a S/P ratio of 0.7 or above in the confirmation test were considered positive. In goats, the sensitivity of ELISA ranged from 63 to 100%, while the specificity of ELISA generally ranged from 92 to 100% [20].

2.2. Welfare and Biosecurity Assessment

On-farm animal welfare and biosecurity assessments were performed using the specific IZSLER-CReNBA protocol for dairy goat farms. This protocol was developed in 2015 in the frame of another research project called "RuminantWelfare", following the method described in Bertocchi et al. [21] and Lorenzi et al. [22]. The protocol included animal-based measures (ABMs) and resource-based indicators (also referred to as non-animal-based measures-N-ABMs) [21]. Briefly, these welfare indicators were selected based on the available scientific literature [23,24], on the European legislation (Council Directive 98/58/EC) [25] and its Italian transposition (Legislative Decree 26 March 2001, n.146), on the AWIN welfare assessment protocol for goats [26] and on an expert knowledge elicitation (EKE) [27]. In particular, the opinion of 14 Italian veterinarians was gathered during an EKE in order to characterize a set of management and housing factors potentially associated with negative or positive welfare outcomes in dairy goats kept in loose housing systems in Italy [27]. Experts were asked to weigh the potential negative or positive impacts of each factor in relation to five welfare categories (udder health, metabolic needs, locomotion and foot health, integument integrity and behavior) by estimating the magnitude (scoring scale from 0–none–to 3–high) and the likelihood (from 0% to 100%) of the negative or positive welfare consequences that could be associated with the exposure of the animals to each of proposed factors and to rate the certainty in relation to the likelihood value they provided (scoring scale: high, medium, low). The data obtained from the EKE were used to set the final protocol and to weight the different indicators.

The welfare protocol includes 36 parameters divided into three areas: "Area A: farm management and staff training" (indicators from 1 to 11), "Area B: housing" (indicators from 12 to 24), and "Area C: animal-based measures" (indicators from 25 to 36) (Table 1).

Additionally, the protocol includes 15 indicators on relevant aspects of biosecurity in ruminant breeding (Table 2) [28,29].

Currently, this protocol is part of the ClassyFarm system of the Italian Ministry of Health (https://www.classyfarm.it/check-list/, accessed on 11 April 2023) and it is voluntarily applied in intensive, semi-intensive and semi-extensive dairy goat farms at national level.

By means of the described protocol, data were collected by six trained veterinarians who have previously attended a specific training course in IZSLER and routinely applied the assessment method in their field activity, with the aim to ensure a high intra- and inter-observer reliability. For each indicator, the evaluator assigned a score based on a 2 or 3-point scale scoring system, where 1 indicated an insufficient status or high level of risk, 2 and 3 indicated, respectively, an acceptable and excellent status of the indicator, or low level of risk. An insufficient assessment corresponds to clear negative evidence or measures below the target levels set by the system, an acceptable assessment corresponds to clear adequate evidence or measures that meet the target levels, whereas an excellent assessment corresponds to clear positive evidence or measures above the target levels. A resulting value for each section was calculated as described in [30,31]. Briefly, each indicator threshold (i.e., 1, 2 or 3) had a different 'weight' according to its potential impact on animal welfare and health, these weights were assigned by means of the previously described EKE. A value for each of the three welfare areas and for the biosecurity section was calculated per farm by summing up the score of each indicator, according to the answer assigned on farm [31]. The final welfare score was calculated as reported in [30], considering a 50% contribution by Areas A and B and 50% by Area C. All these values were expressed as a percentage from 0 to 100, where 0% indicated lack of any welfare/biosecurity measure and 100% indicated their full application. An overall result of less than 60% was considered insufficient.

Table 1. Indicators for welfare assessment (36), divided into "Area A: farm management and staff training", "Area B: housing", "Area C: animal-based measures" and based on a 2 or 3-point scale scoring system (1: insufficient status, 2: acceptable status, 3: excellent status).

	Area A—Management Factors	Scoring Scale
1.	Number of stockpersons	1, 2, 3
2.	Experience and training of stockpersons	1, 2, 3
3.	Animal grouping strategy	1, 2, 3
4.	Inspection of the animals	1, 2, 3
5.	Management of sick or injured animals	1, 2

Table 1. *Cont.*

	Area A—Management Factors	Scoring Scale
6.	Feeding strategy	1, 2, 3
7.	Water provision	1, 2, 3
8.	Cleanliness of water points	1, 2, 3
9.	Cleanliness and hygiene of floor in walking areas and of bedding	1, 2, 3
10.	Hygiene of milking procedures	1, 2, 3
11.	Biosecurity measures	1, 2, 3
	Area B—Housing factors	
12.	Building and animal housing	1, 2
13.	Presence of shelters in outdoor areas	1, 2, 3
14.	Type of housing	1, 2, 3
15.	Space availability in lying area (adult goats)	1, 2, 3
16.	Space availability in lying area (young goats)	1, 2, 3
17.	Space availability in lying area (bucks)	1, 2, 3
18.	Space availability in lying area and feed bunk dimension (kids)	1, 2, 3
19.	Feeding place dimension and accessibility	1, 2, 3
20.	Functioning and number of water points	1, 2, 3
21.	Facilities for sick animals	1, 2, 3
22.	Milking machine or robot maintenance	1, 2, 3
23.	Temperature and humidity	1, 2, 3
24.	Lighting	1, 2
	Area C—Animal based measures	
25.	First-contact latency test	1, 2, 3
26.	Isolated animals, staring into space and estranged from the environment	1, 2, 3
27.	Body condition score	1, 2, 3
28.	Cleanliness of the animals	1, 2, 3
29.	Integument alterations	1, 2, 3
30.	Lameness	1, 2, 3
31.	Overgrown claws	1, 2, 3
32.	Abscesses	1, 2, 3
33.	Udder asymmetries	1, 2, 3
34.	Annual mortality rate (adult goats)	1, 2, 3
35.	Annual mortality rate (kids)	1, 2, 3
36.	Mutilations	1, 2, 3

Table 2. Measures for the biosecurity assessment (15) based on a 2 or 3-point scale scoring system (1: insufficient status, 2: acceptable status, 3: excellent status).

	Biosecurity Plan	Scoring Scale
1.	Measures for rodent and insect fighting	1, 2, 3
2.	Contact with other animal species	1, 2, 3

Table 2. *Cont.*

	Biosecurity Plan	Scoring Scale
3.	Measures for preventing the entrance of strangers	1, 2, 3
4.	Measures for monitoring the entrance of regular visitors	1, 2, 3
5.	Disinfection of vehicles entering the farm	1, 2, 3
6.	Indirect–direct contacts between own animals and outside vehicles	1, 2
7.	Collection disposal of fallen stock (dead livestock)	1, 2
8.	Loading of live animals	1, 2
9.	Purchase and/or movement of animals to outside the farm	1, 2, 3
10.	Quarantine measures	1, 2, 3
11.	Control and prevention of major infectious diseases	1, 2, 3
12.	Health monitoring activities	1, 2
13.	Control and prevention of mammary infections	1, 2, 3
14.	Control and prevention of endo/ectoparasitosis	1, 2, 3
15.	Monitoring and analysis of water sources	1, 2, 3

After obtaining the overall rating for welfare and biosecurity, a statistical evaluation was carried out in order to evaluate if the scores assigned to each indicator had a correlation with the presence of paratuberculosis at farm level.

2.3. Data Analysis

Descriptive analysis was performed for lactating goats, milk production, welfare and biosecurity assessment, and for each indicator. Continuous variables are shown as mean ± standard deviation (SD), and selected percentiles. Categorical variables (scoring for the indicators) are presented as absolute and relative frequency. Statistical analysis using the Chi-Square test was also provided for categorical variables, in order to explore potential association with paratuberculosis status. All the analyses were performed using R 4.0.2 [32].

3. Results

The data relative to number of lactating goats of the flocks, recorded at the moment of the sampling and welfare assessment, as well as their mean year-long milk production (expressed as kilograms per head per day) are summarised in Table 3.

Table 3. Data about number of lactating goats and mean year-long milk production (expressed as kilograms per head per day) in seropositive (19) and seronegative (14) flocks.

		Mean	SD	Min	1st Quartile	Median	3rd Quartile	Max
Lactating goats		100	64	20	46	80	133	280
	pos	111	75	20	47	83	138	280
	neg	85	44	34	43	71	131	153
Milk production (Kg/head/day)		2.84	0.72	1.60	2.25	2.90	3.22	4.70
	pos	2.69	0.7	1.6	2.1	2.9	3.26	4
	neg	3.06	0.72	2	2.48	3	3.33	4.7

An intensive type of management was adopted by 25 farms (76%), whereas eight farms (24%) adopted a semi-intensive management. The represented breeds were Chamois Coloured goat, Saanen, Roccaverano goat, Murciana, Nubian, Nera Verzasca and crossbreed (Table S1).

3.1. Serological Assay

A total of 164 out of 4431 analysed samples tested positive for paratuberculosis. A total of 19 out of 33 (58%) farms were infected (with at least one seropositive animal), with a mean intra-herd apparent seroprevalence of 7.4% (range: 0.4–17.4%; Figure 1). When distinguishing on the basis of the type of management adopted, 14 out of 25 (56%) intensive farms and 5 out of 8 (62.5%) semi-intensive farms were infected.

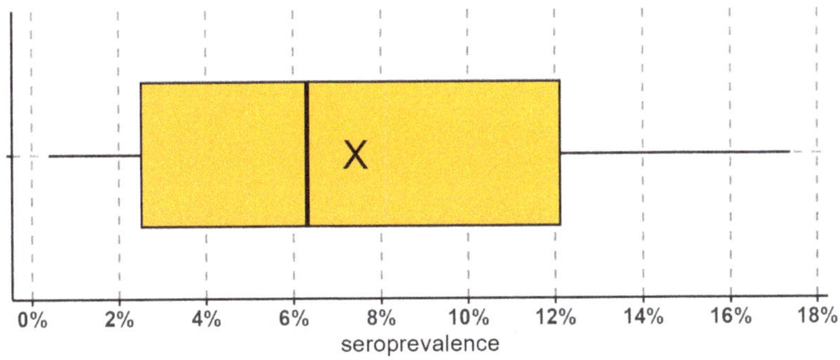

Figure 1. Boxplot of the intra-herd paratuberculosis seroprevalence. The mean value is represented by the "X".

3.2. Welfare and Biosecurity Assessment

The results obtained from welfare and biosecurity assessment in the sampled farms are summarized in Table 4. Overall, total welfare ranged from 39.56 to 90.7% (mean = 68.64%, SD = 12.61%); in detail, data ranged from 41.29 to 96.16% (mean = 67.92 %, SD = 15.25%) for Area A, from 37.21 to 93.59% (mean = 60.36%; SD = 13.65%) for Area B, and from 30 to 100% (mean = 70.41%; SD = 15.57%) for Area C. Biosecurity data ranged from 10.04 to 90.01% (mean = 57.57%, SD = 19.25%).

Table 4. Results obtained from welfare (Total, Area A, Area B and Area C) and biosecurity assessment in the 33 farms by paratuberculosis positive (19 farms) or negative (14 farms) serological status.

		Mean	Min.	1st Quartile	Median	3rd Quartile	Max.
Total welfare		68.64	39.56	60.16	66.14	78.88	90.70
	pos	64.66	39.56	53.54	65.41	76.59	88.89
	neg	74.03	58.50	63.16	77.21	83.63	90.70
Area A [1]		67.92	41.29	56.14	67.20	78.72	96.16
	pos	61.23	41.29	50.50	62.84	73.00	92.05
	neg	76.99	54.73	66.20	78.72	88.34	96.16
Area B [2]		60.36	37.21	51.24	57.08	68.43	93.59
	pos	56.91	37.21	47.25	53.03	62.75	89.36
	neg	65.03	46.99	51.78	63.19	78.11	93.59
Area C [3]		70.41	30.00	56.50	69.21	82.93	100.00
	pos	67.11	30.00	56.21	66.21	82.79	94.07
	neg	74.89	52.7	64.75	76.11	84.36	100.00
Biosecurity		57.57	10.04	44.60	55.68	75.19	90.01
	pos	53.08	10.04	40.86	51.99	74.79	89.76
	neg	63.65	29.73	51.89	64.25	76.98	90.01

[1] Farm management and staff training, [2] Housing factors, [3] Animal based measures.

From the compilation of the welfare checklist, regarding "total welfare", 24% of the farms (8 out of 33) obtained an insufficient score (below 60%), 61% (20/33) a medium score (between 60 and 80%), and 15% (5/33) a high score (over 80%; Figure 2). From the compilation of the biosecurity checklist, 58% (19/33) of the enrolled flock showed insufficient measures of biosecurity, 27% (9/33) obtained a medium score, and 15% (5/33) a high score (Figure 3).

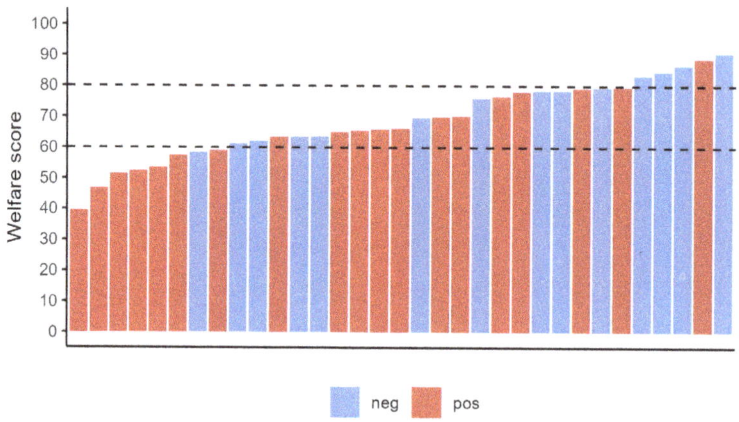

Figure 2. Welfare scores in paratuberculosis positive and negative flocks. The dotted lines indicate the thresholds among insufficient (below 60%), medium (61–80%), and high score (over 80%).

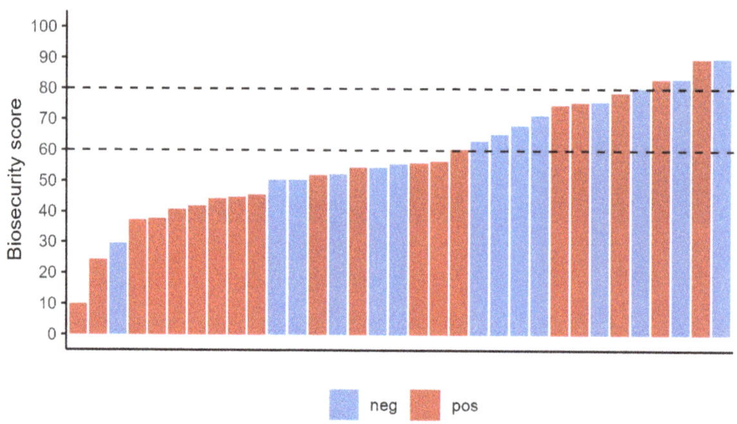

Figure 3. Biosecurity scores in paratuberculosis positive and negative flocks. The dotted lines indicate the thresholds among insufficient (below 60%), medium (61–80%), and high score (over 80%).

In Figure 4 we reported the percentage distribution of the scores assigned for each indicator, according to the presence or absence of the disease in the farms.

Table 5 shows the result of statistical analysis (Chi-square test). In general, there is no strong evidence of an association between the indicators and paratuberculosis status. The analysis suggests a potential association with paratuberculosis status (p-value < 0.10) for the indicators "Animal grouping strategy", "Inspection of the animals", "Cleanliness and hygiene of floor in walking areas and of bedding", "Space availability in lying area (young goats)", and "Annual mortality rate (adult goats)" from the animal welfare assessment, "Contact with other animal species" and "Measures for preventing the entrance of strangers" from the biosecurity assessment.

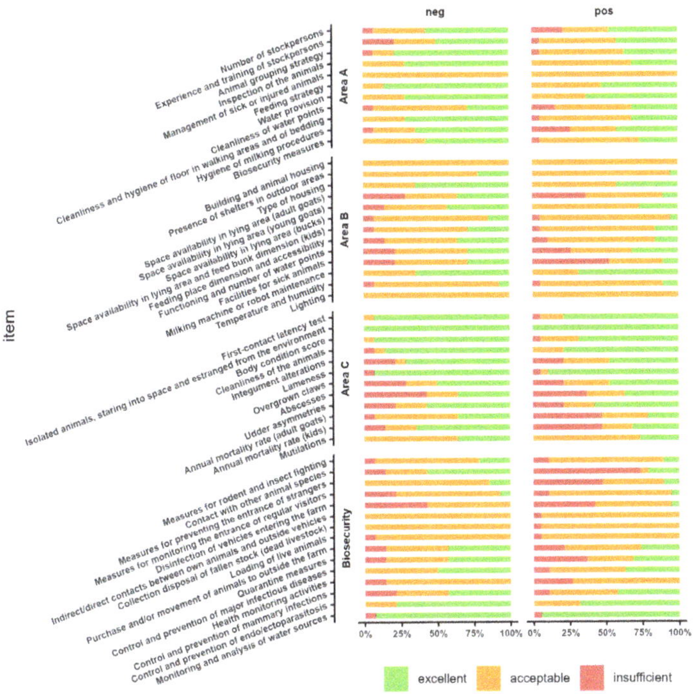

Figure 4. Percentage distribution of the scores assigned to each indicator in paratuberculosis positive and negative flocks.

Table 5. Statistical analysis of the indicators from the animal welfare and biosecurity assessment that showed a correlation with paratuberculosis seropositivity (p-value < 0.10).

Indicator	Score	Flocks (N)	Positive Flocks	Statistical Analysis
Welfare				
Animal grouping strategy	excellent	18	7	$\chi^2 = 6.51$ p-value = 0.039
	acceptable	13	11	
	insufficient	2	1	
Inspection of the animals	excellent	16	6	$\chi^2 = 3.65$ p-value = 0.056
	acceptable	17	13	
	insufficient	0	0	
Cleanliness and hygiene of floor in walking areas and of bedding	excellent	16	6	$\chi^2 = 5.37$ p-value = 0.068
	acceptable	16	12	
	insufficient	1	1	
Space availability in lying area (young goats)	excellent	11	5	$\chi^2 = 4.64$ p-value = 0.098
	acceptable	20	14	
	insufficient	2	0	
Annual mortality rate (adult goats)	excellent	9	4	$\chi^2 = 6.18$ p-value = 0.045
	acceptable	14	6	
	insufficient	10	9	

Table 5. *Cont.*

Indicator	Score	Flocks (N)	Positive Flocks	Statistical Analysis
Biosecurity				
Contact with other animal species	excellent	12	4	$\chi^2 = 11.64$ *p*-value = 0.003
	acceptable	5	1	
	insufficient	16	14	
Measures for preventing the entrance of strangers	excellent	4	2	$\chi^2 = 9.25$ *p*-value = 0.01
	acceptable	20	8	
	insufficient	9	9	

4. Discussion

In the present study, we carried out a serological survey for paratuberculosis on 33 dairy goat farms, located in four regions of northern Italy. Contextually, animal welfare and biosecurity assessment were carried out on the same farms, using a protocol developed by IZSLER-CReNBA and based on the use of animal-based measures (ABMs) and resource-based indicators.

More than half (58%) of the investigated farms were seropositive for paratuberculosis, with mean intra-herd seroprevalence of 7.4%. In all the seropositive farms, at least one seropositive animal was confirmed to be infected by testing a fecal sample by a qPCR targeting the IS900 sequence of MAP [33]. Considering the limitations of the serological test, with a diagnostic sensitivity varying according to the stage of the disease [20], the prevalence of seropositive animals has been probably underestimated. The data shown confirmed, in agreement with those worldwide reported, the high diffusion of the disease in the northern Italy dairy goat farms. In 2023 the Italian Ministry of Health, in the frame of application of the Regulation (EU) 2016/429 [28] ("Animal health law"), included goat in the "National guidelines for the control of paratuberculosis" [34].

Regarding the welfare assessment, favorable average scores were found both in the overall score and for the three areas individually taken (Table 4). Out of the 33 farms considered in the present study, 25 (76%) obtained a positive score from the overall welfare assessment while eight farms did not reach the threshold score of 60%. In infected flocks, seven out of 19 (37%) showed an insufficient total welfare score (Figure 2). Considering the mean score of the three areas, the "housing" (Area B) had a mean insufficient score (56.91%), while "farm management and staff training" (Area A) and ABMs (Area C) showed, respectively a mean score of 61.23 and 67.11%, above the acceptability threshold. In non-infected flocks, registered mean scores for the three areas were higher: 76.99%, 65.03 and 74.89% for Area A, B and C, respectively.

In order to explore potential association between paratuberculosis status and the 36 welfare indicators provided by the protocol, we compared the scores assigned to paratuberculosis-positive farms with the negative ones. According to the statistical analysis (Table 5), five indicators showed an association with paratuberculosis positivity: (i) "Animal grouping strategy", (ii) "Inspection of the animals", (iii) "Cleanliness and hygiene of floor in walking areas and of bedding", (iv) "Space availability in lying area (young goats)"and (v) "Annual mortality rate (adult goats)" from the animal welfare assessment.

Regarding animal grouping, this indicator evaluates the separation into homogeneous groups from six months of age. Although this represents a good practice for the management of infectious diseases [35], the association highlighted may be misleading as the indicator does not take into account the early separation of the kids from the mother, a cornerstone of paratuberculosis management in the farm [7]. A regular inspection of the animals by the stockpersons is important for the detection of symptomatic animals and therefore allows the improvement of biosecurity and management practices. Cleanliness of facilities is one of the most important factors related to paratuberculosis, due to its fecal–oral transmission [35,36]. Regarding the space availability in lying area, the higher density of animals may increase horizontal transmission, mainly due to closer contact between goats. The statistical analysis underlined an association for young goats and not for adults, even if this category is the major source of the infection. This is especially true during the lambing season, since goats usually give birth all together in one pen and kids can easily come into contact with feces of adults before being separated from them. With respect to the annual mortality rate, even if not all the deaths can be referred to paratuberculosis, it is important evidence of the health problems and mismanagement inside the farms [5].

Concerning the biosecurity assessment, more critical issues were found compared to welfare. Nineteen out of 33 farms (58%) obtained biosecurity scores below the acceptability threshold. The average score was 57.57% (Table 4). Not surprisingly, most of paratuberculosis infected flocks showed insufficient biosecurity assessments results: 13 out of 19 (68%) infected farms had biosecurity values below 60% (Figure 3).

The adoption of proper biosecurity measures is a key tool to prevent the introduction of MAP into the farm and to tackle its spread [5,7]. Therefore, also for the 15 biosecurity indicators a statistical analysis was performed, in order to explore possible association with paratuberculosis status. For two indicators a correlation was demonstrated: (i) "Contact with other animal species" and (ii) "Measures for preventing the entrance of strangers".

Contact prevention with other animal species represents a general biosecurity measure on farms, which also applies to paratuberculosis, although the check-list indicator considers all species and is not focused on those that may be a source of MAP (e.g., ruminants). Similarly, preventing the entrance of strangers in the farm is a general biosecurity measure, therefore a correlation with this indicator and infected farms is not surprising.

Correlation with the other biosecurity indicators could not be demonstrated, probably due to the diffuse low level of biosecurity in all assessed flocks, a critical issue already reported in cattle farming [37].

Moreover, the assessment through checklists provides a "picture" at the moment of the visit to the farm. So, since paratuberculosis is a chronic disease, conditions that could have led to the entry and spread of the infection in the flocks could may not been highlighted at the moment of the assessment. For paratuberculosis control, the general biosecurity measures included by the checklist should be implemented with specific management procedures, such as feed milk replacers or pasteurized milk in infected farms, ensure at least the separation of test-positive and test-negative animals, organizing positive and negative groups, prevent manure contamination of feed and water.

Our findings may be indicative of how some deficiencies in livestock management and biosecurity may favor the introduction and spread of communicable diseases, which, in turn, may affect some animal welfare parameters. In this exploratory study, the execution of a multiple model was dispensed with, since the study design, characterized by opportunistic sampling, is not suitable for this type of analysis.

Finally, in accordance with data reported in Italy in cattle [38], we observed that the mean year-long milk production is higher in MAP negative flocks (Table 3). Nevertheless, the high SD values suggest that this observation may not be significant.

5. Conclusions

The results of this study confirm that paratuberculosis is present in northern Italy dairy goat farms. Because of possible biases related to low representativeness of the sample considered, this investigation was presented as exploratory.

Knowledge and surveillance by veterinarians and farmers are essential in order to counteract paratuberculosis impact on the economy of the farm, on the animal welfare and on public health.

The animal welfare and biosecurity assessments through the described checklists proved to be an accurate and easy-to-use tool in the field, capable of identifying critical points and providing the farmers with indications to improve farm management.

Our survey highlighted that, on the investigated dairy goat farms, the animal welfare level is on average acceptable, while more effort should be directed toward improving biosecurity levels.

Supplementary Materials: The following supporting information can be downloaded at: https://www.mdpi.com/article/10.3390/ani13142346/s1: Table S1: Details of breeds, management, number of lactating goats and mean year-long milk production. Table S2: Statistical analysis of the indicators from the animal welfare and biosecurity assessment associated with paratuberculosis seropositivity. Statistical analysis was not applicable when the same score was assigned to all the flocks.

Author Contributions: Conceptualization, A.G. and C.G.; methodology, A.G. and C.G.; software, G.G.; validation, N.A., M.R., F.F. and V.L.; formal analysis, G.G.; investigation, A.G., C.G., E.M., L.D.P., A.P., S.B. and D.D.; resources, A.G., C.G., E.M., L.D.P., A.P., S.B. and D.D.; data curation, F.B. and G.G.; writing—original draft preparation, F.B.; writing—review and editing, F.B., A.G. and C.G.; visualization, A.G., F.B., G.G. and C.G.; supervision, N.A., M.R., F.F. and V.L.; project administration, A.G.; funding acquisition, A.G. All authors have read and agreed to the published version of the manuscript.

Funding: This research was funded by Ministero della Salute, grant number CUP E56C18001750001 (IZS LER 005/18 RC) "Analisi di parametri produttivi e loro correlazione con lo stato di benessere in allevamenti di capra da latte".

Institutional Review Board Statement: Not applicable.

Informed Consent Statement: Informed consent was made for the farmers to sign.

Data Availability Statement: The data presented in this study are available on request from the corresponding author. The data are not publicly available due to privacy restrictions.

Acknowledgments: We gratefully acknowledge the veterinary practitioners Fabio Bencetti, Claudio Gelmini, Marzia Monaci, Bastianino Mossa, Mario Villa, Adriana Vismara and the farmers for their collaboration.

Conflicts of Interest: The authors declare no conflict of interest. The funders had no role in the design of the study; in the collection, analyses, or interpretation of data; in the writing of the manuscript; or in the decision to publish the results.

References

1. WOAH. Chapter 3.1.16 Paratuberculosis (Johne's Disease) (version adopted in May 2021). In *Manual of Diagnostic Tests and Vaccines for Terrestrial Animals 2022*; WOAH: Paris, France, 2022; Available online: https://www.woah.org/en/what-we-do/standards/codes-and-manuals/terrestrial-manual-online-access/ (accessed on 11 April 2023).
2. Whittington, R.; Donat, K.; Weber, M.F.; Kelton, D.; Nielsen, S.S.; Eisenberg, S.; Arrigoni, N.; Juste, R.; Sáez, J.L.; Dhand, N.; et al. Control of Paratuberculosis: Who, Why and How. A Review of 48 Countries. *BMC Vet. Res.* **2019**, *15*, 1–29. [CrossRef]
3. Robbe-Austerman, S. Control of Paratuberculosis in Small Ruminants. *Vet. Clin. N. Am. Food Anim. Pract.* **2011**, *27*, 609–620. [CrossRef]
4. Stewart, D.J.; Vaughan, J.A.; Stiles, P.L.; Noske, P.J.; Tizard, M.L.V.; Prowse, S.J.; Michalski, W.P.; Butler, K.L.; Jones, S.L. A Long-Term Bacteriological and Immunological Study in Holstein-Friesian Cattle Experimentally Infected with Mycobacterium Avium Subsp. Paratuberculosis and Necropsy Culture Results for Holstein-Friesian Cattle, Merino Sheep and Angora Goats. *Vet. Microbiol.* **2007**, *122*, 83–96. [CrossRef] [PubMed]
5. Jiménez-Martín, D.; García-Bocanegra, I.; Risalde, M.A.; Fernández-Molera, V.; Jiménez-Ruiz, S.; Isla, J.; Cano-Terriza, D. Epidemiology of Paratuberculosis in Sheep and Goats in Southern Spain. *Prev. Vet. Med.* **2022**, *202*, 105637. [CrossRef] [PubMed]
6. Gavin, W.G.; Porter, C.A.; Hawkins, N.; Schofield, M.J.; Pollock, J.M. Johne's Disease: A Successful Eradication Programme in a Dairy Goat Herd. *Vet. Rec.* **2018**, *182*, 483. [CrossRef]
7. Windsor, P.A. Paratuberculosis in Sheep and Goats. *Vet. Microbiol.* **2015**, *181*, 161–169. [CrossRef]
8. Sardaro, R.; Pieragostini, E.; Rubino, G.; Petazzi, F. Impact of Mycobacterium Avium Subspecies Paratuberculosis on Profit Efficiency in Semi-Extensive Dairy Sheep and Goat Farms of Apulia, Southern Italy. *Prev. Vet. Med.* **2017**, *136*, 56–64. [CrossRef]
9. Idris, S.M.; Eltom, K.H.; Okuni, J.B.; Ojok, L.; Elmagzoub, W.A.; El Wahed, A.A.; Eltayeb, E.; Gameel, A.A. Paratuberculosis: The Hidden Killer of Small Ruminants. *Animals* **2021**, *12*, 12. [CrossRef]
10. Iarussi, F.; Paradies, P.; Sardaro, R.; Rubino, G.; Scaltrito, D.; Pieragostini, E.; Petazzi, F. Epidemiology and Risk Factors of Mycobacterium Avium Subspecies Paratuberculosis in Semi-Extensive Dairy Sheep and Goat Farms of Apulia, Southern Italy. *Small Rumin. Res.* **2019**, *177*, 89–96. [CrossRef]

11. EFSA. Assessment of Listing and Categorisation of Animal Diseases within the Framework of the Animal Health Law (Regulation (EU) No 2016/429): Paratuberculosis. *EFSA J.* **2017**, *15*, 1–47. [CrossRef]
12. Ikonomopoulos, J.; Pavlik, I.; Bartos, M.; Svastova, P.; Ayele, W.Y.; Roubal, P.; Lukas, J.; Cook, N.; Gazouli, M. Detection of Mycobacterium Avium Subsp. Paratuberculosis in Retail Cheeses from Greece and the Czech Republic. *Appl. Environ. Microbiol.* **2005**, *71*, 8934–8936. [CrossRef] [PubMed]
13. Botsaris, G.; Slana, I.; Liapi, M.; Dodd, C.; Economides, C.; Rees, C.; Pavlik, I. Rapid Detection Methods for Viable Mycobacterium Avium Subspecies Paratuberculosis in Milk and Cheese. *Int. J. Food Microbiol.* **2010**, *141*, S87–S90. [CrossRef]
14. Williams, A.G.; Withers, S.E. Microbiological Characterisation of Artisanal Farmhouse Cheeses Manufactured in Scotland. *Int. J. Dairy Technol.* **2010**, *63*, 356–369. [CrossRef]
15. Galiero, A.; Fratini, F.; Mataragka, A.; Turchi, B.; Nuvoloni, R.; Ikonomopoulos, J.; Cerri, D. Detection of Mycobacterium Avium Subsp. Paratuberculosis in Cheeses from Small Ruminants in Tuscany. *Int. J. Food Microbiol.* **2016**, *217*, 195–199. [CrossRef] [PubMed]
16. European Union. Commission Implementing Regulation (EU) 2018/1882 of 3 December 2018 on the Application of Certain Disease Prevention and Control Rules to Categories of Listed Diseases and Establishing a List of Species and Groups of Species Posing a Considerable Risk for the Spread of Those Listed Diseases. *Off. J. Eur. Union.* **2018**, *L 308*, 21–29. Available online: http://data.europa.eu/eli/reg_impl/2018/1882/oj (accessed on 11 April 2023).
17. Nielsen, S.S.; Toft, N. A Review of Prevalences of Paratuberculosis in Farmed Animals in Europe. *Prev. Vet. Med.* **2009**, *88*, 1–14. [CrossRef] [PubMed]
18. Galiero, A.; Turchi, B.; Pedonese, F.; Nuvoloni, R.; Cantile, C.; Colombani, G.; Forzan, M.; Cerri, D.; Bandecchi, P.; Fratini, F. Serological, Culture and Molecular Survey of Mycobacterium Avium Paratuberculosis in a Goat Flock in Tuscany. *Folia Microbiol.* **2017**, *62*, 471–477. [CrossRef]
19. Muri, K.; Leine, N.; Valle, P.S. Welfare Effects of a Disease Eradication Programme for Dairy Goats. *Animal* **2016**, *10*, 333–341. [CrossRef]
20. Nielsen, S.S.; Toft, N. Ante Mortem Diagnosis of Paratuberculosis: A Review of Accuracies of ELISA, Interferon-Gamma Assay and Faecal Culture Techniques. *Vet. Microbiol.* **2008**, *129*, 217–235. [CrossRef]
21. Bertocchi, L.; Fusi, F.; Angelucci, A.; Bolzoni, L.; Pongolini, S.; Strano, R.M.; Ginestreti, J.; Riuzzi, G.; Moroni, P.; Lorenzi, V. Characterization of Hazards, Welfare Promoters and Animal-Based Measures for the Welfare Assessment of Dairy Cows: Elicitation of Expert Opinion. *Prev. Vet. Med.* **2018**, *150*, 8–18. [CrossRef]
22. Lorenzi, V.; Sgoifo Rossi, C.A.; Compiani, R.; Grossi, S.; Bolzoni, L.; Mazza, F.; Clemente, G.A.; Fusi, F.; Bertocchi, L. Using Expert Elicitation for Ranking Hazards, Promoters and Animal-Based Measures for on-Farm Welfare Assessment of Indoor Reared Beef Cattle: An Italian Experience. *Vet. Res. Commun.* **2023**, *47*, 141–158. [CrossRef]
23. Anzuino, K.; Bell, N.J.; Bazeley, K.J.; Nicol, C.J. Assessment of Welfare on 24 Commercial UK Dairy Goat Farms Based on Direct Observations. *Vet. Rec.* **2010**, *167*, 774–780. [CrossRef] [PubMed]
24. Muri, K.; Stubsjøen, S.; Valle, P. Development and Testing of an On-Farm Welfare Assessment Protocol for Dairy Goats. *Anim. Welf.* **2013**, *22*, 385–400. [CrossRef]
25. European Union. Council Directive 98/58/EC of 20 July 1998 Concerning the Protection of Animals Kept for Farming Purposes. *Off. J. Eur. Union.* **1998**, *L 221*, 23–27. Available online: http://data.europa.eu/eli/dir/1998/58/oj (accessed on 11 April 2023).
26. Mattiello, S.; Battini, M.; Vieira, A.; Stilwell, G. AWIN Welfare Assessment Protocol for Goats 2015. Available online: https://air.unimi.it/handle/2434/269102 (accessed on 11 April 2023).
27. Bertocchi, L.; Gaffuri, A.; Angelucci, A.; Fusi, F.; Strano, R.M.; Ginestreti, J.; Ferrara, G.; Bolzoni, L.; Zanardi, G.; Lorenzi, V. Expert Knowledge Elicitation for Ranking Hazards Affecting Dairy Goat Welfare in Italian Farms. In Proceedings of the The 30th World Buiatrics Congress, Sapporo, Japan, 28 August–1 September 2018; pp. 345–346.
28. European Union. Regulation (EU) 2016/429 of the European Parliament and of the Council of 9 March 2016 on Transmissible Animal Diseases and Amending and Repealing Certain Acts in the Area of Animal Health ('Animal Health Law'). *Off. J. Eur. Union.* **2016**, 1–208.
29. Fusi, F.; Lorenzi, V.; Franceschini, G.; Compiani, R.; Harper, V.; Ginestreti, J.; Ferrara, G.; Sgoifo Rossi, C.A.; Bertocchi, L. Animal Welfare and Biosecurity Assessment: A Comparison between Italian and Irish Beef Cattle Rearing Systems. *Anim. Prod. Sci.* **2021**, *61*, 55–63. [CrossRef]
30. Ginestreti, J.; Lorenzi, V.; Fusi, F.; Ferrara, G.; Scali, F.; Alborali, G.L.; Bolzoni, L.; Bertocchi, L. Consumo Di Antimicrobici, Benessere Animale e Biosicurezza in 16 Aziende Di Bovine Da Latte in Lombardia. *Large Anim. Rev.* **2020**, *26*, 3–11.
31. Diana, A.; Lorenzi, V.; Penasa, M.; Magni, E.; Alborali, G.L.; Bertocchi, L.; De Marchi, M. Effect of Welfare Standards and Biosecurity Practices on Antimicrobial Use in Beef Cattle. *Sci. Rep.* **2020**, *10*, 20939. [CrossRef]
32. Stonos, N.; Bauman, C.; Menzies, P.; Wootton, S.K.; Karrow, N.A. Prevalence of Small Ruminant Lentivirus and Mycobacterium Avium Subsp. Paratuberculosis Co-Infection in Ontario Dairy Sheep and Dairy Goats. *Can. J. Vet. Res.* **2017**, *81*, 155–159.
33. Barrero-Domínguez, B.; Luque, I.; Huerta, B.; Gomez-Laguna, J.; Galán-Relaño, Á.; Gómez-Gascón, L.; Sánchez, M.; Astorga, R.J. Paratuberculosis in Dairy Goat Flocks from Southern Spain: Risk Factors Associated with Seroprevalence. *Vet. Rec.* **2019**, *185*, 600. [CrossRef]
34. R Core Team. *A Language and Environment for Statistical Computing 2020*; R Foundation for Statistical Computing: Vienna, Austria, 2020; Available online: https://www.R-project.org/ (accessed on 11 April 2023).

35. Russo, S.; Galletti, G.; Leo, S.; Arrigoni, N.; Garbarino, C.; Ricchi, M. Validation of IS900- qPCR assay to assess the presence of *Mycobacterium avium* subs. *paratuberculosis* in faecal samples according to the OIE procedure. *Prev. Vet. Med.* **2022**, *208*, 105732 [CrossRef]
36. Conferenza permanente per i rapporti tra lo Stato, le Regioni e le Province Autonome di Trento e Bolzano. Accordo, Ai Sensi Dell'articolo 4, Comma 1, Del Decreto Legislativo 28 Agosto 1997, n. 281, Tra Il Governo, Le Regioni e Le Province Autonome Di Trento e Bolzano Concernente: «Linee Guida per l'adozione Dei Piani Di Controllo e per l'assegnazione Della Qualifica Sanitaria Agli Allevamenti Di Specie Sensibili (Bovini, Bufalini, Ovini, Caprini) Nei Confronti Della Paratubercolosi». (Repertorio atti n 230/CSR del 30 novembre 2022). (23A00089). *GU Ser. Gen.* **2023**, *10*, 44–49.
37. Renault, V.; Humblet, M.-F.; Pham, P.N.; Saegerman, C. Biosecurity at Cattle Farms: Strengths, Weaknesses, Opportunities and Threats. *Pathogens* **2021**, *10*, 1315. [CrossRef] [PubMed]
38. Luini, M.; Foglia, E.; Andreoli, G.; Tamba, M.; Arrigoni, N. Piano Di Controllo e Certificazione per La Paratubercolosi Bovina: Criteri e Analisi Costo-Beneficio. *Large Anim. Rev.* **2013**, *19*, 257–265.

animals

MDPI

Article

Mycobacterium avium subsp. *paratuberculosis* in Sheep and Goats in Austria: Seroprevalence, Risk Factors and Detection from Boot Swab Samples

Juliane Schrott [1], Eva Sodoma [1,2], Michael Dünser [1,2], Alexander Tichy [3] and Johannes Lorenz Khol [4,*]

1 Austrian Agency for Health and Food Safety (AGES), Institute for Veterinary Disease Control Innsbruck, Technikerstraße 70, 6020 Innsbruck, Austria; juliane.schrott@ages.at (J.S.); eva.sodoma@ages.at (E.S.); michael.duenser@ages.at (M.D.)
2 Austrian Agency for Health and Food Safety (AGES), Institute for Veterinary Disease Control Linz, Wieningerstraße 8, 4020 Linz, Austria
3 Platform for Bioinformatics and Biostatistics, Department of Biomedical Sciences, University of Veterinary Medicine Vienna, Veterinärplatz 1, 1210 Wien, Austria; alexander.tichy@vetmeduni.ac.at
4 University Clinic for Ruminants, Department for Farm Animals and Veterinary Public Health, University of Veterinary Medicine Vienna, Veterinärplatz 1, 1210 Wien, Austria
* Correspondence: johannes.khol@vetmeduni.ac.at

Simple Summary: This study aimed to investigate the occurrence of paratuberculosis (Johne's disease), caused by *Mycobacterium avium* subsp. *paratuberculosis* (MAP), in Austrian sheep and goats by testing 22,019 blood samples. Furthermore, detailed investigations in five MAP-infected goat herds were carried out to detect the disease in blood and faecal samples. The detected animal MAP seroprevalence was 2.0% for goats and 0.7% for sheep (calculated true prevalence 3.5% and 1.2%, respectively). Herd-level apparent MAP seroprevalence was 11.1% for goat herds and 8.9% for sheep flocks. Herds with a more intensive production system had a significantly higher risk of being infected, as well as farms with frequent trading of animals or where other ruminant species were kept on the same premise. In the five goat farms investigated, 21.8% (11.7%–28.0%, calculated true seroprevalence 38.6%) of the animals were found to be infected and 12.3% (5.0%–24.7%) of the animals were shedding the bacterium with their faeces. It was further possible to identify the bacterium using boot swab samples from the stable environment in each of the five herds. The results indicated a moderate paratuberculosis infection rate in small ruminants in Austria.

Abstract: This study aimed to investigate the prevalence of *Mycobacterium avium* subsp. *paratuberculosis* (MAP) in small ruminants in Austria by testing 22,019 serum samples with ELISA for the presence of specific antibodies. Furthermore, detailed investigations in five MAP-infected goat herds were carried out by ELISA, qPCR and bacterial culture. The found animal-level apparent MAP seroprevalence was 2.0% for goats and 0.7% for sheep (calculated true prevalence 3.5% and 1.2%, respectively). Herd-level apparent MAP seroprevalence was 11.1% for goat herds and 8.9% for sheep flocks. Significant risk factors for seropositivity in goat herds were: herd size, animal trading, farmed as a dairy herd, Animal Health Service membership and cohabitation with farmed game. For sheep flocks, seroprevalence was significantly higher in flocks with animal trading and where cattle or goats were kept in the flock, respectively. The overall apparent within-herd MAP seroprevalence in the five goat farms investigated was 21.8% (11.7%–28.0%, calculated true seroprevalence 38.6%) and an overall rate of MAP shedding of 12.3% was detected (5.0%–24.7%). It was possible to identify MAP by culture using boot swab samples in each herd. The results indicated a moderate MAP infection rate in small ruminants in Austria.

Keywords: paratuberculosis; goat; sheep; risk factors; seroprevalence; boot swab samples

1. Introduction

Paratuberculosis (Johne's disease, JD) is a chronic intestinal disease caused by *Mycobacterium avium* subsp. *paratuberculosis* (MAP), affecting ruminants with a global distribution [1]. Infections in sheep and goats are mainly acquired at an early age by faecal–oral transmission of MAP, followed by a long subclinical period. The main clinical sign of JD in small ruminants is progressive weight loss [2]. Caprine and ovine paratuberculosis is known to cause considerable economic losses in affected herds [3].

In contrast to cattle, the MAP prevalence in sheep and goats is unknown in many countries and the prevalence appears to be underestimated [1]. Additionally, prevalence studies carried out among sheep and goats in Europe are scarce. Due to different study design, sampling frames and the uncertainty of the diagnostic tests, MAP prevalence among small ruminants is difficult to estimate [4]. A few studies describe several risk factors associated with MAP prevalence in small ruminants [5–11], but further research is needed to confirm their validity for the local small ruminant population.

The enzyme-linked immunosorbent assay (ELISA) is a suitable cost-effective and commonly used diagnostic test for MAP prevalence studies [12]. For herd-level diagnostics and control programs in dairy herds, culture and/or polymerase chain reaction (PCR) of boot swab (BS) samples and pooled faecal samples have been applied successfully [13,14], but, to the authors' knowledge, the application of BS sampling has not been verified for sheep and goats so far.

In the past, cattle in Austria were predominantly affected by JD. However, MAP was also isolated from goat, sheep and wild ruminants (red deer, roe deer and mouflon) [15]. The herd-level MAP seroprevalence of Austrian cattle was observed to have increased over the years to 19.1% [16,17], but more recent studies utilising BS sampling in a region in western Austria showed a much lower MAP prevalence in dairy cattle of 0.97% [18]. In 2006, clinical JD became a notifiable disease in cattle, sheep, goat and farmed deer in Austria. Based on the compulsory national paratuberculosis control program, suspicious animals that test MAP positive have to be culled. Moreover, preventive hygienic measures have to be implemented in affected farms [19].

Sheep and goat are minor species in Austria and the production is characterised by smallholders. The Austrian national goat herd comprises 92,800 goats distributed in approximately 10,000 herds, with an average herd size of 9 animals (only 3% of the herds have more than 50 animals). About 16,000 sheep flocks hold 394,000 sheep, corresponding to an average sheep flock size of 25 animals; 13% of the flocks hold more than 50 sheep. Since the year 2011, the number of goats kept in Austria increased by about 28% and the number of sheep by about 9%, respectively [20]. According to the national animal health information system, goats are mainly kept for milk production (33%), whereas sheep are predominantly kept for meat production. The high number of animals grazing on common mountain pastures during summer months (112,000 sheep (28%) and 13,000 goats (14%) in 2020) underlines the importance of small ruminants for alpine landscape preservation in Austria [21].

The aim of the present study was to investigate the current herd-level MAP seroprevalence in Austrian sheep and goat holdings and to identify risk factors associated with a positive MAP status. Furthermore, subsequent investigations in five infected dairy goat herds were carried out to gain information about the within-herd MAP seroprevalence and bacterial shedding and whether boot swab or pooled faecal sampling may be suitable for MAP herd-level detection in goats.

2. Materials and Methods

2.1. Study Design, Animal Sampling and Data Collection

2.1.1. Serological Screening

In total, serum samples from 6434 goats in 638 herds and from 15,585 sheep in 1032 flocks were analysed in this study (Table 1). Serum sample numbers were consistent with the totality of the 2020 national annual risk-based sampling plan for the maintenance of the *Brucella melitensis*-free status in Austria [22]. In this sampling plan, herds were selected using stratified random sampling, with sample sizes proportional to the number of herds in the respective provinces (see Figures 1 and 2). A sampling probability proportional to the natural logarithm of the herd size was used to select farms. In addition, dairy herds, farms with increased domestic or international livestock trade, farms using common pastures and farms with previous brucellosis history were assigned a higher selection probability. The number of samples tested for each selected herd depended on the herd size to ensure a minimum level of herd sensitivity: In herds with fewer than 12 animals, all animals were sampled. In larger herds, the sample size increased slightly. A maximum sample size of 22 animals was achieved in herds with more than 164 animals. Only sheep and goats older than 6 months were included in the sample scheme. The median herd size of the study population was 42 animals per sheep flock and 11 animals per goat herd, respectively, while the median sample size per herd was 16 samples for sheep and 8 samples for goats. Further production parameters, such as information about cohabitation with other ruminant species, common alpine pastures (yes/no), organic farming (yes/no) and Animal Health Service membership (yes/no) were obtained from the national animal health information system for the herds participating in the study (Table 2). The Animal Health Service supports its members (livestock owners) regarding animal health and welfare, providing consulting, diagnostic services, training courses and control programs.

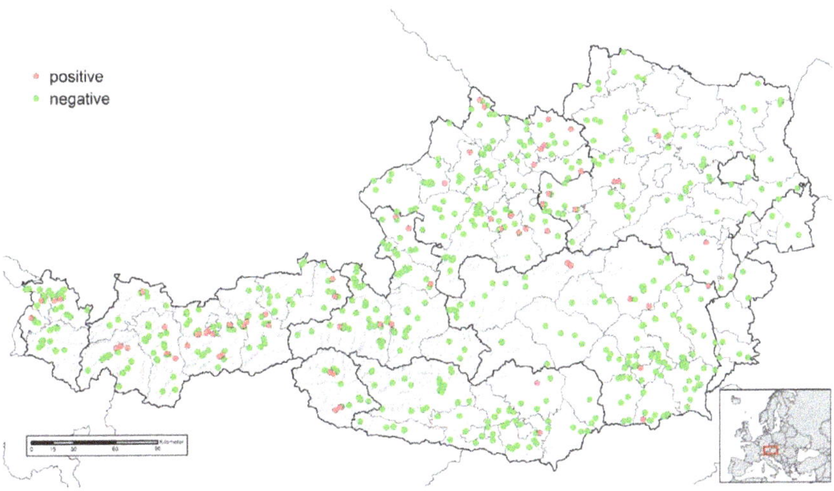

Figure 1. Location of goat herds investigated for the presence of *Mycobacterium avium* subsp. *paratuberculosis* specific antibodies in Austria and test results (positive, n = 71; negative, n = 567).

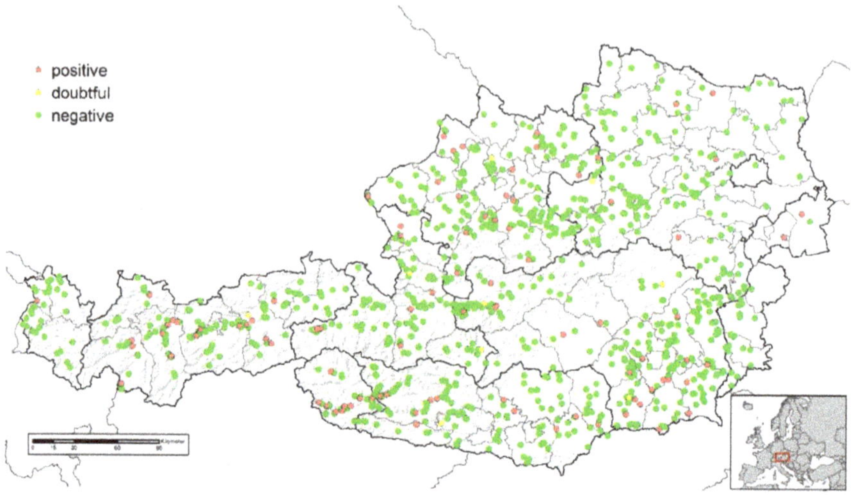

Figure 2. Location of sheep flocks investigated for the presence of *Mycobacterium avium* subsp. *paratuberculosis* specific antibodies in Austria and test results (positive, n = 92; doubtful, n = 10; negative, n = 930).

2.1.2. Herd-Level Examination in Selected Dairy Goat Herds

In five dairy goat herds, either detected as seropositive during serological screening or during previous investigations by the National Reference Laboratory (NRL) for Paratuberculosis, detailed herd-level examinations were carried out. Herds were selected based on herd size and the willingness of the owner and the responsible veterinarian to participate in the study. Dairy goats were housed in stalls and at pasture (n = 4) or housed indoors only (n = 1). In two holdings, the herd was divided into two physically separated groups. The predominant breed was the Saanen Goat, other breeds were the German Improved Fawn, the German Improved White and the Chamois Coloured Goat. Animals did not have access to common pastures, but billy goats originating from external herds were introduced periodically to the herds. A separated kidding area was provided in three herds; moreover, kids were separated from their does after ingestion of colostrum during the first days postpartum (n = 4) or kids were separated immediately after birth from their does and fed with artificial colostrum (n = 1). In addition, two herds had dairy cattle kept on the farm.

Individual blood and faecal samples were taken from all animals older than 12 months in three herds and from a representative random sample in two herds. Blood samples were taken from the external jugular vein, according to Baumgartner and Wittek [23], using a vacuum system with serum tubes containing a clot activator (Greiner bio-one, Kremsmünster, Austria). Faecal samples were taken from the rectum; in 8 animals, no faecal sample could be collected, because faecal material was absent in the rectum at the time of sampling. Moreover, environmental BS samples and pooled faecal samples (5 animals each) were collected. BS samples were collected mainly in stable locations with high animal traffic and manure concentration in a meandering pattern, as described for cattle elsewhere [24]. Individual faecal samples were tested by qPCR, whereas BS samples and pooled faecal samples were tested by culture and qPCR, as described below. Herd size and numbers of samples per herd are shown in Table 3. For each animal, the age was documented in years, according to owners' records, and the body condition score (BCS) was determined by a trained veterinarian, as described elsewhere [25]. Samples were transported to the laboratory for further processing immediately after collection. At the Austrian NRL for Paratuberculosis, samples were stored and refrigerated and processing started on the same working day.

2.2. Serological Analyses

Blood samples were first centrifuged for 5 min at 2325 g. Serum samples were tested for the presence of MAP specific antibodies, using a commercial ELISA test kit (ID Screen® Paratuberculosis Indirect Screening test, ID.Vet, Grabels, France) following the manufacturer's instructions, using the short incubation protocol. According to the data provided by the technical datasheet, the kit detects antibodies in cattle sera with a sensitivity (Se) of 56.86% and a specificity (Sp) of 99.63%; these values were applied for true prevalence (TP) calculation. Samples with positive and questionable test results were confirmed by ID Screen® Paratuberculosis Indirect Confirmation test (ID.Vet, Grabels, France).

2.3. qPCR of Individual Faecal Samples

According to the prescribed method in the Austrian NRL for Paratuberculosis, faecal samples underwent a pre-treatment: 3 g of faeces each was put into an EZ-Drop bottle (ID Gene® Easy preparation for faeces samples, ID.vet, Grabels, France), pressed carefully with a spatula and mixed for 30 s in a vortex mixer. A total of 1.5 mL of the homogenate was transferred to a 2 mL tube and centrifuged for 10 min at 13,000× *g*. Then, the supernatant was discarded and 1 mL of PowerBead Solution (Qiagen GmbH, Hilden, Germany) and 1 µL IC-DNA (bactotype® MAP PCR Kit, Indical Bioscience GmbH, Leipzig, Germany) were added to the sediment. The entire suspension was transferred to a 2 mL tube with a screw cap, containing 300 mg glass beads, size 0.1 mm (Retsch GmbH, Haan, Germany), and homogenised twice for 30 s at 6800 rpm on a Precellys 24 (Bertin Technologies, Montigny Le Bretonneux, France). Then, the tube was centrifuged for 1 min at 20,000× *g* and 200 µL of the supernatant was used for the extraction. DNA was extracted using the IndiMag® Pathogen Kit (Indical Bioscience GmbH, Leipzig, Germany) and the KingFisher™ Flex Purification System (Thermo Fisher Scientific, Waltham, MA, USA) or IDEAL™ 32 Extraction Robot (ID.vet, Grabels, France), depending on the number of samples, following the manufacturer's instructions. Extracts were subjected to a commercial qPCR protocol targeting the IS900 DNA fragment of MAP (bactotype® MAP PCR Kit, Indical Bioscience GmbH, Leipzig, Germany) using the following detection systems: CFX96 Touch Real-Time PCR Detection System and CFX Maestro V1-1 Software (Bio-Rad Laboratories, California, USA), Applied Biosystems 7500 Fast Real-Time PCR System and 7500 v2.3 Software (Thermo Fisher Scientific, Waltham, MA, USA), Quantstudio 6 Pro Real-Time Systems and Software 1.3.1 (Thermo Fisher Scientific, Waltham, MA, USA).

2.4. Culture and qPCR of Boot Swab and Pooled Faecal Samples

For analysis of the BS, a pre-treatment was applied: BS samples were homogenised for 1 min using a Laboratory Blender Stomacher 400 (Seward LTD, Worthing, UK) together with 50 mL bi-distilled water. The dissolved faeces were transferred to a 50 mL tube, centrifuged for 15 min at 2938× *g* and the supernatant was discarded. Thereafter, 3 g of pooled faecal sample or the BS manure sediment was mixed together with 30 mL 0.75% HCP solution for 30 min at 150 rpm on a laboratory shaker. Subsequently, gross particles were sedimented for 5 min. A total of 15 mL of the supernatant was transferred to another tube and incubated in the dark for 48 h at room temperature. Then, after centrifugation for 15 min at 2938× *g*, the supernatant was discarded and the sediment was re-suspended with 1 mL HPC solution. Finally, four HEYM tubes (Becton-Dickinson, Heidelberg, Germany), one M7H10 agar slope (Sigma-Aldrich, Vienna, Austria) and one liquid culture medium M7H9C, manufactured in-house as described elsewhere (Australian and New Zealand Standard Diagnostic Procedure, July 2015), respectively, were inoculated with 200 µL of each sample and incubated at 37 °C. Tubes were incubated for 12 weeks and monitored periodically for bacterial growth and fungal contamination. After four weeks of incubation, one HEYM tube of each sample was rinsed with 400 µL bi-distilled water and the liquid obtained was examined by MAP qPCR. DNA was extracted using the IndiMag® Pathogen Kit and the KingFisher™ Flex Purification System and qPCR was carried out using the bactotype® MAP PCR Kit, as described above. After 12 weeks, the remaining HEYM and M7H10 tubes underwent the visual evaluation. Colonies suspected to be MAP were confirmed by PCR, as described elsewhere [26]. The liquid culture samples were evaluated after 12 weeks by MAP qPCR.

2.5. Statistical Analyses

Statistical data analysis was performed using the IBM SPSS Statistics v27 (IBM, New York, NY, USA) software. Potential risk factors associated with herd-level MAP seropositivity were estimated using chi-square test and logistic regression analyses and are listed in Table 2. The level of significance was set at $p < 0.05$. Operating chi-square test, the age and BCS parameters were dichotomized: BCS regular to poor were grades 1–2; BCS good to optimal were grades 3–5. Ages were also merged into two groups (age 1–4 years and 5 years or older). For the analysis of serological screening data, 10 sheep flocks with doubtful ELISA results were excluded from the dataset.

3. Results

3.1. Serological Screening

The investigation revealed an animal-level apparent MAP seroprevalence of 2.0% for goats and 0.7% for sheep (TP 3.5% and 1.2%), respectively. In 71 goat herds and 92 sheep flocks, at least one seropositive animal was detected, leading to a herd-level apparent MAP seroprevalence of 11.1% for goat herds and 8.9% for sheep flocks. In 24 goat herds, two or more samples tested positive (34%), whereas in sheep flocks 13 (14%) flocks had more than one positive sample (Table 1).

From the serological screening (ID Screen® Paratuberculosis Indirect Screening test), 33 sheep and 5 goat samples showed a doubtful result. Out of these samples, 7 samples originating from sheep were confirmed positive, 11 samples remained doubtful (10 sheep and 1 goat sample) and 20 tested negative (16 sheep and 4 goat samples) by the second confirmatory testing (by ID Screen® Paratuberculosis Indirect-Confirmation test). All samples testing positive at the serological screening were confirmed by the second test method.

Table 1. Results of the serological screening for *Mycobacterium avium* subsp. *paratuberculosis* using ID Screen® Paratuberculosis Indirect Screening and Confirmation test (ID.Vet, Grabels, France).

	Animals (*n*)	Seropositive Animals (*n*)	Animal-Level Apparent Seroprevalence (%)	Animal-Level Calculated True Seroprevalence (%)	Herds (*n*)	Seropositive Herds (*n*)	Herd-Level Apparent Seroprevalence (%)
Goat	6434	126	2.0 (95% CI = 1.6–2.3)	3.5 (95% CI = 3.1–4.0)	638	71	11.1 (95% CI = 8.9–13.8)
Sheep	15,585	110	0.7 (95% CI = 0.6–0.9)	1.2 (95% CI = 1.0–1.4)	1032	92	8.9 (95% CI = 7.3–10.8)
Total	22,019	236	1.1 (95% CI = 0.9–1.2)	1.9 (95% CI = 1.7–2.1)	1670	163	9.8 (95% CI = 8.4–11.3)

Logistic regression showed significant association between herd size and MAP seropositivity for goat herds ($p < 0.001$), but not for sheep flocks ($p = 0.895$). For goat herds, seropositivity rate was significantly higher in herds with animal trading; in dairy herds, if herd owners were Animal Health Service members and if farmed game was kept alongside goats in the herd (Table 2). For sheep flocks, seroprevalence was significantly higher in flocks with animal trading and if cattle or goat were kept together in the flock, respectively (Table 2). Depending on the region, seroprevalence ranged from 2.9 to 20.9% for goat herds and from 3.1 to 14.6% for sheep flocks, respectively (two regions with an unrepresentative sample size, each <22 herds, were not included). The prevalence in the region with the highest prevalence in both goat and sheep herds was significantly higher when compared to the prevalence in the other regions together ($p < 0.001$ and $p < 0.004$).

Table 2. Description of data and chi-square test *p*-values of potential risk factors for *Mycobacterium avium* subsp. *paratuberculosis* seropositivity in goat and sheep holdings in Austria.

Variable [a]		Goat Herds Apparent seroprevalence in %, absolute numbers in brackets [b]	*p*-value	Sheep Flocks Apparent seroprevalence in %, absolute numbers in brackets	*p*-value
Animal trading [c]	Yes No	16.8 (25/149) 9.4 (46/489)	0.012 *	11.8 (48/406) 7.1 (44/616)	0.011 *
Dairy herd	Yes No	18.6 (32/172) 8.4 (39/466)	<0.001 *	6.1 (4/66) 9.2 (88/956)	0.388
Grazing on common pastures	Yes No	10.5 (9/86) 11.2 (62/552)	0.833	10.6 (32/302) 8.3 (60/720)	0.249
Grazing on alpine pastures	Yes No	9.9 (11/111) 11.4 (60/527)	0.653	10.7 (34/319) 8.3 (58/703)	0.213
Animal Health Service membership	Yes No	18.9 (25/132) 9.1 (46/506)	0.001 *	8.9 (24/270) 9.0 (68/752)	0.940
Organic farming	Yes No	13.6 (23/169) 10.2 (48/469)	0.232	8.0 (22/274) 9.4 (70/748)	0.511
Cohabitation with other animal species					
Cattle	Yes No	8.7 (23/265) 12.9 (48/373)	0.097	11.8 (34/287) 7.9 (58/735)	0.047 *
Goat	Yes No	-	-	13.0 (40/308) 7.3 (52/714)	0.003 *
Sheep	Yes No	10.1 (28/278) 11.9 (43/360)	0.456	-	-
Farmed game	Yes No	33.3 (7/21) 10.4 (64/617)	0.001 *	0.0 (0/10) 9.1 (92/1012)	0.318
South American camelids	Yes No	20.8 (5/24) 10.7 (66/614)	0.123	10.0 (2/20) 9.0 (90/1002)	0.875

[a] variables were chosen based on their availability in the national animal health information system. [b] (number of seropositive herds/total number of herds). [c] herd entry of one or more animals in the year 2020 for replacement or breeding purposes. * $p < 0.05$ = significant.

3.2. Herd-Level Examination

Detailed investigations in five infected dairy goat herds (432 animals) revealed an overall apparent within-herd MAP seroprevalence of 21.8% (ranging from 11.7 to 28.0%); the calculated overall TP was 38.6%, with an overall rate of MAP shedding of 12.3% (ranging from 5.0 to 24.7%). The results are shown for each herd in Table 3 in detail. In 6 out of the 52 faecal PCR positive animals (11.5%), no seroconversion could be determined by ELISA.

The potential risk factors of body condition score and age were analysed in terms of seropositivity and MAP shedding (data shown in Table 4). Both factors, BCS and age, were not identified as significant risk factors in the logistic regression model nor in the chi-square test (dichotomized variables as described above). However, for animals of three years old, a high frequency of MAP seropositivity (35.5%) and shedding (31.1%) was observed.

It was possible to identify MAP by culture from BS in each of the five investigated MAP positive herds. In total, 15 out of 16 BS samples tested positive (94%) on HEYM agar. Using pooled faecal samples (five animals each), 7 out of 15 samples tested positive (47%) on HEYM agar (Table 3). A minimum of one pooled faecal sample tested positive in every herd. Because of an insufficient sample amount, M7H10 was carried out for 13 BS samples (9 tested positive, 2 were contaminated) and 12 pooled faecal samples (5 resulted positive) only; M7H9C was conducted for five boot swab samples (all five positive) and nine pooled faecal samples (four with positive results; of those, two tested negative on HEYM and M7H10).

Table 3. Herd-level testing for *Mycobacterium avium* subsp. *paratuberculosis* in five dairy goat herds by ELISA on serum samples, individual faecal sample qPCR and culture and qPCR of boot swab and pooled faecal samples.

Herd Number	Herd Size (*n*)	Serum Sample Size (*n*)	Faecal Sample Size (*n*)	ELISA Positive [a]	PCR Positive [a]	Both ELISA and PCR Positive [a]	Boot Swab Samples (HEYM) [d]	Pooled Faecal Samples (HEYM) [d]
1	114	114	112 [c]	20.2 (23)	10.7 (12)	8.9 (10)	3/3	1/3
2	126	126 [b]	122 [c]	24.0 (30)	10.7 (13)	10.7 (13)	4/4	1/3
3	57	57	57	22.8 (13)	10.5 (6)	8.8 (5)	3/3	2/3
4	600	75	73 [c]	28.0 (21)	24.7 (18)	23.3 (17)	2/3	2/3
5	550	60	60	11.7 (7)	5.0 (3)	1.7 (1)	3/3	1/3
Total		432	424	21.8 (94)	12.3 (52)	10.9 (46)	15/16	7/15

Abbreviation: HEYM—Herrold's Egg Yolk agar slants with Mycobactin J. [a] results in %, absolute numbers in brackets. [b] 1 serum sample resulted doubtful. [c] in 8 animals, no faecal sample could be collected. [d] positive samples/number of samples.

Table 4. *Mycobacterium avium* subsp. *paratuberculosis* ELISA and PCR results of dairy goats in different BCS classes and age groups in %, absolute numbers in brackets.

	Number of Animals	ELISA Positive	PCR Positive
Age, years			
1	20.5 (88)	18.2 (16)	5.7 (5)
2	19.6 (84)	13.1 (11)	8.3 (7)
3	14.5 (62)	35.5 (22)	30.6 (19)
4	10.0 (43)	16.3 (7)	18.6 (8)
5 or older	35.4 (152)	25.0 (38)	8.6 (13)
Total	100 (429)	21.9 (94)	12.1 (52)
BCS			
1	0.9 (4)	75.0 (3)	75.0 (3)
2	17.4 (75)	25.3 (19)	12.0 (9)
3	57.4 (248)	20.6 (51)	11.7 (29)
4	23.8 (103)	20.4 (21)	10.7 (11)
5	0.5 (2)	0.0 (0)	0.0 (0)
Total	100 (432)	21.8 (94)	12.3 (52)

4. Discussion

This study gives an overview about the current epidemiological situation of MAP in the Austrian small ruminant population. The investigation revealed a herd-level apparent MAP seroprevalence of 11% for goat herds and 9% for sheep flocks, respectively. The animal-level apparent MAP seroprevalence was 2.0% for goats and 0.7% for sheep. In contrast, a study conducted in Austria in 2003 reported an animal-level apparent prevalence of 0.6% in sheep and 0.0% in goats, concluding that MAP was not widespread in the small ruminant population [27]. In that study, serum and tissue samples from 169 sheep and 80 goats were investigated for MAP by ELISA and culture. One sheep was positive by ELISA and a second sheep showed positive results for MAP in PCR and culture; all the goats tested negative. The considerable difference in MAP prevalence may be partially explicable by differences in study design (the study population in the present study is more representative) and variation in the diagnostic tests used. Nevertheless, the results of our study show, that the MAP prevalence has increased in small ruminants over the last years. In Germany, the herd-level seroprevalence was estimated to be 65% in sheep flocks and 71% in goat herds [8]. According to a global review, MAP herd-level prevalence estimation determined by laboratory testing was higher than 10% in 5 out of 11 countries for sheep and in 7 out of 12 countries for goats, respectively. More than 40% of herds being affected by MAP were reported in two countries for sheep and four countries for goats [1]. However, as the number of prevalence studies carried out among sheep and goats is low and as they differ in study design and diagnostic tests used, MAP prevalence in small ruminants is still unknown in many countries and the prevalence appears to be underestimated [4]. The results of our study indicate a moderate MAP seroprevalence and compared to prevalence reported in other countries, the Austrian MAP prevalence seems to be relatively low.

The large serological screening sample size of the present study allows flocks to be selected randomly, considering a proportional provincial distribution. As the samples originated from the national *Brucella melitensis* sampling scheme, it was not required to take blood samples for this prevalence study, which implied a benefit for animal welfare. However, large flocks, dairy herds and herds with increased animal movement were selected more frequently. As the statistical analysis showed significant association between MAP seropositivity and the risk factors' herd size, dairy herd and animal trading in goat herds, prevalence of caprine paratuberculosis might have been overestimated in this study.

The serological tests used in our study are validated for small ruminant sera. However, since no particular test accuracy data for sheep and goats are available for these tests, Se/Sp indicated for cattle sera had to be applied to calculate TP. The calculated TP therefore should be interpreted cautiously. MAP diagnosis using ELISA is challenging, especially in young stock, since the immune response varies between the different stages of infection [28] and between individuals [29]. Nielsen and Toft [30] reviewed test accuracies of several ELISAs for serological MAP diagnostic and concluded that test Se/Sp values differ between species. It appears that ELISA is more accurate in the detection of infection in goats than in cattle (Se ranging from 0.63 to 0.84). In contrast, Se in sheep appears to be poor (0.16–0.44). Consequently, MAP prevalence in sheep is likely to be underestimated in this study. Furthermore, test accuracy may only be valid for the target population in which a test has been validated [31]. Therefore, more research on validation of serological tests for small ruminant species is needed. The fact that a representative number of samples was investigated per herd (the median sample size per herd was 16 samples for sheep and 8 samples for goats) enabled the identification of infected flocks despite the low test sensitivity, assuming a common within-herd MAP seroprevalence of >10% in infected flocks [1]. However, infected flocks with a very low seroprevalence might have been classified as false negative.

The present study revealed a significant association between herd size and MAP seropositivity for goat herds; the same effect was not seen in sheep flocks. This finding is in contrast to what was described previously, where flock size was found to be associated with increased MAP prevalence in sheep only [8,9]. It might be assumed that association between herd size and seropositivity in goats is related also to the fact that larger goat herds in Austria are also applying an intensive production system. For sheep, the overall moderate MAP seroprevalence in Austrian flocks and the general poor test sensitivity in sheep may circumvent the detection of a potential association. We could demonstrate significant differences in flock-level MAP prevalence between regions for both species too, as described elsewhere [5,7]. Prevalence differences between regions in Austrian goat herds may be associated with an increased percentage of herds applying a more intensive production system in some geographic areas, whereas prevalence differences between regions in sheep could probably be related to a variable regional incidence of cohabitation with cattle and/or goats. Therefore, region might be considered a confounding variable. To answer this question, further studies, with a sampling scheme specifically designed for the detection of MAP, should be performed. Goat herds were significantly more often affected by MAP if they were categorised as flocks with animal trading, as a dairy herd and if herd owners were Animal Health Service members, all factors synonymous with elements of an intensive production system. These results are consistent with current knowledge: contact with other flocks [10] and intensive production systems [5,11] were reported as risk factors associated with an increased MAP seroprevalence. Furthermore, seropositivity rate was significantly higher in the present study in goat herds with cohabitation with farmed game. To the authors' knowledge, a similar correlation has not been described before, though Verdugo et al. [32] found high flock prevalence of 46% in deer herds. For sheep flocks, the seropositivity rate was significantly higher in flocks with animal trading and if cattle or goats were kept together on the farm, respectively. These findings enable the suggestion of a possible cross-species transmission of MAP between cattle, sheep, goat and farmed game in Austria. An important tool to reconstruct infection chains and confirm MAP cross-species transmission is the bacterial strain genotyping of MAP isolates, which should be a consideration for future studies [33]. Cross species infection is suspected to occur between ruminant species [34–38]. Consequently, the infection of small ruminants should also be considered as a possible risk factor in the epidemiology of MAP in cattle herds. Furthermore, there is evidence for interspecies transmission between wild and domestic ruminants [15,39,40]. In an Austrian study, four out of seven MAP genotypes detected in wild ruminants were also present in local cattle and sheep herds [15]. In another study, 15 MAP genotypes across multiple wild and domestic ruminant host species were identified, leading to the conclusion that wild ruminants can act as reservoirs in the transmission of MAP to ruminant livestock [40]. Several risk factors, such as breed, kidding area, management by batches, ventilation and lactation, found to bias MAP prevalence in other studies were not considered in this study because of absent data due to the retrospective study design [5,6,10,11]. Furthermore, widespread MAP infection has been reported in goat flocks seropositive to CAEv (caprine arthritis encephalitis virus) [5]. As control programs for CAE have been implemented in Austria for many years and most herds are free of the infection, no attempts were made to survey the CAE status in participating herds in the course of our study. Both factors of age and BCS were not significant in terms of seropositivity and MAP shedding in goats in the present study, although weight loss was the main clinical sign of paratuberculosis. However, the number of investigated herds and individual animals was not sufficient to gain consolidated findings.

In the present study, it was possible to isolate MAP by culture using BS samples in all five investigated herds (15/16 samples tested positive). Using pooled faecal samples, the detection rate was comparatively low (9/15), but all herds were detected as MAP positive by a minimum of one pooled faecal sample. According to Eamens et al. [41], pooled faecal culture is suitable for herd diagnosis if the animals are moderate or high shedders of MAP but not in herds with a few low shedders only. As two samples were positive on M7H9C, but not on HEYM in this study, it should also be considered for future investigations to use both media types for caprine samples, as goats can be infected by both MAP-S strains (sheep-type) and MAP-C strains (cattle-type) [42]. Environmental sampling is mainly used in cattle holdings as an effective and cost-efficient surveillance method [43]. Combined culture and PCR of BS samples has been applied successfully for large-scale herd-level testing in a few control programs for dairy cattle herds [13,44]. The results of the present study indicate that BS sampling may be suitable to detect MAP infection in goat herds. However, further studies are needed to validate the sensitivity of this diagnostic tool in goat holdings.

To prevent a future spread of MAP among herds and between species, a nationwide control programme should be established for sheep and goat holdings, including herd-level monitoring (e.g., by repeated environmental sample testing) and on-farm control and biosecurity measures. Some regions in Germany and Austria are currently applying such an approach in cattle [18]. Moreover, the purchase or movement of livestock should rely on herd status certification combined with an individual animal test result in order to reduce the risk of between-herd transmission of MAP by infected livestock. Goat production in Austria is a growing sector (since the year 2011 the number of goats kept has increased by about 28%), with the trend for larger flocks and more intensive production systems. Furthermore, according to experimental infection, goats have been found to be more susceptible to paratuberculosis than cattle and sheep [45]. As the prevalence was quite high in Austrian dairy goat herds, it could be judicious to focus on caprine paratuberculosis control in this production group. Windsor [2] mentioned three main approaches to control ovine and caprine paratuberculosis: management changes to decrease transmission, test and cull strategies and vaccination. Few countries are also using vaccination for disease control [1] and a meta-analysis showed vaccination reduces MAP shedding, production losses and pathology [46]. On the other hand, MAP vaccination can interfere with testing for bovine tuberculosis (TB). Austria is officially free of bovine TB, but a wildlife reservoir for *Mycobacterium caprae* in red deer exists in some geographic areas in western Austria and serves as source of infections for domestic cattle grazing on Alpine pastures during the summer [47]. In addition, vaccination may interfere with MAP control based on test-and-cull programmes using immunological tests, as ELISA is not able to differentiate between antibodies produced by infection and by vaccination [12]. Therefore, vaccination against paratuberculosis is prohibited for all ruminant species in Austria.

5. Conclusions

Mycobacterium avium subsp. *paratuberculosis* infection is present in Austrian sheep and goat holdings, with a herd-level apparent seroprevalence of 11.1% for goat herds and 8.9% for sheep flocks. Identified risk factors for MAP seropositivity in goat herds were herd size, animal trading, dairy farms, Animal Health Service membership and cohabitation of farmed game; for sheep flocks, identified risk factors were animal trading and cohabitation with cattle or goat, respectively. Furthermore, prevalence differences were observed between different regions for both species. At the herd level, average apparent seroprevalence was 21.8% (calculated true prevalence 38.6%) and MAP shedding was 12.3% in dairy goats. It was possible to identify MAP by culture from boot swab samples in each of five MAP infected herds.

It can be concluded, that the seroprevalence of MAP in the Austrian small ruminant population is moderate and below reported prevalence in other countries. Goat herds with an intensive production system were identified to have an increased risk to be affected by MAP; moreover, interspecies transmission seemed to occur. The use of BS sampling to identify MAP positive goat flocks may be promising but needs further investigation. Based on these results, a close monitoring of the MAP prevalence in Austrian small ruminants as well as a national control program, especially for dairy goats, seems to be advisable.

Author Contributions: Conceptualization, J.S, M.D. and J.L.K.; Methodology, J.S., E.S., M.D., A.T. and J.L.K.; Investigation, J.S., E.S. and J.L.K.; Resources, M.D. and J.L.K.; Data Curation, J.S.; Writing—Original Draft Preparation, J.S. and J.L.K.; Writing—Review and Editing, J.S., E.S., M.D., A.T. and J.L.K.; Visualization, J.S. and A.T.; Supervision, M.D. and J.L.K. All authors have read and agreed to the published version of the manuscript.

Funding: This study was supported and funded by the Austrian Agency for Health and Food Safety (AGES) and the University Clinic for Ruminants, University of Veterinary Medicine Vienna.

Institutional Review Board Statement: This study was discussed and approved by the institutional ethics and animal welfare committee in accordance with GSP guidelines and national legislation.

Data Availability Statement: The data presented in this study are available in the article.

Acknowledgments: Open Access Funding by the University of Veterinary Medicine Vienna.

Conflicts of Interest: The authors declare no conflict of interest.

References

1. Whittington, R.; Donat, K.; Weber, M.F.; Kelton, D.; Nielsen, S.S.; Eisenberg, S.; Arrigoni, N.; Juste, R.; Sáez, J.L.; Dhand, N.; et al. Control of paratuberculosis: Who, why and how. A review of 48 countries. *BMC Vet. Res.* **2019**, *15*, 198. [CrossRef] [PubMed]
2. Windsor, P.A. Paratuberculosis in sheep and goats. *Vet. Microbiol.* **2015**, *181*, 161–169. [CrossRef] [PubMed]
3. Sardaro, R.; Pieragostini, E.; Rubino, G.; Petazzi, F. Impact of *Mycobacterium avium* subspecies paratuberculosis on profit efficiency in semi-extensive dairy sheep and goat farms of Apulia, southern Italy. *Prev. Vet. Med.* **2017**, *136*, 56–64. [CrossRef] [PubMed]
4. Nielsen, S.S.; Toft, N. A review of prevalences of paratuberculosis in farmed animals in Europe. *Prev. Vet. Med.* **2009**, *88*, 1–14. [CrossRef]
5. Barrero-Domínguez, B.; Luque, I.; Huerta, B.; Gomez-Laguna, J.; Galán-Relaño, Á.; Gómez-Gascón, L.; Sánchez, M.; Astorga, R.J. Paratuberculosis in dairy goat flocks from southern Spain: Risk factors associated with seroprevalence. *Vet. Rec.* **2019**, *185*, 600. [CrossRef]
6. Begg, D.J.; Purdie, A.C.; de Silva, K.; Dhand, N.K.; Plain, K.M.; Whittington, R.J. Variation in susceptibility of different breeds of sheep to Mycobacterium avium subspecies paratuberculosis following experimental inoculation. *Vet. Res.* **2017**, *48*, 36. [CrossRef]

7. ESG Sergeant. Ovine Johne's disease in Australia—The first 20 years. *Aust. Vet. J.* **2001**, *79*, 484–491. [CrossRef]
8. Stau, A.; Seelig, B.; Walter, D.; Schroeder, C.; Ganter, M. Seroprevalence of Mycobacterium avium subsp. paratuberculosis in small ruminants in Germany. *Small Rumin. Res.* **2012**, *105*, 361–365. [CrossRef]
9. Morales-Pablos, M.I.; Mejía-Sánchez, P.; Díaz-Aparicio, E.; Palomares-Resendiz, E.G.; Gutiérrez-Hernández, J.L.; Reyna-Granados, J.R.; Luna-Nevárez, P.; Munguía-Xóchihua, J.A.; Segura-Correa, J.C.; Leyva-Corona, J.C. Risk factors associated with the seroprevalence of paratuberculosis in sheep flocks in the hot-arid region of Sonora, México. *Trop. Anim. Health Prod.* **2020**, *52*, 1357–1363. [CrossRef]
10. Angelidou, E.; Kostoulas, P.; Leontides, L. Flock-level factors associated with the risk of Mycobacterium avium subsp. paratuberculosis (MAP) infection in Greek dairy goat flocks. *Prev. Vet. Med.* **2014**, *117*, 233–241. [CrossRef]
11. Freitas, T.D.; de Azevedo, S.S.; Silva, M.L.C.R.; Júnior, F.G.; Santos, C.d.S.A.B.; Clementino, I.J.; Amaral, F.R.-C.; Alves, C.J. Epidemiological characterization and risk factors associated with Mycobacterium avium subsp. paratuberculosis infection in dairy goats in the Brazilian semiarid region. *Semin. Ciências Agrárias* **2015**, *36*, 267–275. [CrossRef]
12. OIE. Chapter 3.1.15. Paratuberculosis. In *Manual of Diagnostic Tests and Vaccines for Terrestrial Animals*; OIE: Paris, France, 2021.
13. Donat, K.; Hahn, N.; Eisenberg, T.; Schlez, K.; KÖHler, H.; Wolter, W.; Rohde, M.; Pützschel, R.; RöSler, U.; Failing, K.; et al. Within-herd prevalence thresholds for the detection of *Mycobacterium avium* subspecies paratuberculosis-positive dairy herds using boot swabs and liquid manure samples. *Epidemiol. Infect.* **2016**, *144*, 413–424. [CrossRef]
14. Sodoma, E.; Mitterhuemer, S.; Altmann, M.; Mader, C.; Kössler, J.; Ortner, P.; Vill, M.; Duenser, M. Results and experiences of laboratory diagnostic analysis within the voluntary paratuberculosis control program in Tyrol. *Berl. Und Munch. Tierarztl. Wochenschr.* **2021**, *134*. [CrossRef]
15. Sodoma, E.; Altmann, M.; Mitterhuemer, S.; Moebius, P.; Duenser, M. First comprehensive study on molecular diversity of Austrian *Mycobacterium avium* subspecies paratuberculosis isolates from domestic and wild ruminants. *Berl. Und Munch. Tierarztl. Wochenschr.* **2018**, *131*, 2–11.
16. Dreier, S.; Khol, J.L.; Stein, B.; Fuchs, K.; Gütler, S.; Baumgartner, W. Serological, Bacteriological and Molecularbiological Survey of Paratuberculosis (Johne's Disease) in Austrian Cattle. *J. Vet. Med. Ser. B* **2006**, *53*, 477–481. [CrossRef]
17. Baumgartner, W.; Damoser, J.; Khol, J. Comparison of two studies concerning the prevalence of bovine paratuberculosis (Johne's Disease) in Austrian cattle in the years 1995-97 and 2002/03. *Wien. Tierarztl. Mon.* **2005**, *92*, 274–277.
18. Khol, J.L.; Eisenberg, S.; Noll, I.; Zschöck, M.; Eisenberg, T.; Donat, K. Two-stage control of paratuberculosis: Herd-status surveillance as the basis for operational measures to reduce the prevalence. Experiences from Lower Saxony, Hesse, Thuringia and Tyrol [Article in German]. *Tierarztl Prax Ausg G Grosstiere Nutztiere* **2019**, *47*, 171–183. [CrossRef]
19. Khol, J.L.; Damoser, J.; Dünser, M.; Baumgartner, W. Paratuberculosis, a notifiable disease in Austria—Current status, compulsory measures and first experiences. *Prev. Vet. Med.* **2007**, *82*, 302–307. [CrossRef]
20. Statistics Austria. Sheep and Goat Livestock. 1 December 2020. Available online: https://www.statistik.at/web_en/statistics/Economy/agriculture_and_forestry/livestock_animal_production/livestock/index.html (accessed on 21 December 2021).
21. AgrarMarktAustria. Sheep and Goat Grazing on Alpine Pastures 2020. Available online: https://www.ama.at/ (accessed on 9 February 2022).
22. *Regulation(EU)2020/689*; Commission Delegated Regulation (EU) 2020/689 of 17 December 2019 Supplementing Regulation (EU) 2016/429 of the European Parliament and of the Council as Regards Rules for Surveillance, Eradication Programmes, and Disease-Free Status for Certain Listed and Emerging Diseases. European Parliament: Strasbourg, France, 2019.
23. Baumgartner, W.; Wittek, T. *Propaedeutics of Domestic Animals*; Enke: Stuttgart, Germany, 2017; Volume 9.
24. Eisenberg, T.; Wolter, W.; Lenz, M.; Schlez, K.; Zschöck, M. Boot swabs to collect environmental samples from common locations in dairy herds for Mycobacterium avium ssp. paratuberculosis (MAP) detection. *J. Dairy Res.* **2013**, *80*, 485–489. [CrossRef]
25. Hervieu, J.; Morand-Fehr, P.; Schmidely, P.; Fedele, V.; Delfa, R. Measures anatomiques permettant d'expliquer les variations des notes sternales, lombaires et caudales utilisées pour estimer l'état corporel des chèvres laitières. *Options Méditerranéennes* **1991**, *13*, 43–56.
26. Englund, S.; Ballagi-Pordány, A.; Bölske, G.; Johansson, K.-E. Single PCR and nested PCR with a mimic molecule for detection of Mycobacterium avium subsp. paratuberculosis. *Diagn. Microbiol. Infect. Dis.* **1999**, *33*, 163–171. [CrossRef] [PubMed]
27. Khol, J.; Stein, B.; Dreier, S.; Baumgartner, W. Paratuberculosis (JOHNE's disease) in small ruminants in Austria. *Slov. Vet. Res.* **2006**, *43*, 129–130.
28. Gumber, S.; Eamens, G.; Whittington, R.J. Evaluation of a Pourquier ELISA kit in relation to agar gel immunodiffusion (AGID) test for assessment of the humoral immune response in sheep and goats with and without *Mycobacterium paratuberculosis* infection. *Vet. Microbiol.* **2006**, *115*, 91–101. [CrossRef] [PubMed]
29. Köhler, H.; Soschinka, A.; Meyer, M.; Kather, A.; Reinhold, P.; Liebler-Tenorio, E. Characterization of a caprine model for the subclinical initial phase of Mycobacterium avium subsp. paratuberculosis infection. *BMC Vet. Res.* **2015**, *11*, 74. [CrossRef] [PubMed]
30. Nielsen, S.S.; Toft, N. Ante mortem diagnosis of paratuberculosis: A review of accuracies of ELISA, interferon-γ assay and faecal culture techniques. *Vet. Microbiol.* **2008**, *129*, 217–235. [CrossRef]
31. Greiner, M.; Gardner, I.A. Epidemiologic issues in the validation of veterinary diagnostic tests. *Prev. Vet. Med.* **2000**, *45*, 3–22. [CrossRef]

32. Verdugo, C.; Jones, G.; Johnson, W.; Wilson, P.; Stringer, L.; Heuer, C. Estimation of flock/herd-level true Mycobacterium avium subspecies paratuberculosis prevalence on sheep, beef cattle and deer farms in New Zealand using a novel Bayesian model. *Prev. Vet. Med.* **2014**, *117*, 447–455. [CrossRef]
33. Kao, R.R.; Haydon, D.T.; Lycett, S.J.; Murcia, P.R. Supersize me: How whole-genome sequencing and big data are transforming epidemiology. *Trends Microbiol.* **2014**, *22*, 282–291. [CrossRef]
34. Holstad, G.; Sigurðardóttir, Ó.G.; Storset, A.K.; Tharaldsen, J.; Nyberg, O.; Schönheit, J.; Djønne, B. Description of the Infection Status in a Norwegian Cattle Herd Naturally Infected by Mycobacterium avium subsp. paratuberculosis. *Acta Vet. Scand.* **2005**, *46*, 45. [CrossRef]
35. Fridriksdottir, V.; Gunnarsson, E.; Sigurdarson, S.; Gudmundsdottir, K.B. Paratuberculosis in Iceland: Epidemiology and control measures, past and present. *Vet. Microbiol.* **2000**, *77*, 263–267. [CrossRef]
36. Muskens, J.; Bakker, D.; Boer, J.d.; Keulen, L.v. Paratuberculosis in sheep: Its possible role in the epidemiology of paratuberculosis in cattle. *Vet. Microbiol.* **2001**, *78*, 101–109. [CrossRef] [PubMed]
37. Verdugo, C.; Pleydell, E.; Price-Carter, M.; Prattley, D.; Collins, D.; de Lisle, G.; Vogue, H.; Wilson, P.; Heuer, C. Molecular epidemiology of Mycobacterium avium subsp. paratuberculosis isolated from sheep, cattle and deer on New Zealand pastoral farms. *Prev. Vet. Med.* **2014**, *117*, 436–446. [CrossRef] [PubMed]
38. Djønne, B.; Pavlik, I.; Svastova, P.; Bartos, M.; Holstad, G. IS900 Restriction Fragment Length Polymorphism (RFLP) Analysis of Mycobacterium avium subsp. paratuberculosisIsolates from Goats and Cattle in Norway. *Acta Vet. Scand.* **2005**, *46*, 13. [CrossRef] [PubMed]
39. Pavlik, I.; Bartl, J.; Dvorska, L.; Svastova, P.; du Maine, R.; Machackova, M.; Yayo Ayele, W.; Horvathova, A. Epidemiology of paratuberculosis in wild ruminants studied by restriction fragment length polymorphism in the Czech Republic during the period 1995–1998. *Vet. Microbiol.* **2000**, *77*, 231–251. [CrossRef]
40. Gerritsmann, H.; Stalder, G.L.; Spergser, J.; Hoelzl, F.; Deutz, A.; Kuebber-Heiss, A.; Walzer, C.; Smith, S. Multiple strain infections and high genotypic diversity among Mycobacterium avium subsp. paratuberculosis field isolates from diseased wild and domestic ruminant species in the eastern Alpine region of Austria. *Infect. Genet. Evol.* **2014**, *21*, 244–251. [CrossRef]
41. Eamens, G.; Walker, D.; Porter, N.; Fell, S. Pooled faecal culture for the detection of Mycobacterium avium subsp paratuberculosis in goats. *Aust. Vet. J.* **2007**, *85*, 243–251. [CrossRef]
42. de Juan, L.; Mateos, A.; Domínguez, L.; Sharp, J.M.; Stevenson, K. Genetic diversity of Mycobacterium avium subspecies paratuberculosis isolates from goats detected by pulsed-field gel electrophoresis. *Vet. Microbiol.* **2005**, *106*, 249–257. [CrossRef]
43. Sergeant, E.S.G.; McAloon, C.G.; Tratalos, J.A.; Citer, L.R.; Graham, D.A.; More, S.J. Evaluation of national surveillance methods for detection of Irish dairy herds infected with Mycobacterium avium ssp. paratuberculosis. *J. Dairy Sci.* **2019**, *102*, 2525–2538. [CrossRef]
44. Sodoma, E. Ergebnisse und Erfahrungen aus den labordiagnostischen Analysen im freiwilligen Paratuberkulose-Bekämpfungsprogramm in Tirol. *Berl. Munch. Tierarztl. Wochenschr.* **2021**, *134*, 1–10. [CrossRef]
45. Stewart, D.J.; Vaughan, J.A.; Stiles, P.L.; Noske, P.J.; Tizard, M.L.V.; Prowse, S.J.; Michalski, W.P.; Butler, K.L.; Jones, S.L. A long-term bacteriological and immunological study in Holstein-Friesian cattle experimentally infected with Mycobacterium avium subsp. paratuberculosis and necropsy culture results for Holstein-Friesian cattle, Merino sheep and Angora goats. *Vet. Microbiol.* **2007**, *122*, 83–96. [CrossRef]
46. Bastida, F.; Juste, R.A. Paratuberculosis control: A review with a focus on vaccination. *J. Immune Based Ther. Vaccines* **2011**, *9*, 8. [CrossRef] [PubMed]
47. Fink, M.; Schleicher, C.; Gonano, M.; Prodinger, W.M.; Pacciarini, M.; Glawischnig, W.; Ryser-Degiorgis, M.-P.; Walzer, C.; Stalder, G.L.; Lombardo, D.; et al. Red deer as maintenance host for bovine tuberculosis, Alpine region. *Emerg. Infect. Dis.* **2015**, *21*, 464–467. [CrossRef] [PubMed]

animals

MDPI

Article

Molecular Diversity of *Mycobacterium avium* subsp. *paratuberculosis* in Four Dairy Goat Herds from Thuringia (Germany)

Chris Pickrodt [1], Heike Köhler [1], Udo Moog [2], Elisabeth M. Liebler-Tenorio [1] and Petra Möbius [1,*]

[1] Institute of Molecular Pathogenesis, Friedrich-Loeffler-Institut, Federal Research Institute for Animal Health, Naumburger Straße 96a, 07743 Jena, Germany; chris.pickrodt@fli.de (C.P.); elisabeth.liebler-tenorio@fli.de (E.M.L.-T.)

[2] Thuringian Animal Diseases Fund, Sheep and Goat Health Service, Victor-Goerttler-Straße 4, 07745 Jena, Germany; umoog@thtsk.de

* Correspondence: petra.moebius@fli.de; Tel.: +49-3641-8042280

Simple Summary: Paratuberculosis (Johne's disease) is a chronic granulomatous enteritis that affects domestic and wild ruminants worldwide, mainly dairy and beef cattle, but also small ruminants such as goats and sheep. To gain a better understanding of the disease epidemiology and to develop effective control strategies, it is useful to determine the strain diversity of the causative agent, *Mycobacterium avium* subsp. *paratuberculosis* (MAP). The aim of this study was to determine the diversity of MAP isolated from four goat herds affected by paratuberculosis in Thuringia (Germany), as well as the detailed distribution of genotypes among the animals and their environment in one herd (herd 1). A combination of three methods was used to genotype isolates from fecal samples of infected animals, from various intestinal and other tissues of clinically affected animals, and from environmental samples. The six MAP-C genotypes identified could be assigned to five different phylogenetic subgroups. The results suggest individual infection strains within each herd. In herd 1, one predominant strain was found, and two strains occurred sporadically. The genotypes detected were not goat specific. Introduction of MAP into the herd through uncontrolled animal trade or environmental exposure to MAP in barns that were previously used by cattle with unknown paratuberculosis status are suspected as possible sources of infection.

Abstract: This study investigated the intra- and inter-herd diversity of *Mycobacterium avium* subsp. *paratuberculosis* (MAP) isolates from four goat herds in Thuringia (Germany) that were affected by paratuberculosis for several years. The main focus was on the characterization and distribution of genotypes among animals and the environment of goat herd 1. This study included 196 isolates from the feces of 121 infected goats, various tissues from 13 clinically diseased goats, 29 environmental samples from herd 1, and additionally, 22 isolates of different origin from herds 2 to 4. The isolates, sampled between 2018 and 2022, were genotyped using short-sequence-repeat (SSR) analysis, mycobacterial-interspersed repetitive units–variable-number tandem repeat (MIRU–VNTR) analysis, and a single nucleotide polymorphism (SNP)-based assay for phylogenetic grouping. All the isolates belonged to the MAP-C group. In herd 1, one predominant genotype was determined, while two other genotypes were identified very rarely and only in fecal and environmental samples. One of three further genotypes was found in each of herds 2 to 4. The assignment of genotypes to different phylogenetic clades suggested six different infection strains. The results indicated no epidemiological links between the examined herds. Based on the current MAP genotyping data from Germany, possible sources of infection are MAP-contaminated barns previously used by infected cattle and the purchase of sub-clinically infected goats.

Keywords: MAP; SNP-based assay; MIRU–VNTR; SSR analysis; MAP-C genotypes; fecal; tissue; environmental samples; transmission; small ruminants

1. Introduction

Mycobacterium avium subsp. *paratuberculosis* (MAP) is the causative agent of paratuberculosis (Ptb) or Johne's Disease, an incurable chronic enteritis of domestic and wild ruminants as well as camelids and some other non-ruminant species with less clear symptoms [1–4]. The disease is characterized by therapy-resistant diarrhea, progressive emaciation, and a decrease in milk production, finally leading to death [5]. The clinical symptoms appear after an incubation period of up to several years in the final stage of the disease [6].

In small ruminants, softening of feces is observed in individual animals, but severe watery diarrhea, typical for cattle, occurs very rarely [1]. There is evidence that goats are more susceptible to MAP than sheep [7] and have a shorter clinically unapparent interval, faster disease progression compared to sheep and cattle [7,8], and often show bacterial shedding before the peripheral immune response is measurable [9].

During the subclinical stage of Ptb, the pathogen may be shed intermittently in low to high concentrations in the feces [10], thus contaminating the immediate environment of animals in the barn [11,12] and on pasture [13], or at feeding sites of wild animals [14]. MAP has a high tenacity in manure, soil, grass, and water [15,16]. Animals become infected mainly in the first weeks of their life via the fecal–oral route upon ingestion of contaminated milk and other feed, water, or oral contact with contaminated surfaces [5,17,18]. However, adult animals can also be infected with MAP, as described for cattle and sheep [18–21]. There is evidence that some individual animals are inherently susceptible to MAP infection, regardless of age [21].

In Germany, in addition to many cattle herds [22], goat herds are also affected by paratuberculosis [23–25], as has been described for other countries worldwide [3,26]. Over the last few decades, the global dairy goat population has steadily increased, as has global goat milk production, with regional variations [27]. Goats are traditionally managed differently than cattle, and they graze in expansive enclosures or not enclosed at all [28]. The species (goat) and the absence of perimeter fencing for livestock are considered the main risk for MAP exposure in specific regions [29]. In countries with increased goat milk production, dairy goat farming systems are still mainly managed semi-extensively or semi-intensively, but these conditions provide ample opportunities for the transmission of MAP [1]. Several management practices (poor kidding/birthing hygiene, administration of bulk colostrum or milk, deficits in rearing young animals, inappropriate ventilation) and co-infections with other pathogens have been proposed to facilitate the spread of MAP or MAP infection [26,30,31].

Numerous studies have characterized MAP isolates from different host species and regions of the world to reveal possible transmission routes, persistence of infection, the diversity and evolution of MAP, and to find differences in pathogenicity. A variety of single or combined molecular genotyping methods have been used for this purpose, summarized in [32–34], including analyses based on whole genome sequencing (WGS), which provides the highest resolution and can reliably identify phylogenetic relationships [35–37]. Depending on the applied method, previously described or novel distinct genotypes have been identified, or isolates have been assigned to defined or novel phylogenetic groups and subgroups by comparing WGS-based single nucleotide polymorphisms (SNPs) [37,38]. Recently, Fawzy et al. [39] recommended a hierarchical genotyping approach including a combination of mycobacterial interspersed repetitive unit variable number tandem repeat (MIRU-VNTR) analysis [40] and an SNP-based assay [38], resulting in fine discrimination of MAP from animals within limited geographical regions.

MAP genomes have been divided into two main groups with different designations in the literature: C-type/Type II (MAP-C, cattle type), with the main subgroups Type C and Type B (bison type); and S-type (MAP-S, sheep type), with the main subgroups Type I and III [33,36,38]. Goats are susceptible to both C-type and S-type strains [1,36, 41,42]. In previous studies, goat isolates originating from different herds in Spain or from individual goats in some European countries showed a high diversity [43–45] and belonged to various phylogenetic subgroups, together with cattle or sheep isolates [36].

However, most previous studies investigating the intra-herd or inter-herd diversity of MAP or discussing the possible transmission routes of the disease have focused on cattle, sheep, or deer, but not on goats [39,46–53]. Currently, no genotyping data of goat isolates are available from Germany.

In this study, various MAP isolates from goats in Thuringia (Germany) were characterized using the hierarchical genotyping method recommended above, as well as short sequence repeat (SSR) analysis. The aims of this study were: (1) to investigate the intra-herd diversity of MAP in one goat herd, including the immediate environment in the housing, and to determine which strains caused clinical disease in some of the goats; (2) to reveal inter-herd diversity in paratuberculosis-affected dairy goat herds in Thuriingia and provide evidence for epidemiological associations between some of these herds; and (3) to compare the genotypes identified, including their assignment to phylogenetic groups with genotypes previously determined in other hosts or regions.

2. Materials and Methods

2.1. Study Herds and Animals

Four dairy goat herds, situated in different counties of Thuringia, a federal state of Germany (Figure 1), were included in this study. Herds 1 and 2 kept goats of the breed "Thüringer Wald Ziege", while herds 3 and 4 mainly consisted of the breed "Saanenziege" (Saanen goat). These herds were monitored by the Sheep and Goat Health Service of the Thuringian Animal Diseases Fund.

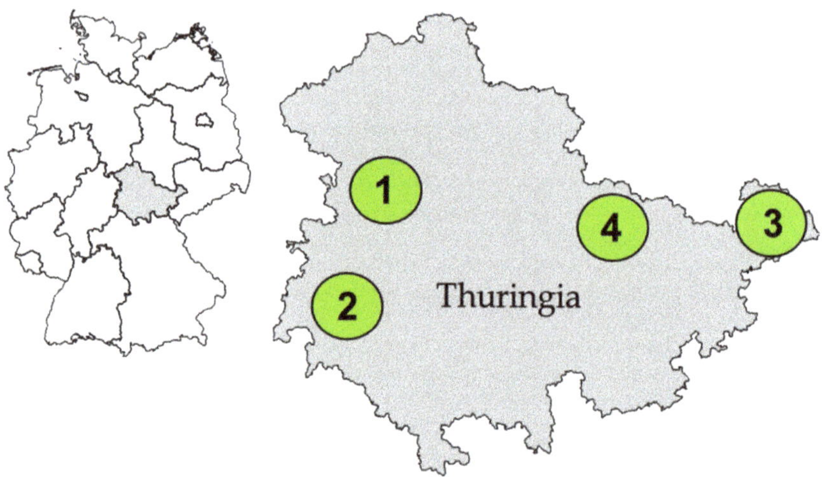

Figure 1. Localizations (1–4) of the examined four goat herds (herd 1 to 4) within the federal state Thuringia in Germany. The maps were generated with © GeoBasis-DE/Bundesamt für Kartographie und Geodäsie (2023).

Most of the isolates analyzed in this study originated from herd 1. This herd was an organic dairy herd with approximately 400 lactating goats. Herd 1 was established as early as 2002, with a mix of goats, sheep, and cattle kept in a former cattle barn from a herd with an unknown Ptb status (Barn A). Later, the focus changed to solely dairy goat husbandry with their own reproduction. Over the years, several goats from different breeders with unknown Ptb statuses were introduced into the herd to increase the herd size and genetic pool. The last time adult goats (female, n = 35) were purchased from a herd with an unknown Ptb status was in 2011. The purchased goat bucks always originated from Ptb non-suspect herds. Between 2014 and 2015, all the goats and kids were moved to another old cattle barn (Barn B) with an unknown Ptb history before the herd was moved to its current housing, which was also a former dairy cattle barn (Barn C), in 2016.

During the data acquisition for this study, it was subsequently revealed that in an initially anonymous screening study in 2011, four out of five of the environmental samples from Barn C had been tested positive for MAP based on their cultures. The corresponding cattle herd was removed in 2012, but eight of the dairy cattle remained on the pasture of the farm until 2016. Barn C was not disinfected after destocking, and the liquid manure remained under the slatted floor in the manure channel. The goat bucks and kids from herd 1 were introduced to this environment in June 2016. In August 2016, the cattle manure was removed, the manure channel was cleaned, and the barn was disinfected, but not using a mycobacterial disinfectant. In early December 2016, the lactating goats were moved into Barn C. Beginning in August 2016, herd 1 had problems with pseudotuberculosis. The last purchase of kids (n = 28) was in 2021 from a Ptb non-suspect dairy goat herd.

MAP was first detected culturally in herd 1 in January 2018 in the tissues of seven clinically affected goats after necropsy. A paratuberculosis control program was implemented shortly after this diagnosis, including vaccination of the entire herd using an inactivated vaccine (Gudair, CZ Vaccines, O Porriño, Spain). In February 2018, MAP was found in 89 of 333 fecal samples from adult female goats based on their cultures, as described further below. In addition to the isolates obtained from the sampling in 2018, fecal isolates recovered from biannual herd samplings in 2020, 2021, and 2022 were also included in this study. Furthermore, the clinically diseased goats from herd 1 were sampled after necropsy in 2018, 2021, and 2022, and MAP was isolated from different tissues. The age of the goats sampled ranged from <1 to 11 years. Most of the goats were two to four years old when MAP was successfully isolated and genotyped. The MAP isolates originated from the goats that were born between 2014 and 2020, with 2014 and 2015 as the focus. Furthermore, environmental samples from various locations in Barn C were sampled in 2020, 2021, and 2022 and cultivated, and if MAP isolation was successful, these isolates were characterized the same way as the isolates from the feces and tissues.

Herd 2 was established in 2016 through purchases from herds of three different breeders. This included 35 female kids (12 weeks old) from herd 1 with an unknown Ptb status at that time, which were born in February and March 2016 and exclusively housed in Barn B. The purchased adult goat bucks originated from two Ptb non-suspect herds. The goats were kept in a newly built barn. However, in previous years, cattle with unknown MAP statuses grazed on some pastures of this farm. MAP was first detected in November 2018 using the fecal culture of samples from several clinically non-suspect goats. Fecal samples from regular whole herd samplings in 2019 and 2022 were examined in this study.

Herd 3 was established in the early 1990s, first in a former sheep barn and then in a renovated dairy cattle barn. The young goats originated from different small goat and sheep herds with unknown Ptb statuses from Thuringia in Germany. In addition, approximately 600 goats were purchased from different breeders in Switzerland (1997) and in France (2014) with unknown Ptb statuses. In 2004, MAP was culturally confirmed for the first time in tissues from a clinically diseased female goat after necropsy.

Herd 4 was established in 1998. The herd was kept in two parts on separate locations. The goats originated from different breeders in The Netherlands, both without consideration of their Ptb status. MAP was first detected in May 2017 by fecal culture of samples from several clinically suspect goats.

MAP isolates from fecal samples collected in 2017 from herds 3 and 4 as part of another research project were included in the study. Furthermore, bedding samples collected in 2017 and fecal bulk samples collected in 2021 and 2022 for monitoring purposes were analyzed. In addition, isolates from tissues originating from a clinically affected goat of herd 4 necropsied in 2020 were examined.

2.2. *Sample Collection and Cultivation of MAP*

2.2.1. Fecal Samples

Fecal samples from herd 1 were collected at nine sampling events between 2018 and 2022. Each sample was collected from the rectum of the goat using a fresh glove, transferred to the laboratory, and stored at 5 °C when processed within the next days or at −20 °C if this was not feasible. Fecal samples from herds 2 to 4 were collected between 2017 and 2022 in the same way.

Bacterial culture was performed on Herrold's Egg Yolk Agar (HEYM, Becton Dickinson, Sparks, MD, USA) supplemented with Mycobactin J and ANV (Amphotericin, Nalidixic acid and Vancomycin) for all fecal samples according to the official diagnostic manual of diagnostic procedures published by the Friedrich-Loeffler-Institut [54], or as described previously [55], but with some modifications. A total of 3 g of feces was decontaminated over 48 h at room temperature (RT) using 30 mL of 0.75% hexadecyl pyridinium chloride (HPC) solution (Sigma Aldrich, Taufkirchen, Germany). After discarding the supernatants, 200 µL of the remainder was transferred to each of the three slopes of the growing medium. The slopes were incubated at 37 °C for up to six months and examined every second week. Visible MAP colonies appeared after 4 to 6 weeks at the earliest. The presence of MAP in characteristic colonies was confirmed using subspecies-specific PCR after DNA isolation from the colony material, as described in Section 2.3.

The culture-positive MAP isolates from the feces used in this study of herd 1 (n = 133) are listed in Table S1, and of herd 2 (n = 7), herd 3 (n = 3), and herd 4 (n = 4) in Table S2. At least one isolate from each animal in herd 1 with successful MAP cultivation during the study period was genotyped.

2.2.2. Tissue Samples after Necropsy

Almost all of the tissue samples originated from animals from herd 1, with the exception of two tissue samples from a goat from herd 4. Thirteen female goats were removed from herd 1 due to clinical symptoms of Ptb, e.g., severe progressive weight loss, intermittent diarrhea, and shedding of MAP. They were euthanized for scientific purposes according to §4, sentence 3 of the German Animal Welfare Act, and they were necropsied for macroscopic examination and sampling for histology and microbiology.

Tissue samples were collected from four representative sites of the jejunum and from the ileum, colon, mesenteric lymph nodes (Lymphonodi [LNN] mesenteriales, LNN ileocolici), colonic lymph nodes (LNN colici), liver, hepatic lymph nodes (LNN hepatici), mammary gland, supramammary lymph nodes (LNN inguinales superficiales), and other parenchyma (spleen, kidney, lung, heart). In all of these goats, lesions were found in the small intestine and mesenteric lymph nodes. Macroscopic inspection revealed thickened intestinal walls and nodular mucosa affecting at least two-thirds of the jejunum. This was confirmed by histology as moderate to severe granulomatous enteritis with few to no acid-fast bacteria (AFB) in five of the goats (paucibacillary) and many AFB in eight of the goats (multibacillary). Multiple calcified granulomas and/or granulomatous infiltrates were present in the mesenteric lymph nodes of all the goats.

The fat and connecting tissue were removed from the lymph nodes and other tissue. The intestine was opened and the ingesta was removed. One gram of sample was cut from different locations, minced with scissors, and transferred into a plastic bag containing 7 mL of 0.9% HPC solution. The samples were homogenized in a stomacher for 6 min, transferred to a 50 mL tube, and agitated on a shaker at 200 rpm for 10 min at RT. Afterwards, they were incubated in an upright position for 24 h at RT in the dark. After centrifugation at 1880× g for 20 min at RT, the supernatants were discarded, and the pellet was re-suspended in 1 mL of sterile phosphate-buffered saline (pH 7.2). A total of 200 µL of the pellet was transferred on each of the four slopes of HEYM. The cultures were further treated as described above.

The MAP isolates used for genotyping originated from the jejunum, ileum, mammary gland, spleen, and different lymph nodes (LNN mesenteriales, LNN ileocolici, LNN colici, LNN hepatici, LNN inguinales superficiales). All of the isolates were from different tissues. Their host origin and the year of isolation are presented in Tables S2 and S3.

2.2.3. Environmental Samples

Environmental samples (n = 256) were collected in the goat barn of herd 1 at eight different timepoints between 2020 and 2022 within the context of another study [12]. Cultivation of bedding, dust, and roughage from the feeding place and trough water was performed as described in detail elsewhere [12]. Briefly, 50 g of bedding and feed samples were soaked in distilled water overnight. The fluid was centrifugated and the supernatants were discarded. The dust was bloated with distilled water overnight and subsequently filtrated. The water was the centrifugated to obtain the sediment. The sediment pellets of all the sample materials were further processed the same way as described for the fecal samples, with the modification that for the decontamination of the dust and water samples, only 5 mL of HPC was used. Cultivation was performed as described above for the fecal samples.

The MAP isolates from all of the culture-positive environmental samples (n = 29) from the goat barn (Barn C) of herd 1 are listed, together with their respective collection sites, in Table S4. These isolates, as well as six isolates originating from the bedding materials of different localizations in the goat barn of herd 4 (see Table S2), were included in this study.

2.3. *DNA Isolation and MAP Identification*

Mycobacterial DNA was obtained using the method previously described in reference [56]. Extraction started with suspending a loopful of bacterial colony material in 100 µL of distilled water, followed by heating for 20 min at 80 °C, ultrasonication (35 Hz) for 10 min, another heating step for 10 min at 100 °C, and then centrifugation for 5 min at 12,000 rpm (15,300× *g*). The supernatant was transferred to another tube and was centrifuged again for 5 min at 12,000 rpm. The concentration of the DNA in the supernatant was determined using a NanoDrop 1000 spectrophotometer (Thermo Fisher Scientific, Wilmington, DE, USA).

MAP was confirmed by the detection of the subspecies-specific insertion element IS900 according to Englund et al. [57]. Mixed isolates with other *Mycobacterium avium* subspecies were excluded by a lack of IS1245 using PCR, described by Guerrero et al. [58]. Furthermore, the subspecies-specific genotyping profiles determined from the MIRU-VNTR analysis verified the subspecies and the MAP type identification (see Section 2.4.1).

2.4. *Genotyping of MAP*

A total of 133 fecal MAP isolates were randomly chosen for genotyping from 121 out of the 186 goats in herd 1 that were detected to be MAP-positive based on their cultures between 2018 and 2022. The MAP isolates from 13 of the goats with clinical Ptb symptoms selected for genotyping originated from one to four different tissues per goat and included those from the small intestine, various lymph nodes, and occasionally from tissues of other organs. All of the isolates, as well as those from all the environmental samples of herd 1 and from all the samples of the other herds, were genotyped using three techniques: MIRU-VNTR analysis, an SNP-based assay, and SSR analysis. For these analyses, the DNA extracts isolated as described in Section 2.3 were used.

2.4.1. MIRU-VNTR Typing

Using this method, differences in the number of tandem repeat sequences were detected using single PCR reactions targeting eight defined mini-satellite loci in the MAP genomes: MIRU-VNTR Loci 292, X3, 25, 47, 3, 7, 10, and 32 [40,59]. Different repeat numbers describe different alleles at these specific loci. These numbers were arranged according to the mentioned loci order, and based on the INMV classification database (http://mac-inmv.tours.inra.fr/ (accessed on 14 June 2023)), the so-called INMV types were determined. This method can be used for isolates belonging to different subspecies of *Mycobacterium avium*, which show subspecies-specific INMV types [60].

The same primers and PCR conditions described by Thibault et al. [40] were used with minor modifications according to Möbius et al. [61] as follows: approximately 50 ng of DNA was amplified in a 20 μL reaction volume containing 2 μM of each primer, 1× reaction buffer containing 1.5 mM $MgCl_2$, 1× Q-Solution (Qiagen, Hilden, Germany), 100 μM of each deoxynucleotide triphosphate (Qiagen, Hilden, Germany), and 0.5 U of HotStarTaq polymerase (Qiagen, Hilden, Germany). The PCR started with an initial denaturation step at 94 °C for 15 min, followed by 35 cycles of 96 °C for 30 s, 58 °C (MIRU 292, VNTR 25) for 1 min, and 72 °C for 30 s, with a final extension step at 72 °C for 10 min. The annealing temperature for MIRU X3 was 62 °C, and for VNTR 10, it was 54 °C. The annealing temperatures of the other target loci were applied according to [40]. Based on the amplicon sizes of the amplified target sequences analyzed using a 1.5% agarose gel, the number of repeats were determined using an allele-calling table published by Radomski et al. [59]. Amplicons presenting as single bands show that there was no mixed genotype.

Additionally, to identify polymorphisms at VNTR locus 7 for isolates with unusual amplicon sizes, the PCR products were purified using the QIAquick PCR purification kit (Qiagen, Hilden, Germany) according to the manufacturer's instructions and sequenced by an external sequencing service. The resulting sequences were compared with a reference sequence, as described in reference [62].

2.4.2. SNP-Based Assay

This PCR-based method with subsequent restriction enzyme digestion or sequencing of the amplified product allows for the assignment of MAP isolates to 14 phylogenetic groups within the MAP population, each defined by a specific SNP [38]. These SNPs are mutually exclusive, i.e., if an isolate possess one particular SNP, it lacks all the other ones [38]. As recommended in [39] (see supplementary Table S3), a MIRU-VNTR analysis was performed first. Based on the detected INMV types, specific PCRs were prioritized for the SNP-based analysis to reduce the number of SNP-PCRs for each sample. According to this work, the SNPs were analyzed stepwise, and analysis was stopped once a phylogenetic group-defining SNP was identified.

For single-plex PCRs, the primers as published were applied, and the SNP-PCRs were named based on their primer designation according to reference [38]. A minimum of 50 ng of DNA was amplified in a 50 μL reaction volume containing 2 μM of each primer, 1× reaction buffer containing 1.5 mM $MgCl_2$, 1× Q-Solution (Qiagen), 100 μM of each deoxynucleotide triphosphate (Qiagen), and 0.5 U of HotStarTaq polymerase (Qiagen). The PCR started with an initial denaturation step at 94 °C for 15 min, followed by 35 cycles of 96 °C for 30 s, 64 °C for 1 min, and 72 °C for 1 min, with a final extension step at 72 °C for 10 min. The original method was modified, with no restriction endonuclease digestion applied, and all the PCR products were sequenced. Approximately 40 μL of the amplified product was purified using a QIAquick PCR Purification Kit according to the manufacturer's instructions. Sequencing was carried out by Eurofins Genomics GmbH (Ebersberg, Germany). The presence or absence of a specific SNP was analyzed by comparing the sequence result with the genome sequence of MAP K10 (GenBank accession number NC_002944.2) at the defined position according to reference [38] using Geneious Prime 2021.0.1. (https://www.geneious.com (accessed on 11 May 2023)). The original sequence chromatograms were checked for the absence of mixed peaks to exclude mixed genotypes.

2.4.3. SSR Typing

The SSR sequencing method was applied according to [63]. In deviation from the original method, only micro-satellite loci SSR1 (g repeats), SSR8 (ggt repeats), and SSR9 (tgc repeats) were chosen because these loci are among the most informative SSR loci [60,63,64] and the results are stable and reliable [63]. As a result, the number of specific repeats at these three loci was given as a numerical code. The primers were used as published [63]. The PCR mixture and the cycler program were slightly modified as follows: The 50 μL PCR reaction mixture for SSR8 and SSR9 comprised 0.6 μM of each primer, 1× reaction buffer containing 1.5 mM $MgCl_2$, 5% dimethyl sulfoxide, 100 μM of each deoxynucleotide triphosphate, 1 U of HotStarTaq polymerase, and 50 ng of DNA. For SSR1, 2.5 U of Herculase II Fusion DNA polymerase (Agilent Technologies Deutschland GmbH, Waldbronn, Germany) was used to minimize the risk of polymerase slippage at the poly(G) motif, and no dimethyl sulfoxide was used. The PCR for SSR8 and SSR9 started with an initial denaturation step at 94 °C for 15 min, followed by 35 cycles of 96 °C for 15 s, 60 °C for 1 min, and 72 °C for 30 s, with a final extension step at 72 °C for 3 min. For SSR1, the PCR started with 98 °C for 2 min, followed by 30 cycles of 98 °C for 20 s, 60 °C for 20 s, and 72 °C for 30 s, with a final extension step at 72 °C for 3 min. The PCR products were purified using a QIAquick PCR Purification Kit according to the manufacturer's instructions, and sequencing was performed by an external sequencing company. The resulting sequences were analyzed in silico, and the original sequence chromatograms were checked for mixed peaks to exclude mixed genotypes.

2.4.4. Calculation of the Discriminatory Power of Typing

As a numerical index with the discriminatory power of a combination of MIRU-VNTR, an SNP-based assay, and SSR typing, the Simpson's Index of Diversity (SID) was calculated, as described by Hunter and Gaston [65]. Only epidemiologically unrelated strains were included in the calculation, i.e., only isolates with different genotypes from the same herd.

3. Results

3.1. *Determined Combined MAP Genotypes*

Using MIRU-VNTR typing, five different INMV types were determined for the isolates included in this study: INMV1, 2, 3, 16*, and 33. The amplicons always showed clear individual bands for each MIRU-VNTR locus in the agarose gels, indicating that none of the isolates were mixed isolate with different genotypes. All the isolates were assigned to the C-type group based on the identified MAP-specific INMV types. In addition, none of the isolates belonged to any other subspecies of *Mycobacterium avium*. Furthermore, SNP-PCRs were prioritized, as described in Appendix A. The specific SNPs identified within the MAP genome using the SNP-based assay were used to assign the isolates to the five different phylogenetic groups: Clades 1, 3, 4, and 9 in Subgroup A and Subgroup B (SGB). Using SSR analysis, two different profiles were found: 7g-4ggt-4tgc (7-4-4) and 7g-4ggt-5tgc (7-4-5) at SSR loci 1, 8, and 9. The combination of the individual methods resulted in six combined genotypes for the tested isolates, designated as T1 to T6 (see Table 1).

Table 1. Combined genotypes for *Mycobacterium avium* subsp. *paratuberculosis* isolates from goats and their environment in this study.

Genotype	Phylo-Group [a]	INMV Type [b]	INMV Code [c]	SSR Profile [d]
T1	9	1	42332228	7-4-4
T2	SGB	33	32522228	7-4-4
T3	1	2	32332228	7-4-4
T4	4	2	32332228	7-4-5
T5	4	3	32332218	7-4-4
T6	3	16*	323325*28	7-4-4

[a] The phylogenetic groups based on SNP analysis using SNP-based assay according to reference [38] were designated also as Clades (here, Clades 1, 3, 4, and 9) belonging to a larger group named subgroup A; or designated as SGB—subgroup B. [b] INMV type according to the INMV database (http://mac-inmv.tours.inra.fr/ (accessed on 14 June 2023)). [c] The INMV code comprises the number of tandem repeats (TR) at MIRU-VNTR loci 292, X3, 25, 47, 3, 7, 10, and 32 according to reference [40]. Underlined are repeat numbers different from type INMV2. [d] The short sequence repeat (SSR) profile is a combination of the number of SSRs at [G] locus 1, [GGT] locus 8, and [TGC] locus 9 according to reference [63]. * At VNTR 7, in addition to three repeats, an additional sequence region of 34 bp between repeats 2 and 3 was identified. The amplicon size of 259 bp looked like almost five copies; therefore, this result was highlighted with 5*, and the respective code (323325*28) was designated as type INMV16*.

Genotype T6 included a MIRU-VNTR profile with an unusual sequence at MIRU-VNTR locus 7. The PCR amplicon size at the VNTR 7 locus of several isolates in the agarose gel was slightly below 269 bp, corresponding to five repeats according to Table S3 in reference [59]. Sequencing and in silico analysis revealed three perfect copies of the 22 bp repeat (CGAAATATTCGCCGTGAGAACA) at VNTR 7, but an additional sequence region of 34 bp (CGTGCGGCGAAGGCTGGCCGGCCCGAAAAGCCA) was identified between repeats 2 and 3. The resulting amplicon size of 259 bp determined through sequencing in the present study was visually almost identical to the amplicon size for the isolates with five repeats. The assignment of isolates with such an unusual sequence at VNTR locus 7 to a specific INMV type using the database (http://mac-inmv.tours.inra.fr (accessed on 14 June 2023)) is not yet clear. Therefore, this result was highlighted with an asterisk (5*), and the respective INMV16 type (32332528) was designated as INMV16* (323325*28).

3.2. *MAP Strain Diversity in Herd 1*

In herd 1, three combined genotypes were identified with different frequencies in isolates from 121 different goats (Table 2). In total, 56 out of 57 fecal MAP isolates of individual animals from 2018 harbored the combined genotype T1 (INMV1/7-4-4/Clade 9). The remaining isolate from 2018 showed genotype T2 (INMV33/7-4-4/Subgroup B), which was isolated again from another goat in 2020 and 2021. Furthermore, genotype T1 was recovered in 36 of the 39 fecal isolates in 2020 (including one isolate from a buck), in 27 of the 29 fecal isolates in 2021, and in 8 of the 8 fecal isolates in 2022 (Table S1). Fecal isolates from twelve goats originating from two different samplings were genotyped with the same T1 result. In addition, genotype T1 was determined in all the tissue isolates (n = 34) from 13 clinically diseased goats necropsied and sampled in 2018, 2020, and 2022 (Table S3) and in 28 of the 29 environmental MAP isolates (Figure 2, Table S4). The third genotype, T3 (INMV2/7-4-4/Clade 1), was isolated in 2020 from the feces of one goat (4 years old), an environmental sample from the barn entrance in 2020, and from the feces of another goat (one year old) in 2021 (Tables S1and S4, Figure 2).

Table 2. Thuringian goat farms tested for *Mycobacterium avium* subsp. *paratuberculosis* in this study: identified genotypes and origin of isolates (n = 220).

Farm	Genotype *	Origin	Isolates (n)	Goats (n)	Year
1	T1	Feces	127	117 **	2018, 2020, 2021, 2022
		Tissue sample	34	13	2018, 2020, 2022
		Environment	28	-	2020, 2021, 2022
	T2	Feces	3	2	2018, <u>2020</u>, <u>2021</u>
	T3	Feces	3	2	<u>2020</u>, 2021
		Environment	1	-	2020
2	T4	Feces	5	5	2019
		Feces	2	2	2022
3	T5	Feces	2	2	2017
		Feces, bulk sample	1		2022
4	T6	Feces	3	3	2017
		Feces, bulk sample	1	-	2021
		Tissue sample	2	1	2020
		Environment	6	-	2017

* Description of combined genotypes T1–T6, see Table 1. ** including one buck. LNN—Lymph node. Underlined—identical genotypes isolated from the same animal in different years (T2) or twice in the same year (T3).

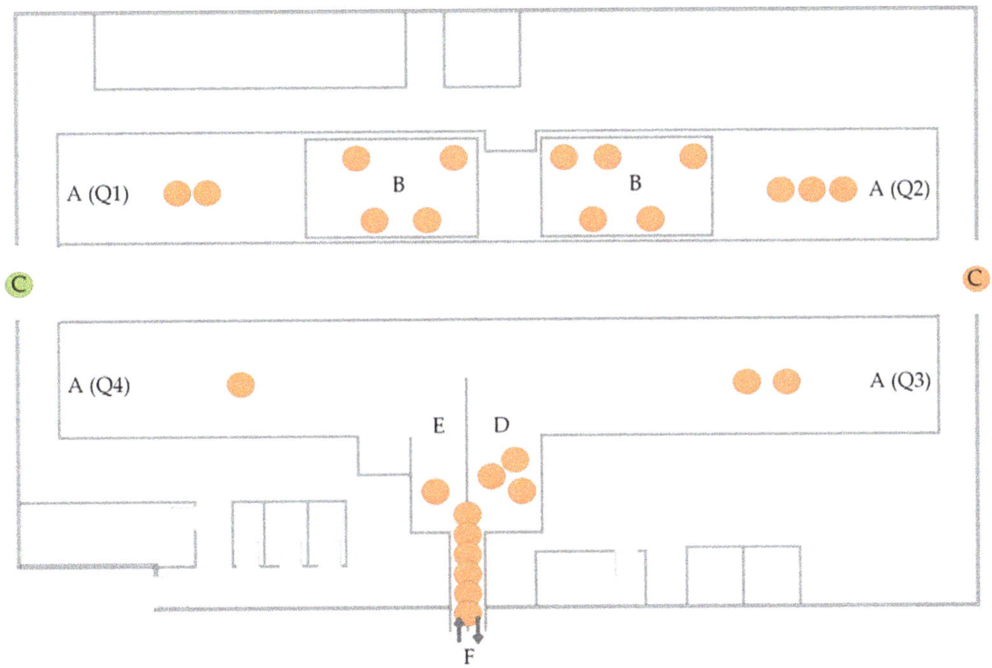

Figure 2. Origins of culture-positive environmental samples within the goat barn of herd 1 ($n = 29$) and determined *Mycobacterium avium* subsp. *paratuberculosis* genotypes. A (Q1–4)—Lactating goat area (stable quadrants 1–4); B—kidding area; C—stable entrance; D—waiting area milking parlor; E—exit milking parlor; F—milking parlor; orange points: genotype T1, green point: genotype T3.

3.3. MAP Genotypes in Other Goat Herds in Thuringia

The other three Thuringian goat herds investigated harbored three other genotypes (Table 2). In herd 2, seven fecal isolates from 2019 and 2022 all showed the same genotype: T4. In herd 3, isolates originating from two fecal samples of one goat in 2017 and one pooled fecal sample in 2022 revealed genotype T5. Furthermore, in herd 4, isolates from the feces, mesenteric lymph nodes, and jejunum of a total of 4 goats in 2017 and 2020, from six environmental samples in 2017 and from a pooled fecal sample in 2021, all showed genotype T6.

3.4. Discriminatory Index of the Genotyping Approach Used

In each goat herd, a different genotype was found in addition the two genotypes in herd 1. Based on these six epidemiologically unrelated strains (Table 2), the discriminatory power of typing by using the combination of MIRU-VNTR, SSR analysis, and an SNP-based assay was calculated and resulted in a discriminatory index of 1.

4. Discussion

This study provides the first insights into the molecular diversity of MAP isolated from naturally infected goats in one federal state in Germany, including 203 isolates from 147 dairy goats and 35 environmental isolates from 4 herds in the federal state Thuringia. Three genotyping methods were used in combination. The previously published stepwise hierarchical typing workflow, combining MIRU-VNTR typing and an SNP-based assay [39], was successfully applied, and the typing effort of the SNP-based assay was reduced. The main findings of this study are as follows: (1) the genotypes of the isolates were different in the four goat herds tested, indicating that they were not epidemiologically linked.

Altogether, six combined MAP-C genotypes were identified. (2) In herd 1, where most of the isolates originated from, one dominant genotype, and sporadically, two additional genotypes, were found. (3) MIRU-VNTR typing resulted in types INMV1, 2, 3, 33, and additionally, a type with a specific structure at locus VNTR 7, designated as INMV16*. No mixed MAP genotypes and no other subspecies of the *Mycobacterium avium* complex were identified. (4) Using SSR analysis, two profiles were detected (7-4-4, 7-4-5). (5) Using the SNP-based assay, the isolates were assigned to five different phylogenetic groups (Clades 1, 3, 4, and 9 in Subgroup A, and Subgroup B). In previous studies, these subgroups and phylogenetic clades included isolates from different hosts, regions, and countries worldwide.

4.1. Genotyping Results

The sampled Thuringian goat herds, all infected by C-type (Type II) strains, were kept solitary; not together with sheep or cattle. In Germany, S-type (Type III) strain isolation so far only has been described in a few sheep [66]. The current study focused on C-type strains using solid HEYM medium with Mycobactin J for MAP isolation, which is less sensitive for isolation of S-type strains [67,68]. In studies characterizing MAP isolates or MAP-DNA from goats in Spain, Mexico, Canada, and Switzerland, the detected predominant type was also the C-type; however, S-type strains/genomes (Type I/III) were also identified [41,69–71]. The goat herds tested were infected either by C-type or S-type strain(s) only [41], or both types were identified in one herd [70]. In Spain, many farms with native goat breeds were infected with S-type strains, and a larger number of these goats were in contact with sheep that showed no signs of infection [41].

Worldwide, only a few previous studies have been carried out in goats investigating the diversity of MAP at the herd level [43,70,72]. The first common methods for genotyping of MAP isolates were pulsed-field gel electrophoresis (PFGE) and restriction fragment length polymorphism coupled with hybridization to IS900 (IS900-RFLP) analysis. De Juan et al. [43] identified 15 combined genotypes (14 × Type II, 1 × Type III) in 44 caprine isolates from Spain, Scotland, and Norway, including high intra-herd diversity with six genotypes in a single herd from Spain, detected using PFGE in combination with IS900-RFLP. In contrast, only one genotype (Type II) could be found in 25 caprine isolates from one individual herd in Germany using IS900-RFLP [72]. In the meantime, PFGE and IS900-RFLP have been replaced by PCR- and sequence-based methods (MIRU-VNTR typing, SSR-analysis), which are faster and less labor-intensive. In addition to one S-type isolate, Bauman et al. [70] found three genotypes in 21 C-type isolates from 11 dairy goat farms in Canada using a combination of MIRU-VNTR and SSR-typing (analyzing only the ggt SSR locus 8). Almost all of the isolates belonged to the combined subtypes INMV1/4ggt and INMV2/5ggt. This low diversity was attributed to the limited discriminatory power of the typing approach used, which is particularly true for isolates with INMV1 and INMV2 [50]. On the other hand, it could also be due to the fact that there were only two main infection lineages in these Canadian goat herds. More target regions, as in the present study, might have led to a higher discriminatory power of typing and thus to more genotypes. Apart from this, WGS-based data provided the highest resolution for distinguishing MAP as a monomorphic pathogen on a regional and global scale and allowed for the detection of phylogenetic relationships between isolates within the MAP population [50,73,74]. However, a study on Canadian cattle isolates showed that SNPs alone were not sufficient to track MAP transmission due to the slow growth rate of MAP and because the extensive movements of affected animals between different provinces were not well documented [35]. Furthermore, using WGS data, a comparison of genotypes between different studies is complicated, as it would require renewed analysis of the raw data of all included genomes. Alternatively, the recently established PCR-based SNP assay [38] allows strains to be assigned to previously defined known phylogenetic groups and subgroups without the need for WGS of isolates. Combining this assay with MIRU-

VNTR typing and the SSR analysis in this study allowed for successful discrimination between isolates from different goat herds in Thuringia (Germany).

MIRU-VNTR types INMV1 und INMV2 are predominant MAP genotypes isolated from cattle and goats in Europe, Canada, the United States, and several countries in Latin America [36,40,50,70,75,76]. Consistent with this, in the present study, INMV1 was found in almost all the samples from herd 1, and INMV2 was found in all the isolates from herd 2, in four isolates from two goats, and in one environmental sample in herd 1. In addition, the goat isolates from Thuringia harbored INMV3, INMV16*, and INMV33. INMV3 and INMV33 are commonly found in Europe and South America [39,60,77–81]. Type INMV16* was a special case, because sequencing of the VNTR 7 locus amplicon did not reveal the expected five repeats (based on amplicon size) but instead an unusual sequence comprising two repeats and an additional 56 bp sequence. This additional sequence was described before as a second repeat sequence, including 34 bp and a sequence identical to the 22 bp VNTR 7 repeat [62]. The respective complete sequence at the VNTR 7 locus was designated as profile E [62]. Possibly, type INMV16, previously found in bovine isolates from France, Argentina, and Germany [39,77,79,82], corresponds to type INMV16* in the present study because the PCR products were similar. Unfortunately, it was not possible to assign this MIRU-VNTR profile with an unusual amplicon size and the underlying sequence differences at specific loci to defined INMV types using the database or previous studies.

Furthermore, the SSR analysis at SSR loci 1, 8, and 9 revealed two out of 13 profiles/alleles (based on these three loci) that were previously identified in MAP isolates originating from different hosts in the United States [63]. The predominant SSR profile (7-4-4) was found in combination with MIRU-VNTR types INMV1, 2, 3, 16*, and 33; the profile 7-4-5 was found with INMV2 only. In one previous study including 12 goat isolates, the profile 7-4-4 was also found to be predominant. It was detected in nine goat isolates from France in combination with INMV1 ($n = 7$) and INMV2 ($n = 2$) and in one goat isolate from the Netherlands in combination with INMV33 [60], as in this study. Two additional goat isolates from France harbored the 7-4-5 profile, identical to our second SSR profile, but in combination with INMV1 and not with INMV2. The combination of the 7-4-4 profile with INMV1, 2, 3, or 33, found in the present study was also detected in several cattle isolates from France, the Netherlands, Argentina, Czech Republic, Slovenia, and Germany, and in combination with INMV1 or 2 in a few deer isolates [46,60]. Regardless of their assignment to phylogenetic clades, these results suggest that the majority of MAP genotypes from the studied goat herds in Thuringia belonged to the MAP population that is common in infected cattle and other ruminants in Central European countries.

Previously, based on WGS–SNP analysis, a set of MAP isolates originating from different hosts and countries worldwide was compared. The isolates were distributed within a phylogenetic tree to the known major S-type (Type I and III) and C-type (Type II) groups and various subgroups [36]. Type II included Subgroup A with 10 clades, Subgroup B (SGB), and Type B (bison type) [36,38]. These defined assignments of the MAP isolates to phylogenetic groups, subgroups, and clades were used for establishing the SNP-based assay based on 14 discriminative SNPs by Leão et al. [38], and this assay was applied in the present study. In contrast to the misidentification of Australian S-type isolates and the incorrect differentiation of C-type isolates into subgroups A or B [83], this method works well for German C-type isolates. Overall, the goat isolates from Thuringia could be assigned to five of these phylogenetic subgroups: Clades 1, 3, 4, and 9 in Subgroup A, and SGB. In the global MAP set [36,38], these subgroups included isolates from goat (Clades 1, 9, and SGB), cattle (all subgroups), sheep (Clades 1, 3, 4, and 9), deer (Clades 1 and 9), and humans (Clades 3 and 4) from different regions worldwide. The present results indicate a high genetic similarity of the goat isolates from Thuringia with those of the global MAP population.

Most isolates of the global set also showed INMV1 and INMV2 distributed to Clades 1, 2, 8, 9, and 10 (INMV1) or Clades 1, 3, 4, 5, and 10 (INMV2) [36,38]. The assignments of isolates with INMV1 or INMV2 to different phylogenetic clades revealed that the MIRU-

VNTR types (especially these two most common types, which differ only at VNTR 292) do not reflect the phylogenetic relationships [36,37,50]. In comparison with WGS-based phylogeny, instances of MIRU-VNTR homoplasy have been observed; MIRU-VNTR typing over- and underestimates the genetic diversity of MAP [36]. Recent studies of modern *Mycobacterium tuberculosis* lineages have shown that the homoplasy of MIRU-VNTR repeats has arisen through stochastic processes in the absence of natural selection [84], which has not yet been investigated in this way for MAP. On the other hand, based on the WGS phylogeny of French animal isolates, a strong association between phylogeny and INMV1 and evidence of an association between phylogeny and INMV2 was found, but not for SSR typing results [73]. All these results suggest that the informative value of MIRU-VNTR typing for phylogeny compared to WGS-based SNP analysis varies depending on the phylogenetic clade to which the isolates belong, on the diversity of the regional origins of isolates, on the time period of sample collection and possibly the phylogenetic analysis methods, including different cut-offs (number of SNPs) for closely related strains and the phylogenetic distance of the included isolates.

German goat isolates from Thuringia with INMV1 were assigned to Clade 9, and isolates with INMV2 were assigned to Clades 1 and 4. Furthermore, goat isolates with INMV3 were also assigned to Clade 4 (this study), as previously reported for cattle isolates from Hesse and Thuringia in Germany [39]. In contrast, isolates (from cattle) with INMV3 in the global set were assigned to Clades 1 and 5 [36,38]. Additionally, isolates harboring INMV33, originating from goats (this study) or from cattle in Spain or Germany [36,38,39], were all assigned to SGB. Furthermore, MAP isolates with the rare genotype INMV16* and belonging to Clade 3 (this study) have not been previously described; however, cattle and human isolates with rare and particularly high repeat numbers at the VNTR 7 locus (INMV16, INMV75, INMV127) have also been assigned to Clade 3 [36,38,39]. However, there are other goat isolates of Type II from different countries worldwide belonging to Clade 7, Clade 10, and Type B [36,38]. These results suggest that within the C-type (Type II) group, there is no host association with phylogenetic clades, and until now, there has been no goat-specific combined genotype.

4.2. Epidemiological Analysis and MAP Transmission

The results of combined genotyping targeting different genome regions showed that Ptb in the four Thuringian goat herds was caused by different MAP strains and that herd 1 was infected by three strains. In addition to the predominant genotype T1 (INMV1/Clade 9/7-4-4), few fecal and environmental isolates carried genotype T2 (INMV33/SGB/7-4-4) or T3 (INMV2/Clade 1/7-4-4). Various studies have already shown that in certain goat and cattle herds, multiple MAP infections with different genotypes circulate within the herd [43,46,61,73,85]. As mentioned above, herd 1 was identified as being highly affected by Ptb in early 2018 and was included in a Ptb control program. Ptb-infected goats, shedding the pathogen in large numbers or showing clinical signs, were removed from the herd. This led to a reduction in MAP culture-positive fecal samples from 30.0% in 2018 to 5.2% in 10/2020 and to 1.7% in 05/2022 [12]. In the isolates from the various tissues of the clinically affected goats in herd 1 (born between 2014 and 2018), sampled in 2018, 2020, and 2022, only one genotype (T1) was identified. It is assumed that individual animals have been infected with T1 several times.. The pathogen was disseminated throughout the body and was found not only in the different sections of the intestine, but also in the mammary gland, mammary lymph nodes, spleen, and in the hepatic lymph nodes of individual animals. The spread of MAP to tissues outside the gut has already been described for naturally infected cattle and a goat [71,86] and in experimentally infected goats [55]. The other two genotypes in herd 1 (T2 and T3) were not able to spread further or even to displace the predominant strain, although infected animals remain susceptible to co-infection with other MAP genotypes [20]. Based on the WGS-SNP analysis in another study, some MAP strains were found to be more successful in spreading than others; these were designated as dominant strains [20]. However, there is no evidence that individual strains within the

MAP-C type group are more virulent than others. SNPs in various virulence-associated genes in individual strains may be associated with differences in virulence [74,87], but this would need to be demonstrated by in vitro and in vivo experiments. The predominance of T1 could be due to the continuous ingestion of very large amounts of this strain by kids, juvenile, and adult goats over a long period of time, acquired from an environment that is highly contaminated by the previous occupants of the barns and later from high shedders in the goat herd. The high stress induced by the move of the goats to Barn C and the pre-existing problems with pseudotuberculosis may have accelerated the infection processes and the development of clinical Ptb in some animals by the end of 2017. In addition to the detection of MAP in two- to five-year-old goats, genotype T1 was also detected in fecal samples from two seven-months-old goats in 10/2020, born in herd 1. This was possibly caused by the ingestion of MAP during the first hours of life in the highly contaminated kidding area [12]. Furthermore, not only infection of kids but also adult-to-adult contact was assumed to be an important route of transmission, as previously described for dairy cattle [20]. This option is supported by the repeated detection of MAP, particularly at sites with high animal traffic, as previously published [12].

Genotype T2 was only found in the fecal samples of two adult goats, once in 2018 from goat 84,924 (born and purchased in 2011, left the herd March 2018), and twice in 2020 and 2021, from goat 81,024 (born in 2014 in herd 1, left the herd July 2021). It is suspected that either only these two goats were infected with T2 or not all of the T2-infected goats could be identified due to the lack of MAP shedding by 2022. Goat 84,924 was purchased as a kid in 2011 together with other goats with unknown Ptb statuses from a breeder. Some of these other goats, as well as goats born in 2009 and later in herd 1, were found to be infected with genotype T1 in 2018, suggesting different infection events and sources of T1 and T2. Strains with a similar genotype to T2 (but with an unknown SSR profile) have also been previously found in bovine isolates from Thuringia, Hesse, and Saxony in Germany [39].

Furthermore, genotype T3 was detected in 2020 and 2021 in the feces of two goats (born in 2014 and 2020 in herd 1), and in 2020, in an environmental sample collected at the barn entrance. This entrance was utilized twice a day during the grazing season for driving the goats to the pasture. It is possible that the soil was contaminated by the feces of the goats. It is likely that T3 was already present in herd 1 in 2018, as the older goat already tested cultural positive for MAP in the first conducted herd sampling at that time. The younger goat may have been infected via environmental contamination in the kidding box, as this goat in 2020 was born three days after the kidding of the older goat, which was infected with T3. The T1 genotype was predominant in herd 1; however, the true origin of the three found MAP strains could not be determined based on the analyses performed in this study and the available data.

Herd 2, which also kept the breed "Thüringer Wald Ziege", was infected with a different strain than herd 1: genotype T4 (INMV2/Clade 4/7-4-5). The origin of this strain is unknown. It is possible that the goats were infected by environmental contamination with MAP from cattle on the pastures of herd 2. Herds 3 and 4 kept a different breed of goat ("Saanenziege", Saanen goats) than herds 1 and 2 and were infected with T5 (INMV3/Clade 4/7-4-4) in herd 3 and T6 (INMV16*/Clade 3/7-4-4) in herd 4. The animals in herd 3 were purchased from several breeders, and the herd had recently used a former dairy cattle barn from a herd with an unknown Ptb status. Only three isolates from herd 3 were tested, but an identical genotype to T5 in 2017 and 2022 suggests that this was the only or the predominant genotype, with an unknown origin. In herd 4, infection strain T6 with the rare MIRU-VNTR profile INMV16* and assignment to clade 3 was identified in all the fecal, tissue, and environmental samples from 2017, 2020, and 2021, suggesting this was the predominant infection strain in this herd. Ptb may have been introduced through the purchase of sub-clinically infected goats.

Furthermore, the presence of different ruminant species with unknown Ptb statuses has been described as another risk factor for MAP infection [26,30]. As described before, herd 1 was established with a mixture of goats, sheep, and cattle in 2002, and herd 3 was first established in a former sheep barn in the early 1990s. This also raises the possibility that there might be an epidemiological link to sheep, in addition to the already discussed one to cattle. Due to the time that has elapsed since the concerned ruminants left the facilities or herds, no further research can be conducted to address this issue. Highly discriminatory genotyping of MAP isolates from a herd of interest, or from different herds with the same origin of kids, combined with appropriate information on the origin of animals, may allow for the origin of an infection to be determined [88]. However, due to the relatively long subclinical phase of Ptb, this is only possible with a time delay. Using WGS-based SNP analysis with a higher resolution for differentiation could identify mixed isolates in individual animals or additional genotypes, as shown in recent studies for individual dairy cattle herds [20,89]. On the other hand, the WGS approach could potentially reveal additional SNPs as a result of evolution in individual herds [90]. However, as the molecular clocks of the specific target regions used for genotyping are different, a combination of methods could still help to identify the different infection lineages and transmission routes of MAP within a limited region and time period.

5. Summary and Conclusions

The applied combination of MIRU-VNTR, an SNP-based assay, and SSR-typing was well suited to differentiate MAP isolates from different dairy goat herds in Thuringia (Germany). Each of these goat herds had been infected predominantly clonally with a dominant MAP-C genotype for several years, and herd 1 was additionally infected with two other MAP-C genotypes. These three different strains in herd 1 were most likely introduced at different times. There was no evidence for the transmission of MAP between the goat farms studied. The purchase of sub-clinically infected animals and a MAP-contaminated environment of the goats are suspected as possible sources of the initial MAP introduction in the herds. The exact sources of the MAP introduction and transmission routes could not be determined. This would require knowledge of the Ptb history of all the breeders or farms from which the goats were purchased, the place of birth of the goats, and mother–daughter relationships. Furthermore, isolates from the Ptb-positive herds and from the contaminated environment of any previously used barns or pastures would have had to be genotyped.

The goats were probably not infected with unique or regional MAP strains. The infecting strains were assigned to five different phylogenetic clades, which also included isolates from other countries, different regions, and ruminant species of the global MAP population. The identified combined MIRU-VNTR and SSR profiles have also previously been found in different hosts and regions worldwide. The specific combination of the typing results of the three used methods was compiled for the first time. Further studies could characterize and compare these isolates based on WGS-SNP analyses, reveal the relationship of isolates representing the six identified genotypes, and elucidate the possible evolutionary processes of the individual infection strains within the herds.

Supplementary Materials: The following supporting information can be downloaded at: https://www.mdpi.com/article/10.3390/ani13223542/s1, Table S1: *Mycobacterium avium* subsp. *paratuberculosis* isolates (n = 133) from the feces of 120 female goats and one buck belonging to herd 1 in Thuringia (Germany) and their respective genotyping results. Table S2: *Mycobacterium avium* subsp. *paratuberculosis* isolates of goats and their environment from different Thuringian herds (herd 2 to 4) and their respective genotypes. Table S3: *Mycobacterium avium* subsp. *paratuberculosis* isolates (n = 34) from different tissues of 13 goats belonging to herd 1 in Thuringia (Germany) and their respective genotyping results. Table S4: *Mycobacterium avium* subsp. *paratuberculosis* isolates (n = 29) from environmental samples of herd 1 in Thuringia (Germany) and their respective genotyping results.

Author Contributions: Conceptualization, P.M.; methodology, C.P., E.M.L.-T., H.K., U.M. and P.M.; formal analysis, P.M.; investigation, C.P., E.M.L.-T., U.M. and P.M.; resources, C.P., E.M.L.-T., H.K., U.M. and P.M.; data curation, P.M.; writing—original draft preparation, C.P. and P.M.; writing—review and editing, C.P., H.K. and P.M.; visualization, C.P. and P.M.; supervision, P.M.; project administration, P.M.; All authors have read and agreed to the published version of the manuscript.

Funding: The project was partially funded by the European Agricultural Fund for Rural Development (EAFRD) via the Thüringer Aufbaubank, grant number 2019 LFE 0017.

Institutional Review Board Statement: All samplings were carried out in strict accordance with European and National Law for the Care and Use of Animals, and every effort was made to minimize suffering. The individual fecal samples from herd 2, bulk fecal samples from herds 3 and 4, and bedding from herd 4 were sampled during regular herd health monitoring by the Sheep and Goat Health Service of the Thuringian Animal Diseases Fund. Tissue samples were obtained in the course of dissections carried out at the Friedrich-Loeffler-Institut in accordance with §4(3) of the German Animal Welfare Act. Individual fecal samples from herds 1, 3 and 4 were obtained within two different scientific studies. These studies were approved by the competent authority, the Animal Health and Welfare Unit of the Thuringian State Office for Consumer Protection (file references: 2684-04-04-BFI-20-103, date of approval 01.07.2020 and 22-2684-04-04-102/16, date of approval 20.04.2016).

Data Availability Statement: All data are included in the supplemented material.

Acknowledgments: The authors would like to thank Kerstin Steger and Sandy Werner (Friedrich-Loeffler-Institut, Jena) for their excellent technical assistance and the dairy goat farmers involved in this study for their kind support and cooperation.

Conflicts of Interest: The authors declare no conflict of interest.

Appendix A

Use of hierarchical typing workflow

The prioritization of specific SNP-PCRs was based on the recommended hierarchical typing workflow according to supplemented Table S3 in [39] and noticeably reduced the typing effort in the present study. For isolates with INMV1 and INMV33, only one specific SNP reaction each was needed to assign the isolates to Clade 9 and Subgroup B, respectively. For isolates with type INMV2, three and two SNP-PCRs were performed, and then Clade 1 and Clade 4 were determined, respectively. In the case of isolates with type INMV3, only the fourth PCR resulted in an assignment to Clade 4, and in the case of isolates with an unusual number of repeats and an additional sequence at VNTR 7, Clade 3 was determined with a single PCR.

References

1. Windsor, P.A. *paratuberculosis* in sheep and goats. *Vet. Microbiol.* **2015**, *181*, 161–169. [CrossRef]
2. Stanitznig, A.; Khol, J.L.; Lambacher, B.; Franz, S.; Wittek, T.; Kralik, P.; Slana, I.; Vasickova, P. Prevalence of *Mycobacterium avium* subspecies *paratuberculosis* and hepatitis E in New World camelids in Austria. *Vet. Rec.* **2017**, *181*, 46. [CrossRef] [PubMed]
3. Whittington, R.; Donat, K.; Weber, M.F.; Kelton, D.; Nielsen, S.S.; Eisenberg, S.; Arrigoni, N.; Juste, R.; Saez, J.L.; Dhand, N.; et al. Control of paratuberculosis: Who, why and how. A review of 48 countries. *BMC Vet. Res.* **2019**, *15*, 198. [CrossRef] [PubMed]
4. Roller, M.; Hansen, S.; Knauf-Witzens, T.; Oelemann, W.M.R.; Czerny, C.P.; Abd El Wahed, A.; Goethe, R. *Mycobacterium avium* Subspecies *paratuberculosis* Infection in Zoo Animals: A Review of Susceptibility and Disease Process. *Front. Vet. Sci.* **2020**, *7*, 572724. [CrossRef] [PubMed]
5. Lombard, J.E. Epidemiology and economics of paratuberculosis. *Vet. Clin. N. Am. Food Anim. Pract.* **2011**, *27*, 525–535. [CrossRef]
6. Whitlock, R.H.; Buergelt, C. Preclinical and clinical manifestations of paratuberculosis (including pathology). *Vet. Clin. N. Am. Food Anim. Pract.* **1996**, *12*, 345–356. [CrossRef]
7. Stewart, D.J.; Vaughan, J.A.; Stiles, P.L.; Noske, P.J.; Tizard, M.L.V.; Prowse, S.J.; Michalski, W.P.; Butler, K.L.; Jones, S.L. A long-term bacteriological and immunological study in Holstein-Friesian cattle experimentally infected with *Mycobacterium avium* subsp. *paratuberculosis* and necropsy culture results for Holstein-Friesian cattle, Merino sheep and Angora goats. *Vet. Microbiol.* **2007**, *122*, 83–96. [CrossRef]
8. Stewart, D.J.; Vaughan, J.A.; Stiles, P.L.; Noske, P.J.; Tizard, M.L.V.; Prowse, S.J.; Michalski, W.P.; Butler, K.L.; Jones, S.L. A long-term study in Angora goats experimentally infected with *Mycobacterium avium* subsp. *paratuberculosis*: Clinical disease, faecal culture and immunological studies. *Vet. Microbiol.* **2006**, *113*, 13–24. [CrossRef]
9. Lybeck, K.R.; Storset, A.K.; Djonne, B.; Valheim, M.; Olsen, I. Faecal shedding detected earlier than immune responses in goats naturally infected with *Mycobacterium avium* subsp. *paratuberculosis*. *Res. Vet. Sci.* **2011**, *91*, 32–39. [CrossRef]
10. Mitchell, R.M.; Medley, G.F.; Collins, M.T.; Schukken, Y.H. A meta-analysis of the effect of dose and age at exposure on shedding of *Mycobacterium avium* subspecies *paratuberculosis* (MAP) in experimentally infected calves and cows. *Epidemiol. Infect.* **2012**, *140*, 231–246. [CrossRef]

11. Eisenberg, S.W.; Nielen, M.; Santema, W.; Houwers, D.J.; Heederik, D.; Koets, A.P. Detection of spatial and temporal spread of *Mycobacterium avium* subsp. *paratuberculosis* in the environment of a cattle farm through bio-aerosols. *Vet. Microbiol.* **2010**, *143*, 284–292. [CrossRef] [PubMed]
12. Pickrodt, C.; Donat, K.; Moog, U.; Kohler, H. *Mycobacterium avium* subsp. *paratuberculosis* in Different Environmental Samples from a Dairy Goat Barn-Implications for Sampling Strategies for paratuberculosis Diagnostic and Prevention. *Animals* **2023**, *13*, 1688. [CrossRef]
13. Fecteau, M.E.; Whitlock, R.H.; Buergelt, C.D.; Sweeney, R.W. Exposure of young dairy cattle to *Mycobacterium avium* subsp. *paratuberculosis* (MAP) through intensive grazing of contaminated pastures in a herd positive for Johne's disease. *Can. Vet. J.* **2010**, *51*, 198–200. [PubMed]
14. Glawischnig, W.; Steineck, T.; Spergser, J. Infections caused by *Mycobacterium avium* subspecies *avium*, hominissuis, and *paratuberculosis* in free-ranging red deer (*Cervus elaphus* hippelaphus) in Austria, 2001–2004. *J. Wildl. Dis.* **2006**, *42*, 724–731. [CrossRef]
15. Whittington, R.J.; Marshall, D.J.; Nicholls, P.J.; Marsh, I.B.; Reddacliff, L.A. Survival and dormancy of *Mycobacterium avium* subsp. *paratuberculosis* in the environment. *Appl. Environ. Microbiol.* **2004**, *70*, 2989–3004. [CrossRef] [PubMed]
16. Eppleston, J.; Begg, D.J.; Dhand, N.K.; Watt, B.; Whittington, R.J. Environmental Survival of *Mycobacterium avium* subsp. *paratuberculosis* in Different Climatic Zones of Eastern Australia. *Appl. Environ. Microb.* **2014**, *80*, 2337–2342. [CrossRef]
17. Chiodini, R.J.; Van Kruiningen, H.J.; Merkal, R.S. Ruminant paratuberculosis (Johne's disease): The current status and future prospects. *Cornell Vet.* **1984**, *74*, 218–262.
18. Sweeney, R.W. Transmission of paratuberculosis. *Vet. Clin. N. Am. Food Anim. Pract.* **1996**, *12*, 305–312. [CrossRef]
19. Windsor, P.A.; Whittington, R.J. Evidence for age susceptibility of cattle to Johne's disease. *Vet. J.* **2010**, *184*, 37–44. [CrossRef]
20. Nigsch, A.; Robbe-Austerman, S.; Stuber, T.P.; Pavinski Bitar, P.D.; Grohn, Y.T.; Schukken, Y.H. Who infects whom?-Reconstructing infection chains of *Mycobacterium avium* ssp. *paratuberculosis* in an endemically infected dairy herd by use of genomic data. *PLoS ONE* **2021**, *16*, e0246983. [CrossRef]
21. McGregor, H.; Dhand, N.K.; Dhungyel, O.P.; Whittington, R.J. Transmission of *Mycobacterium avium* subsp. *paratuberculosis*: Dose-response and age-based susceptibility in a sheep model. *Prev. Vet. Med.* **2012**, *107*, 76–84. [CrossRef] [PubMed]
22. Eisenberg, S.; Krieger, M.; Campe, A.; Lorenz, I.; Einax, E.; Donat, K. Herd Prevalence Estimation of *Mycobacterium avium* Subspecies *paratuberculosis* Burden in the Three Main Dairy Production Regions of Germany (PraeMAP). *Animals* **2022**, *12*, 447. [CrossRef]
23. Stau, A.; Seelig, B.; Walter, D.; Schroeder, C.; Ganter, M. Seroprevalence of *Mycobacterium avium* subsp. *paratuberculosis* in small ruminants in Germany. *Small Rumin. Res.* **2012**, *105*, 361–365. [CrossRef]
24. Sieber, P.L. Status-quo-Erhebung des Gesundheitszustandes Bayrischer Milchziegenherden Anhand Epidemiologischer Untersuchungen. Ph.D. Thesis, Ludwig-Maximilian-Universität München, Munich, Germany, 2014.
25. Moog, U.; Wehrle, F.; Donat, K.; Köhler, H. Screening of sheep- and goat herds for *Mycobacterium avium* ssp. *paratuberculosis* using both environmental samples and specific antibodies in blood serum. In Proceedings of the 10th International Sheep Veterinary Congress (ISVC 2023), Seville, Spain, 6–10 March 2023; pp. 86–87.
26. Schrott, J.; Sodoma, E.; Dunser, M.; Tichy, A.; Khol, J.L. *Mycobacterium avium* subsp. *paratuberculosis* in Sheep and Goats in Austria: Seroprevalence, Risk Factors and Detection from Boot Swab Samples. *Animals* **2023**, *13*, 1517. [CrossRef]
27. Miller, B.A.; Lu, C.D. Current status of global dairy goat production: An overview. *Asian-Australas. J. Anim. Sci.* **2019**, *32*, 1219–1232. [CrossRef]
28. Utaaker, K.S.; Chaudhary, S.; Kifleyohannes, T.; Robertson, L.J. Global Goat! Is the Expanding Goat Population an Important Reservoir of Cryptosporidium? *Front. Vet. Sci.* **2021**, *8*, 648500. [CrossRef]
29. Jimenez-Martin, D.; Garcia-Bocanegra, I.; Risalde, M.A.; Fernandez-Molera, V.; Jimenez-Ruiz, S.; Isla, J.; Cano-Terriza, D. Epidemiology of paratuberculosis in sheep and goats in southern Spain. *Prev. Vet. Med.* **2022**, *202*, 105637. [CrossRef] [PubMed]
30. Iarussi, F.; Paradies, P.; Sardaro, R.; Rubino, G.; Scaltrito, D.; Pieragostini, E.; Petazzi, F. Epidemiology and risk factors of *Mycobacterium avium* subspecies *paratuberculosis* in semi-extensive dairy sheep and goat farms of Apulia, southern Italy. *Small Ruminant Res.* **2019**, *177*, 89–96. [CrossRef]
31. Barrero-Dominguez, B.; Luque, I.; Huerta, B.; Gomez-Laguna, J.; Galan-Relano, A.; Gomez-Gascon, L.; Sanchez, M.; Astorga, R.J. Paratuberculosis in dairy goat flocks from southern Spain: Risk factors associated with seroprevalence. *Vet. Rec.* **2019**, *185*, 600. [CrossRef] [PubMed]
32. Castellanos, E.; de Juan, L.; Dominguez, L.; Aranaz, A. Progress in molecular typing of *Mycobacterium avium* subspecies *paratuberculosis*. *Res. Vet. Sci.* **2012**, *92*, 169–179. [CrossRef]
33. Stevenson, K. Genetic diversity of *Mycobacterium avium* subspecies *paratuberculosis* and the influence of strain type on infection and pathogenesis: A review. *Vet. Res.* **2015**, *46*, 64. [CrossRef]
34. Fawzy, A.; Zschock, M.; Ewers, C.; Eisenberg, T. Genotyping methods and molecular epidemiology of *Mycobacterium avium* subsp. *paratuberculosis* (MAP). *Int. J. Vet. Sci. Med.* **2018**, *6*, 258–264. [CrossRef]
35. Ahlstrom, C.; Barkema, H.W.; Stevenson, K.; Zadoks, R.N.; Biek, R.; Kao, R.; Trewby, H.; Haupstein, D.; Kelton, D.F.; Fecteau, G.; et al. Genome-Wide Diversity and Phylogeography of *Mycobacterium avium* subsp. *paratuberculosis* in Canadian Dairy Cattle. *PLoS ONE* **2016**, *11*, e0149017. [CrossRef]

36. Bryant, J.M.; Thibault, V.C.; Smith, D.G.; McLuckie, J.; Heron, I.; Sevilla, I.A.; Biet, F.; Harris, S.R.; Maskell, D.J.; Bentley, S.D.; et al. Phylogenomic exploration of the relationships between strains of *Mycobacterium avium* subspecies *paratuberculosis*. *BMC Genom.* **2016**, *17*, 79. [CrossRef]
37. Perets, V.; Allen, A.; Crispell, J.; Cassidy, S.; O'Connor, A.; Farrell, D.; Browne, J.A.; O'Mahony, J.; Skuce, R.; Kenny, K.; et al. Evidence for local and international spread of *Mycobacterium avium* subspecies *paratuberculosis* through whole genome sequencing of isolates from the island of Ireland. *Vet. Microbiol.* **2022**, *268*, 109416. [CrossRef] [PubMed]
38. Leao, C.; Goldstone, R.J.; Bryant, J.; McLuckie, J.; Inacio, J.; Smith, D.G.; Stevenson, K. Novel Single Nucleotide Polymorphism-Based Assay for Genotyping *Mycobacterium avium* subsp. *paratuberculosis*. *J. Clin. Microbiol.* **2016**, *54*, 556–564. [CrossRef] [PubMed]
39. Fawzy, A.; Zschöck, M.; Ewers, C.; Eisenberg, T. Development of a hierarchical typing approach for *Mycobacterium avium* subsp. *paratuberculosis* (MAP) and characterization of MAP field cultures from Central Germany. *J. Appl. Microbiol.* **2020**, *129*, 1193–1206. [CrossRef]
40. Thibault, V.C.; Grayon, M.; Boschiroli, M.L.; Hubbans, C.; Overduin, P.; Stevenson, K.; Gutierrez, M.C.; Supply, P.; Biet, F. New variable-number tandem-repeat markers for typing *Mycobacterium avium* subsp. *paratuberculosis* and *M. avium* strains: Comparison with IS900 and IS1245 restriction fragment length polymorphism typing. *J. Clin. Microbiol.* **2007**, *45*, 2404–2410. [CrossRef]
41. de Juan, L.; Alvarez, J.; Aranaz, A.; Rodriguez, A.; Romero, B.; Bezos, J.; Mateos, A.; Dominguez, L. Molecular epidemiology of Types I/III strains of *Mycobacterium avium* subspecies *paratuberculosis* isolated from goats and cattle. *Vet. Microbiol.* **2006**, *115*, 102–110. [CrossRef] [PubMed]
42. Sevilla, I.; Li, L.; Amonsin, A.; Garrido, J.M.; Geijo, M.V.; Kapur, V.; Juste, R.A. Comparative analysis of *Mycobacterium avium* subsp. *paratuberculosis* isolates from cattle, sheep and goats by short sequence repeat and pulsed-field gel electrophoresis typing. *BMC Microbiol.* **2008**, *8*, 204. [CrossRef]
43. de Juan, L.; Mateos, A.; Dominguez, L.; Sharp, J.M.; Stevenson, K. Genetic diversity of *Mycobacterium avium* subspecies *paratuberculosis* isolates from goats detected by pulsed-field gel electrophoresis. *Vet. Microbiol.* **2005**, *106*, 249–257. [CrossRef] [PubMed]
44. Sevilla, I.; Garrido, J.M.; Geijo, M.; Juste, R.A. Pulsed-field gel electrophoresis profile homogeneity of *Mycobacterium avium* subsp. *paratuberculosis* isolates from cattle and heterogeneity of those from sheep and goats. *BMC Microbiol.* **2007**, *7*, 18. [CrossRef]
45. Stevenson, K.; Alvarez, J.; Bakker, D.; Biet, F.; de Juan, L.; Denham, S.; Dimareli, Z.; Dohmann, K.; Gerlach, G.F.; Heron, I.; et al. Occurrence of *Mycobacterium avium* subspecies *paratuberculosis* across host species and European countries with evidence for transmission between wildlife and domestic ruminants. *BMC Microbiol.* **2009**, *9*, 212. [CrossRef] [PubMed]
46. Fritsch, I.; Luyven, G.; Kohler, H.; Lutz, W.; Mobius, P. Suspicion of *Mycobacterium avium* subsp. *paratuberculosis* transmission between cattle and wild-living red deer (*Cervus elaphus*) by multitarget genotyping. *Appl. Environ. Microbiol.* **2012**, *78*, 1132–1139. [CrossRef] [PubMed]
47. Oakey, J.; Gavey, L.; Singh, S.V.; Platell, J.; Waltisbuhl, D. Variable-number tandem repeats genotyping used to aid and inform management strategies for a bovine Johne's disease incursion in tropical and subtropical Australia. *J. Vet. Diagn. Investig.* **2014**, *26*, 651–657. [CrossRef]
48. Sohal, J.S.; Arsenault, J.; Labrecque, O.; Fairbrother, J.H.; Roy, J.P.; Fecteau, G.; L'Homme, Y. Genetic structure of *Mycobacterium avium* subsp. *paratuberculosis* population in cattle herds in Quebec as revealed by using a combination of multilocus genomic analyses. *J. Clin. Microbiol.* **2014**, *52*, 2764–2775. [CrossRef] [PubMed]
49. Verdugo, C.; Pleydell, E.; Price-Carter, M.; Prattley, D.; Collins, D.; de Lisle, G.; Vogue, H.; Wilson, P.; Heuer, C. Molecular epidemiology of *Mycobacterium avium* subsp. *paratuberculosis* isolated from sheep, cattle and deer on New Zealand pastoral farms. *Prev. Vet. Med.* **2014**, *117*, 436–446. [CrossRef] [PubMed]
50. Ahlstrom, C.; Barkema, H.W.; Stevenson, K.; Zadoks, R.N.; Biek, R.; Kao, R.; Trewby, H.; Haupstein, D.; Kelton, D.F.; Fecteau, G.; et al. Limitations of variable number of tandem repeat typing identified through whole genome sequencing of *Mycobacterium avium* subsp. *paratuberculosis* on a national and herd level. *BMC Genom.* **2015**, *16*, 161. [CrossRef]
51. Marquetoux, N.; Heuer, C.; Wilson, P.; Ridler, A.; Stevenson, M. Merging DNA typing and network analysis to assess the transmission of paratuberculosis between farms. *Prev. Vet. Med.* **2016**, *134*, 113–121. [CrossRef]
52. de Kruijf, M.; Lesniak, O.N.; Yearsley, D.; Ramovic, E.; Coffey, A.; O'Mahony, J. Low genetic diversity of bovine *Mycobacterium avium* subspecies *paratuberculosis* isolates detected by MIRU-VNTR genotyping. *Vet. Microbiol.* **2017**, *203*, 280–285. [CrossRef]
53. Galiero, A.; Leo, S.; Garbarino, C.; Arrigoni, N.; Russo, S.; Giacomelli, S.; Bianchi, A.; Trevisiol, K.; Idrizi, I.; Daka, G.; et al. *Mycobacterium avium* subsp. *paratuberculosis* isolated from wild red deer (*Cervus elaphus*) in Northern Italy. *Vet. Microbiol.* **2018**, *217*, 167–172. [CrossRef] [PubMed]
54. Friedrich-Loeffler-Institut. Paratuberkulose (*Mycobacterium avium* subsp. *paratuberculosis*). Available online: https://www.openagrar.de/servlets/MCRFileNodeServlet/openagrar_derivate_00028517/TK18-Paratuberkulose-2023-04-26_bf.pdf (accessed on 3 August 2023).
55. Kohler, H.; Soschinka, A.; Meyer, M.; Kather, A.; Reinhold, P.; Liebler-Tenorio, E. Characterization of a caprine model for the subclinical initial phase of *Mycobacterium avium* subsp. *paratuberculosis* infection. *BMC Vet. Res.* **2015**, *11*, 74. [CrossRef]

56. Schmidt, V.; Kohler, H.; Heenemann, K.; Mobius, P. Mycobacteriosis in Various Pet and Wild Birds from Germany: Pathological Findings, Coinfections, and Characterization of Causative Mycobacteria. *Microbiol. Spectr.* **2022**, *10*, e0045222. [CrossRef] [PubMed]
57. Englund, S.; Ballagi-Pordany, A.; Bolske, G.; Johansson, K.E. Single PCR and nested PCR with a mimic molecule for detection of *Mycobacterium avium* subsp. *paratuberculosis*. *Diagn. Microbiol. Infect. Dis.* **1999**, *33*, 163–171. [CrossRef]
58. Guerrero, C.; Bernasconi, C.; Burki, D.; Bodmer, T.; Telenti, A. A novel insertion element from *Mycobacterium avium*, IS1245, is a specific target for analysis of strain relatedness. *J. Clin. Microbiol.* **1995**, *33*, 304–307. [CrossRef]
59. Radomski, N.; Thibault, V.C.; Karoui, C.; de Cruz, K.; Cochard, T.; Gutierrez, C.; Supply, P.; Biet, F.; Boschiroli, M.L. Determination of genotypic diversity of *Mycobacterium avium* subspecies from human and animal origins by mycobacterial interspersed repetitive-unit-variable-number tandem-repeat and IS1311 restriction fragment length polymorphism typing methods. *J. Clin. Microbiol.* **2010**, *48*, 1026–1034. [CrossRef] [PubMed]
60. Thibault, V.C.; Grayon, M.; Boschiroli, M.L.; Willery, E.; Allix-Beguec, C.; Stevenson, K.; Biet, F.; Supply, P. Combined multilocus short-sequence-repeat and mycobacterial interspersed repetitive unit-variable-number tandem-repeat typing of *Mycobacterium avium* subsp. *paratuberculosis* isolates. *J. Clin. Microbiol.* **2008**, *46*, 4091–4094. [CrossRef] [PubMed]
61. Mobius, P.; Luyven, G.; Hotzel, H.; Kohler, H. High genetic diversity among *Mycobacterium avium* subsp. *paratuberculosis* strains from German cattle herds shown by combination of IS900 restriction fragment length polymorphism analysis and mycobacterial interspersed repetitive unit-variable-number tandem-repeat typing. *J. Clin. Microbiol.* **2008**, *46*, 972–981. [CrossRef]
62. Fawzy, A.; Zschock, M.; Ewers, C.; Eisenberg, T. New polymorphisms within the variable number tandem repeat (VNTR) 7 locus of *Mycobacterium avium* subsp. *paratuberculosis*. *Mol. Cell Probes* **2016**, *30*, 132–137. [CrossRef]
63. Amonsin, A.; Li, L.L.; Zhang, Q.; Bannantine, J.P.; Motiwala, A.S.; Sreevatsan, S.; Kapur, V. Multilocus short sequence repeat sequencing approach for differentiating among *Mycobacterium avium* subsp. *paratuberculosis* strains. *J. Clin. Microbiol.* **2004**, *42*, 1694–1702. [CrossRef]
64. Harris, N.B.; Payeur, J.B.; Kapur, V.; Sreevatsan, S. Short-sequence-repeat analysis of *Mycobacterium avium* subsp. *paratuberculosis* and *Mycobacterium avium* subsp. avium isolates collected from animals throughout the United States reveals both stability of loci and extensive diversity. *J. Clin. Microbiol.* **2006**, *44*, 2970–2973. [CrossRef]
65. Hunter, P.R.; Gaston, M.A. Numerical index of the discriminatory ability of typing systems: An application of Simpson's index of diversity. *J. Clin. Microbiol.* **1988**, *26*, 2465–2466. [CrossRef] [PubMed]
66. Mobius, P.; Fritsch, I.; Luyven, G.; Hotzel, H.; Kohler, H. Unique genotypes of *Mycobacterium avium* subsp. *paratuberculosis* strains of Type III. *Vet. Microbiol.* **2009**, *139*, 398–404. [CrossRef] [PubMed]
67. Florou, M.; Leontides, L.; Kostoulas, P.; Billinis, C.; Sofia, M. Strain-specific sensitivity estimates of *Mycobacterium avium* subsp. *paratuberculosis* culture in Greek sheep and goats. *Zoonoses Public Health* **2009**, *56*, 49–52. [CrossRef]
68. Whittington, R.J.; Marsh, I.B.; Saunders, V.; Grant, I.R.; Juste, R.; Sevilla, I.A.; Manning, E.J.; Whitlock, R.H. Culture phenotypes of genomically and geographically diverse *Mycobacterium avium* subsp. *paratuberculosis* isolates from different hosts. *J. Clin. Microbiol.* **2011**, *49*, 1822–1830. [CrossRef]
69. Favila-Humara, L.C.; Chavez-Gris, G.G.; Carrillo-Casas, E.M.; Hernandez-Castro, R. *Mycobacterium avium* subsp. *paratuberculosis* detection in individual and bulk tank milk samples from bovine herds and caprine flocks. *Foodborne Pathog. Dis.* **2010**, *7*, 351–355. [CrossRef]
70. Bauman, C.A.; Jones-Bitton, A.; Ahlstrom, C.; Mutharia, L.; De Buck, J.; Jansen, J.; Kelton, D.; Menzies, P. Identification of *Mycobacterium avium* subspecies *paratuberculosis* strains isolated from dairy goats and dairy sheep in Ontario, Canada. *Can. J. Vet. Res.* **2017**, *81*, 304–307. [PubMed]
71. Scherrer, S.; Stephan, R.; Zumthor, J.P.; Kipar, A.; Seehusen, F. Morphological and Molecular Characterization of a New *Mycobacterium avium* subsp. *paratuberculosis* S-Type Strain Genotype in Goats. *Front. Vet. Sci.* **2019**, *6*, 250. [CrossRef]
72. Bauerfeind, R.; Benazzi, S.; Weiss, R.; Schliesser, T.; Willems, H.; Baljer, G. Molecular characterization of *Mycobacterium paratuberculosis* isolates from sheep, goats, and cattle by hybridization with a DNA probe to insertion element IS900. *J. Clin. Microbiol.* **1996**, *34*, 1617–1621. [CrossRef]
73. Conde, C.; Theze, J.; Cochard, T.; Rossignol, M.N.; Fourichon, C.; Delafosse, A.; Joly, A.; Guatteo, R.; Schibler, L.; Bannantine, J.P.; et al. Genetic Features of *Mycobacterium avium* subsp. *paratuberculosis* Strains Circulating in the West of France Deciphered by Whole-Genome Sequencing. *Microbiol. Spectr.* **2022**, *10*, e0339222. [CrossRef]
74. Byrne, A.; Ollier, S.; Tahlan, K.; Biet, F.; Bissonnette, N. Genomic epidemiology of *Mycobacterium avium* subsp. *paratuberculosis* isolates from Canadian dairy herds provides evidence for multiple infection events. *Front. Genet.* **2023**, *14*, 1043598. [CrossRef]
75. Correa-Valencia, N.M.; Moyano, R.D.; Hernandez-Agudelo, M.; Fernandez-Silva, J.A. *Mycobacterium avium* subsp. *paratuberculosis* (MAP) molecular diversity in cattle, sheep, and goats from Latin America and the Caribbean: A systematic review. *Trop. Anim. Health Prod.* **2021**, *53*, 468. [CrossRef]
76. Moyano, R.D.; Imperiale, B.R.; Romero, M.; Santangelo, M.P.; Alvarado Pinedo, M.F.; Traveria, G.; Romano, M.I. Genetic diversity of *Mycobacterium avium* sp. *paratuberculosis* by mycobacterial interspersed repetitive Unit-Variable number tandem repeat and multi-locus short-sequence repeat one-sentence summary: Genetic diversity of *Mycobacterium avium* sp. *paratuberculosis* isolates. *Int. J. Mycobacteriol.* **2021**, *10*, 51–59. [CrossRef]

77. Biet, F.; Sevilla, I.A.; Cochard, T.; Lefrancois, L.H.; Garrido, J.M.; Heron, I.; Juste, R.A.; McLuckie, J.; Thibault, V.C.; Supply, P.; et al. Inter- and intra-subtype genotypic differences that differentiate *Mycobacterium avium* subspecies *paratuberculosis* strains. *BMC Microbiol.* **2012**, *12*, 264. [CrossRef] [PubMed]
78. Fernandez-Silva, J.A.; Abdulmawjood, A.; Akineden, O.; Bulte, M. Genotypes of *Mycobacterium avium* subsp. *paratuberculosis* from South American countries determined by two methods based on genomic repetitive sequences. *Trop. Anim. Health Prod.* **2012**, *44*, 1123–1126. [CrossRef] [PubMed]
79. Gioffre, A.; Munoz, M.C.; Pinedo, M.F.A.; Vaca, R.; Morsella, C.; Fiorentino, M.A.; Paolicchi, F.; Ruybal, P.; Zumarraga, M.; Traveria, G.E.; et al. Molecular typing of Argentinian *Mycobacterium avium* subsp. *paratuberculosis* isolates by multiple-locus variable number-tandem repeat analysis. *Braz. J. Microbiol.* **2015**, *46*, 557–564. [CrossRef] [PubMed]
80. Vasini, B.; Farace, P.; Ariel, A.; Cirone, K.; Mendez, L.; Morsella, C.; Fresia, P.; Iraola, G.; Gioffre, A.; Paolicchi, F. Phylogenetic and Multiple-Locus Variable number tandem repeat analysis of *Mycobacterium avium* subsp. *paratuberculosis* isolates from Argentina. *Vet. Res. Commun.* **2022**, *46*, 1121–1129. [CrossRef] [PubMed]
81. Rasper-Hossinger, M.; Biggel, M.; Stephan, R.; Seehusen, F.; Scherrer, S. Strain diversity in *Mycobacterium avium* subsp. *paratuberculosis*-positive bovine fecal samples collected in Switzerland. *Front. Vet. Sci.* **2023**, *10*, 1154516. [CrossRef]
82. Olivieri, M.A.C.; Moyano, R.D.; Traveria, G.E.; Pinedo, M.F.A.; Mon, M.L.; Gravisaco, M.J.; Delgado, F.O.; Santangelo, M.P.; Romano, M.I. Protection efficacy of Argentinian isolates of *Mycobacterium avium* subsp. *paratuberculosis* with different genotypes and virulence in a murine model. *Res. Vet. Sci.* **2018**, *121*, 4–11. [CrossRef]
83. Hodgeman, R.; Mann, R.; Savin, K.; Djitro, N.; Rochfort, S.; Rodoni, B. Molecular characterisation of *Mycobacterium avium* subsp. *paratuberculosis* in Australia. *BMC Microbiol.* **2021**, *21*, 101. [CrossRef]
84. Outhred, A.C.; Gurjav, U.; Jelfs, P.; McCallum, N.; Wang, Q.; Hill-Cawthorne, G.A.; Marais, B.J.; Sintchenko, V. Extensive Homoplasy but No Evidence of Convergent Evolution of Repeat Numbers at MIRU Loci in Modern Mycobacterium tuberculosis Lineages. *Front. Public Health* **2020**, *8*, 455. [CrossRef] [PubMed]
85. van Hulzen, K.J.; Heuven, H.C.; Nielen, M.; Hoeboer, J.; Santema, W.J.; Koets, A.P. Different *Mycobacterium avium* subsp. *paratuberculosis* MIRU-VNTR patterns coexist within cattle herds. *Vet. Microbiol.* **2011**, *148*, 419–424. [CrossRef]
86. Hasonova, L.; Trcka, I.; Babak, V.; Rozsypalova, Z.; Pribylova, R.; Pavlik, I. Distribution of *Mycobacterium avium* subsp. *paratuberculosis* in tissues of naturally infected cattle as affected by age. *Vet. Med.-Czech* **2009**, *54*, 257–269. [CrossRef]
87. Wibberg, D.; Price-Carter, M.; Ruckert, C.; Blom, J.; Mobius, P. Complete Genome Sequence of Ovine *Mycobacterium avium* subsp. *paratuberculosis* Strain JIII-386 (MAP-S/type III) and Its Comparison to MAP-S/type I, MAP-C, and *M. avium* Complex Genomes. *Microorganisms* **2020**, *9*, 70. [CrossRef] [PubMed]
88. Sodoma, E.; Altmann, S.; Mitterhuemer, S.; Moebius, P.; Duenser, M. First comprehensive study on molecular diversity of Austrian *Mycobacterium avium* subspecies *paratuberculosis* isolates from domestic and wild ruminants. *Berl. Munch. Tierarztl. Wochenschr.* **2018**, *131*, 2–11. [CrossRef]
89. Davidson, F.W.; Ahlstrom, C.; De Buck, J.; Whitney, H.G.; Tahlan, K. Examination of *Mycobacterium avium* subspecies *paratuberculosis* mixed genotype infections in dairy animals using a whole genome sequencing approach. *PeerJ* **2016**, *4*, e2793. [CrossRef]
90. Bolzoni, L.; Scaltriti, E.; Russo, S.; Pongolini, S.; Garbarino, C.; Leo, S.; Arrigoni, N.; Ricchi, M. Transmission patterns of a *Mycobacterium avium* subsp. *paratuberculosis* clone within a single heard investigated by Whole Genome Sequencing. *Vet. Microbiol.* **2021**, *263*, 109272. [CrossRef]

animals MDPI

Article

Morphological Assessment of Concomitant Lesions Detected in Goat Herds Naturally Infected with Paratuberculosis (Johne's Disease)

Elena Plamenova Stefanova [1,2], Óscar Quesada-Canales [1,2,*], Yania Paz-Sánchez [1,2], María José Caballero [1,2,*], María del Pino Quintana-Montesdeoca [3], Antonio Espinosa de los Monteros [1,2], Miguel Antonio Rivero [1,2], Ayoze Castro [1,2], Valentín Pérez [4,5] and Marisa Andrada [1,2]

[1] Veterinary Histology and Pathology, Institute of Animal Health and Food Safety (IUSA), Veterinary School, University of Las Palmas de Gran Canaria, 35413 Arucas, Gran Canaria, Spain; elena.plamenova101@alu.ulpgc.es (E.P.S.); yania.paz102@gmail.com (Y.P.-S.); antonio.espinosa@ulpgc.es (A.E.d.l.M.); miguel.rivero@ulpgc.es (M.A.R.); ayoze.castro@ulpgc.es (A.C.); marisaana.andrada@ulpgc.es (M.A.)

[2] Departament of Morphology, Veterinary School, University of Las Palmas de Gran Canaria, 35413 Arucas, Gran Canaria, Spain

[3] Departament of Mathematics, University of Las Palmas de Gran Canaria, 35017 Las Palmas de Gran Canaria, Gran Canaria, Spain; mariadelpino.quintana@ulpgc.es

[4] Departamento de Sanidad Animal, Facultad de Veterinaria, Campus de Vegazana, Universidad de León, 24007 León, Spain; valentin.perez@unileon.es

[5] Departamento de Sanidad Animal, Instituto de Ganadería de Montaña (CSIC-ULE), 24346 Grulleros, Spain

* Correspondence: oscar.quesada@ulpgc.es (Ó.Q.-C.); mariajose.caballero@ulpgc.es (M.J.C.)

Simple Summary: Paratuberculosis (PTB) is a well-known disease with considerable financial impact on the farm industry worldwide. Nevertheless, data regarding the assessment of naturally infected goat herds is limited. The present study describes in detail the observed gross and histological lesions detected in 39 necropsies of goats (15 vaccinated and 24 non-vaccinated) from herds with a confirmed history of PTB. PTB microscopic lesions of various grades were detected in all animals in target organs and the presence of the causative agent was confirmed using different laboratory tools. The main inflammatory findings affected the hemolymphatic, respiratory and gastrointestinal systems. The lesions were confirmed microscopically with lesser macroscopically visible alterations. Our result demonstrated that non-vaccinated animals presented more severe PTB intestinal lesions and had respiratory inflammation in all age groups studied. Those also presented a higher prevalence of ileocecal valve PTB lesions. Gastrointestinal non-PTB lesions were detected in higher number in non-vaccinated goats. Thus, histology is a powerful tool for herd diagnosis and assessment. Mainly inflammatory lesions of the respiratory and gastrointestinal tract were detected in the studied PTB-affected herds. Additionally, vaccination against PTB could play a key role in the reduction of lung and gastrointestinal inflammatory processes present in the herd.

Abstract: Paratuberculosis (PTB), caused by *Mycobacterium avium* subspecies *paratuberculosis* (MAP), causes significant financial losses in the ruminant industry. The aim of this study is to describe the concomitant pathological findings as well as PTB-induced lesions in 39 naturally infected goats (15 vaccinated and 24 non-vaccinated). All animals exhibited MAP-induced microscopic lesions affecting target organs, although only 62% of those were detected grossly. Mainly concomitant inflammatory pathologies were recognized affecting the hemolymphatic, respiratory and gastrointestinal systems. Non-vaccinated animals exhibited both moderate and marked granulomatous enteritis in contrast with vaccinated ones which presented mild intestinal affection. Our results demonstrate that non-vaccinated animals presented pneumonia in all age groups studied (from 12 up to >48 months old). A significantly higher prevalence of ileocecal valve PTB lesions was detected in non-vaccinated animals with pneumonic lesions ($p = 0.027$). Furthermore, a reduction of gastrointestinal non-PTB processes was described in vaccinated goats. In conclusion, a PTB infected goat herd can be affected by a wide range of concomitant pathologies, mostly inflammatory in origin. Anatomic pathology is of crucial importance for correct herd diagnosis and histopathology is an indispensable tool for lesion detection. Additionally, anti-MAP vaccination could have a beneficial effect on the reduction of respiratory and gastrointestinal non-PTB diseases.

Keywords: *Mycobacterium avium* subspecies *paratuberculosis*; goat pathology; histopathology; lesions; vaccination; concomitant pathologies

1. Introduction

Paratuberculosis (PTB), also known as Johne's disease, is a progressive emaciating disease caused by *Mycobacterium avium* subspecies *paratuberculosis* (MAP), which affects domestic and wild ruminants worldwide [1–6]. Although young animals are more susceptible to infection due to the high degree of environmental contamination, the subclinical phase is long, and the originated economic losses are significant in both cattle and the small ruminant industry [4,7–9]. Clinical-phase signs are non-specific and include a reduction in daily milk yield, weight loss, poor body condition and diarrhea [10].

PTB is a disease which is impacting European countries due to the considerable economic losses it originates and its high prevalence. Herd seropositivity of up to 65% has been reported in goat and sheep herds in Germany and 62.9% in France [11,12]. In Italy, various reports highlight the need for the implementation of control strategies and viable diagnostic tools for PTB early detection [13,14]. In Spain, goat farming is one of the biggest pillars of the national livestock breeding industry since the country counts for 23% of EU's goat population (EUROSTAT 2021). Goat farms represent the primary livestock breeding sector in the Canary Islands, with 202.887 heads and 1289 dairy farms distributed as follows: 478 on Gran Canaria, 224 on Tenerife, 212 on Fuerteventura, 129 on La Palma, 117 on Lanzarote, 82 on El Hierro and 47 on La Gomera (ISTAC 2020). The archipelago has the fourth-largest goat population in Spain (Eurostat 2020), composed mainly of certified autochthonous endangered breeds (Orden APM/26/2018). In continental Spain, considerable economic losses due to goat PTB have been reported, and it is considered a widespread problem for the whole national territory with herd seroprevalence of 87.5% in southern Spain, 46% in Huelva region, 44% in Madrid and 56% in Avila, although data about the prevalence of the disease in the Canary Islands is limited [8,15].

Gross lesions in animals with PTB observed during necropsy are usually confined to the intestinal mucosa and the draining lymph nodes [6,16,17]. The ileocaecal valve mucosa can appear diffusely thickened and folded into transverse rugae, the crests of which may be congested. The mesenteric (MS) and ileocecal lymph (IC) nodes (LNs) are diffusely enlarged, edematous and pale (lymphadenomegaly, LAM). Lymphangitis can be detected in the mesentery. In some mild cases, lymphangitis is the only gross lesion specific enough to justify a presumptive PTB gross diagnosis [6,16,18]. Other commonly associated macroscopic findings are loss of muscle mass, serous atrophy of fat depots, intermandibular oedema, fluid effusion in cavities, with the presence of mineralization plates and intimal fibrosis in the thoracic aorta [6,16,18].

The main histopathological lesions associated with MAP-infection include transmural granulomatous enteritis, mesenteric and ileocecal lymph nodes granulomatous lymphadenitis and lymphangitis [6,16–18]. Villi can present moderate to marked atrophy with infiltrate of epithelioid macrophages and a variable number of Langhans-type multinucleated giant cells focally or diffusely distributed along the lamina propria, submucosa muscular layer or serosa of the intestine. The lesions observed in lymph nodes can vary from multiple foci of epithelioid macrophages up to well-formed encapsulated granulomas with central necrosis [6,16–18].

Although PTB is a major livestock farming issue of worldwide distribution, there may be a variety of other different diseases co-existing in naturally infected herds. The prevalence of the different diseases varies among the age groups present in the herd, the breed of the animals, the herd size, environmental factors as well as the variations in the pathogens present in each farm [8,15,19–21]. Respiratory, gastrointestinal, lymphatic and hepatic pathologies stand out among the main lesions detected in small ruminant herds worldwide [19,22,23].

Respiratory diseases represent 5.6% of the most common diseases in small ruminants affecting both individuals or groups, resulting in considerable financial losses due to decreased meat, milk and wool production along with a reduced number of offspring [24,25]. The etiology is multifactorial, involving mainly bacteria and viruses, and secondary infections are common. The lung can present lesions varying from acute bronchopneumonia up to severe chronic lung involvement [5,24,25].

Gastrointestinal affection, besides PTB-induced lesions, can be associated with diarrhea and weight loss due to viral, bacterial and parasitic infections. Acute lesions can vary from marked hyperemia of the intestinal mucosa up to fibrino-haemorrhagic and even necrotizing enteritis affecting both the small and large intestines. Subacute and chronic cases can exhibit multifocally distributed lesions, such as abscesses affecting the ileum, caecum and colon [6,19,22,23].

Lymphadenopathy is also considered a major clinical finding in many important infectious goat diseases. LNs are the primary disease location in cases of *C. pseudotuberculosis* induced caseous lymphadenitis [5,26]. Acute lymphadenitis, on the other hand, is a common finding in regional LNs associated with organs affected by bacterial infection [5,19,26].

Specific liver lesions are mainly of inflammatory origin (hepatitis) [5]. In adult animals, abscesses are commonly observed in association with *Trueperella pyogenes* or *Fusobacterium necrophorum* infection. Necrotizing hepatitis caused by *Clostridium* spp. is also a common finding [5,27,28].

Anatomopathological examination (gross and histopathological findings) is crucial to adequate herd diagnosis [16,18,29]. For correct assessment and confirmation of the suspected diagnosis or new insights on the investigated disease problem in the herd, all possible necropsies following a systematic procedure should be performed as a useful adjunct to clinical examination [5,19,22,25].

The present study aims to provide a morphological classification of a variety of MAP-induced lesions in naturally infected animals as well as to recognize the different concomitant pathological findings in 39 female majorera goats with histopathological PTB lesions detected during a routine postmortem examination.

2. Materials and Methods

2.1. Study Design

A total of 39 adult goats with histological PTB lesions were included in the study. All animals were submitted with a presumption of a field PTB diagnosis and were examined by the Anatomic Pathology Diagnostic Service of the Veterinary School of the University of Las Palmas de Gran Canaria. The animals were submitted from 8 farms (F1, F2, F3, F4, F5, F6, F7 and F8) in the Canary Islands with a confirmed history of PTB. The on-farm confirmation protocol included serological ELISA immunoassay (PARACHEK® 2 Kit, Thermo Fisher Scientific, Waltham, MA, USA), bacteriological culture of environmental samples and subsequent PCR confirmation, postmortem examination and compatible clinical signs detection. A total of 15/39 animals from 3 of the farms were vaccinated against MAP (Table 1). All goats selected presented histological lesions compatible with PTB including granulomatous enteritis (GE) with the presence of focal to multifocal/diffuse transmural accumulations of epithelioid cells and lymphocytes typically accompanied by Langhans-type multinucleated giant cells (MGCs) and/or granulomatous lymphadenitis (GLA) of the mesenteric lymph nodes (MS LNs) characterized by the infiltration of epithelioid macrophages and/or Langhans-type MGCs.

Table 1. Number of vaccinated (V) and non-vaccinated (nV) goats included in the study by farm and age range.

Farm	Age Range (Months)						Total
	(0–12]	(12–24]	(24–36]	(36–48]	>48	Unknown	
F1	1V	2V	4V	-	-	1V	8V
F2	-	5V + 1nV	1V	-	-	-	6V + 1nV
F3	-	1V + 7nV	-	-	-	-	1V + 7nV
F4	-	5nV	1nV	-	-	-	6nV
F5	-	-	1nV	-	3nV	-	4nV
F6	-	-	2nV	-	-	-	2nV
F7	-	-	-	1nV	1nV	-	2nV
F8	-	-	1nV	1nV	-	-	2nV
Total (n = 39 [1])	1V	8V + 13nV	5V + 5nV	2nV	4nV	1V	15nV + 24nV

V, vaccinated; nV, non-vaccinated. [1] Data about the age of 38/39 animals was available. All animals included in the study presented histological lesions compatible with PTB, including granulomatous lymphadenitis of the mesenteric lymph nodes and/or granulomatous enteritis.

According to the local legislation (Decree 51/2018 del 23 de abril), caprine vaccination against MAP is only permitted after "tuberculosis-free" status confirmation and PTB diagnostic confirmation since the archipelago was declared "officially free" of bovine tuberculosis (TB) in 2017. The vaccination protocols in the 8 farms studied included commercial vaccines against *E. coli*, *Clostridium* spp. (enterotoxemia), *Chlamydia* spp. (abortion) and *Pasteurella* spp. (bronchopneumonia) which were applied following the manufacturer's recommendations.

2.2. Anti-MAP Vaccine

Gudair® is a commercial heat-inactivated vaccine containing 2.5 mg/mL of MAP strain 316 F with mineral oil adjuvant (CZ Vaccines S.A., O Porriño, Pontevedra, Spain) for use in sheep and goats. It had been administered once subcutaneously as per the manufacturer's instructions in the three farms applying PTB vaccination enrolled in this study.

2.3. Sample Collection and Processing for Histolgical Examination

Necropsies were performed following the protocol proposed by King et al., 2014 [29]. Lymph nodes and intestines were sampled for histology according to the recommendations of the national bovine tuberculosis eradication program protocol 2021 (MAPA, 2021) as follows: mesenteric (MS), ileocecal (IC), retropharyngeal (RPh), prescapular (PE) and mediastinal (MD) lymph nodes (LNs) and ileocecal valve region (ICV). Samples from representative organs (lungs and trachea, liver, kidneys and urinary bladder, heart, LNs, spleen, intestine, forestomaches and abomasum, uterus and ovaries, mammary gland, skeletal muscle and brain) were systematically collected for histological examination.

All tissue samples were fixed in 10% buffered formalin, embedded in paraffin and processed routinely, sectioned at 5 µm and stained with hematoxylin/eosin (HE). Additionally, selected tissue sections of 4 µm were stained with the Ziehl–Neelsen (ZN) technique for detecting acid-fast bacilli (AFB).

2.4. Gross Evaluation

All macroscopic lesions were recorded and described in terms of location, color, size, shape, consistency and number or percent of involvement of the affected organ and a presumed morphological gross diagnosis was established by an experienced pathologist.

2.5. Histological Evaluation

All histopathological findings were detailed and organized according to the system affected.

Two grading systems were applied to evaluate the PTB-compatible histological lesions. MS LNs lesions were classified as 0 (no lesions), I (initial), II (solid), III (minimal necrosis) and IV (necrosis and mineralization) as per Wangoo et al., 2005, grading score described for tuberculosis lymph node granulomas [30]. The lesions observed in the lamina propria (LP) and the Peyer's patches (PPs) of the ICV were graded separately in terms of severity (mild, moderate and marked) and distribution (focal, multifocal and diffuse) using the grading system proposed by Krüger et al., 2015 [18]. Severity was graded as mild when only small circumscribed lymphocytic or granulomatous infiltrates were present with no change of tissue architecture; moderate when granulomatous infiltrates with altered tissue architecture were present; and marked when massive granulomatous infiltrates with partially or completely disrupted tissue architecture were observed [18]. Lesion distribution was considered focal when up to 3 distinct granulomatous infiltrates were observed per section; multifocal when more than 3 granulomatous distinct infiltrates per section were seen; and diffuse when granulomatous infiltrate was present throughout the whole section [18].

2.6. MAP Confirmation by Immunohistochemistry

Selected sections (3 μm) from MS LNs and ICV were immunohistochemically labelled with a polyclonal anti-MAP in-house antibody kindly provided by Dr. V. Perez, University of León, León, Spain. A polymer-based detection system (EnVision®System Labelled Polymer-HRP; Dako, Glostrup, Denmark) was employed, following the manufacturer's instructions. Subsequently, immunolabelling was developed with a commercial solution of 3,3′diaminobenzidine (DAB) (K3468; Dako, Glostrup, Denmark). The slides were counterstained with Harris´ hematoxylin and mounted in a hydrophobic medium.

The number of mycobacteria per section was graded using a scoring system described by Krüger et al., 2015 [18], as follows: none (<2 labelled bacteria per section), single (mycobacteria in <20% epithelioid cells and/or MGCs; single/few bacteria per cell or foci of granular labelling, predominantly <21 mm in diameter), few (mycobacteria in 20% to 50% epithelioid cells and/or MGCs; on average, 1 to 10 bacteria per cell or foci of granular labeling predominantly >21 mm in diameter) and many (mycobacteria in >50% to 75% epithelioid cells and/or MGCs; on average >10 bacteria per cell, with up to 50% of cells containing countless bacteria).

2.7. qPCR and Bacterial Culture for MAP Identification

In order to confirm MAP infection, frozen tissue samples from MS LNs and ICV were submitted to external laboratories for real-time polymerase chain reaction (qPCR) and bacterial culture. Frozen tissue samples from 29/39 animals were available for examination. Molecular biology qPCR targeting IS900 gene was performed on samples 13/29 animals using EXOone qPCR kit (EXOPOL SL, Zaragoza, Spain). Samples were considered positive when Cq $\leq$ 38. Tissues from 16/29 animals were examined by bacterial culture as part of the national tuberculosis eradication and surveillance program. Samples from 3/29 animals were submitted to both qPCR and bacterial culture exams. The culture analysis was carried out by the national infectious diseases reference laboratory of VISAVET, Health Surveillance Centre, Madrid, Spain. The qPCR was performed by a private external laboratory.

2.8. Statistical Analysis

An observational cross-sectional study was carried out. The prevalence of macroscopic and microscopic pathological findings was calculated separately in each organic system affected. Lesion type prevalence was also measured in each system affected analyzing gross and histological findings separately.

Statistical analysis of data was performed by IBM SPSS Statistics 27 (IBM Corp. Released 2020. IBM SPSS Statistics for Windows, Version 27.0. Armonk, NY, USA: IBM Corp). Categorical variables were summarized using percentages and relative or absolute frequencies. The ages of the studied goats were categorized in the following groups: (0–12] months,

(12–24] months, (24–36] months, (36–48] months and >48 months. Chi-square test was used to contrast the association between two categorical variables. To analyze the association between two ordinal scale variables, Kendall's Tau-b test was applied. Numerical variables were summarized using the mean, standard deviation (SD), median and interquartile range (IQR). A Shapiro–Wilk test was applied to analyze the sample normality. A non-parametric Mann–Whitney U test was used to compare two independent samples when the condition for normality of the numerical variable was not met. Agreement between qualitative results obtained by two measurement procedures was calculated using the kappa coefficient (κ). The results were considered statistically significant if the p value < 0.05.

3. Results

3.1. Necropsy Findings

All animals studied exhibited poor body conditions characterized by severe emaciation, protrusion of lumbar vertebrae and easily palpable transverse processes, muscle mass loss and a reduction of visceral fat deposits.

Data about the age of the animals was available in 38/39 cases. The studied animals were between 7 and 121 months of age, with a mean of 29 months, median of 16.50 months, SD of 30.2 months and IQR of 16 months. The ages of the studied goats were categorized in the following groups: (0–12] months, (12–24] months, (24–36] months, (36–48] months and >48 months in which 1, 21, 10, 2 and 4 animals were included, respectively.

PTB granulomatous lesions were observed in 61.5% (24/39) of the animals affecting the MS LNs (Figure 1a and Table 2). PTB intestinal gross lesions were described in 7/39 (17.9%) cases consisting in the thickened, corrugated mucosal surface of the small intestine and the ICV region (Figure 1b and Table 2).

A significant difference was observed between the month of age of the animals and the presence of macroscopic PTB granulomatous lesions affecting the MS LNs (p = 0.022). Once the ages of the vaccinated and non-vaccinated groups were compared in relation with PTB induced lesions in MS LNs, significant variation was observed only in the non-vaccinated group (p = 0.029) in contrast with the vaccinated where the animals were mostly between 12 and 24 months (p = 0.471).

Granulomatous lymphadenitis was also detected in the RPh (2/39; 5.1%) and the mammary (MA) (2/39; 5.1%) LNs. Those LNs appeared firm and enlarged, with a thickened capsule. The cut surface revealed the partial to complete loss of the architecture, replaced by multiple white, firm poorly demarcated areas which often presented central caseous necrosis and mineralization.

Lymph nodes were one of the most affected organs (80%; 31/39). Lymphadenomegaly was detected in 64% (25/39) of the animals affecting mainly the MS (21/25), IC (9/25) and MA (9/25) LNs (Figure 1c). Inflammatory lesions (lymphadenitis) were seen in 57% (22/39), affecting different LNs.

Suppurative lymphadenitis (SLA) was observed in the PE (6/39; 15.4%) and MD (3/39; 7.7%) LNs. Those LNs appeared enlarged, soft and hyperemic with thick fibrous well-formed capsules. On the cut surface, the parenchyma bulged and exudated variable amounts of blood, pus or lymph.

Caseous lymphadenitis (CLA) characterized grossly by the classic "cheesy gland" appearance parenchyma affected the RPh (2/39; 5.1%), MD (3/39; 7.7%) and MA (2/39; 5.1%) LNs. Those LNs were large and firm, exudating variable amounts of pasty yellowish material (pus) which completely substituted the organ's parenchyma (Figure 1d).

Abscess formation characterized by the complete substitution of the normal lymph node parenchyma by large amount of yellow-green pus surrounded by a thick fibrous capsule was seen affecting the RPh (3/39; 7.7%), PE (4/39; 10.3%), MD (2/39; 5.1%) and MA (1/39; 2.6%) LNs.

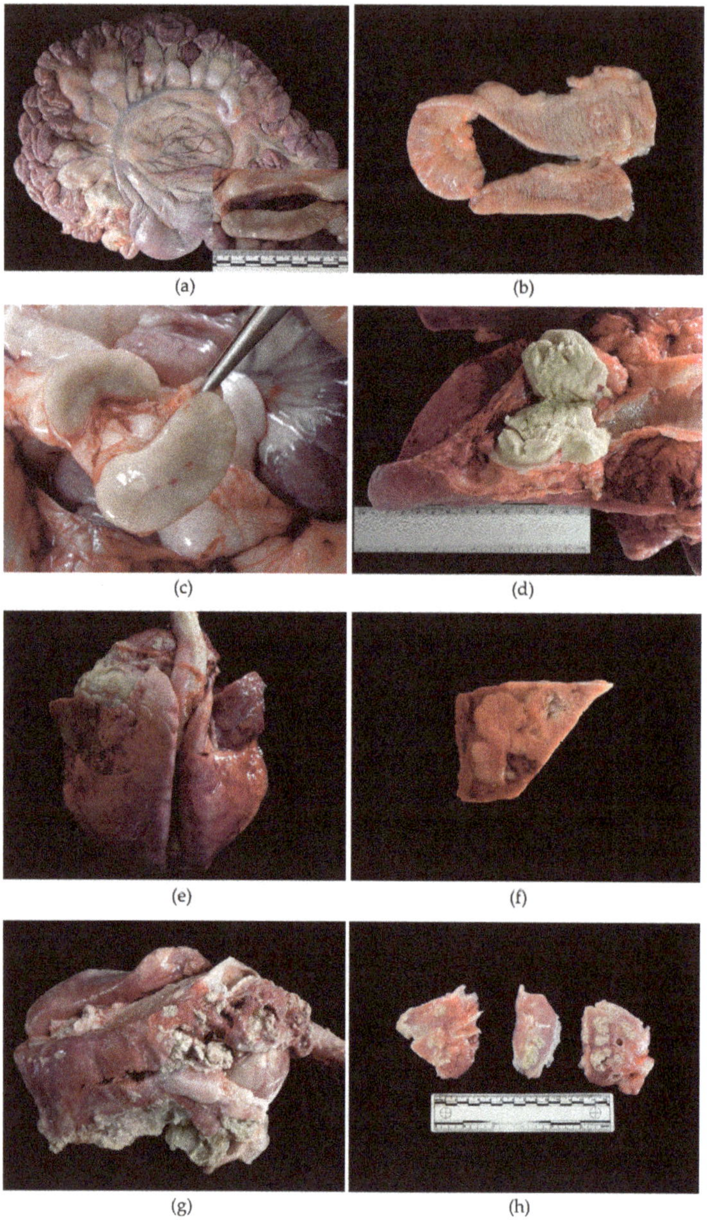

Figure 1. Main gross findings in 39 naturally infected goats with PTB lesions. (**a**) Severe lymphadenomegaly (LAM) affecting the mesenteric lymph nodes (MS LNs); inset: MS LNs with white calcified granulomas on the cut surface. (**b**) Jejunum. Diffuse thickening of the intestinal mucosa folded into transverse rugae, characteristic of PTB. (**c**) Lymphadenomegaly (LAM) in an MS LN. Enlarged lymph node with uniform cut surface. (**d**) Caseous lymphadenitis (CLA) affecting mediastinal lymph nodes (MD LNs) with abundant caseous material replacing the parenchyma completely. (**e**) Cranioventral bronchopneumonia (CBP) characterized by cranioventral dark red consolidation of lung lobes. (**f**) CBP lung section showing whitish areas of necrosis and purulent exudate. (**g**,**h**) Severe CBP affecting 70% of the lung with abundant purulent exudate oozing from airways.

Table 2. Summary of MAP-induced morphological findings and lung pneumonic lesions in 39 naturally infected goats.

			Age Range (Months)								**Total**		
			(0–12]	**(12–24]**		**(24–36]**		**(36–48]**	**>48**	**Unknown**	**V** $n = 15$	**nV** $n = 24$	$n = 39$
Vaccination			**V** $n = 1$	**V** $n = 8$	**nV** $n = 13$	**V** $n = 5$	**nV** $n = 5$	**nV** $n = 2$	**nV** $n = 4$	**V** $n = 1$			
Gross PTB lesions	MS LNs	Yes	1 (100%)	6 (75%)	8 (61.5%)	4 (80%)	2 (40%)	1 (50%)	1 (25%)	1 (100%)	12 (80%)	12 (50%)	24 (61.5%)
		No	-	2 (25%)	5 (38.5%)	1 (20%)	3 (60%)	1 (50%)	3 (75%)	-	3 (20%)	12 (50%)	15 (38.5%)
	ICV	Yes	-	3 (37.5%)	1 (7.7%)	3 (60%)	-	-	-	-	6 (40%)	1 (4.2%)	7 (17.9%)
		No	1 (100%)	5 (62.5%)	12 (92.3%)	2 (40%)	5 (100%)	2 (100%)	4 (100%)	1 (100%)	9 (60%)	23 (95.4%)	32 (82.1%)
Histological PTB lesions	MS LNs (Severity grade)	I	-	-	3 (23.1%)	-	1(20%)	-	3 (75%)	-	-	7 (29.2%)	7 (17.9%)
		III	-	-	2 (15.4%)	-	-	-	-	-	-	2 (8.3%)	2 (5.1%)
		IV	1 (100%)	8 (100%)	7 (53.8%)	5 (100%)	3 (60%)	1 (50%)	1 (25%)	1 (100%)	15 (100%)	12 (50%)	27 (69.2%)
		Subtotal	1 (100%)	8 (100%)	12 (92.3%)	5 (100%)	4 (80%)	1 (50%)	4 (100%)	1 (100%)	15 (100%)	21 (87.5%)	36 (92.3%)
		No	-	-	1 (7.7%)	-	1 (20%)	1 (50%)	-	-	-	3 (12.5%)	3 (7.7%)
	ICV LP (severity)	Mild	-	5 (62.5%)	3 (23.1%)	4 (80%)	-	1 (50%)	1 (25%)	-	9 (60%)	5 (20.8%)	14 (35.9%)
		Moderate	-	-	1 (7.7%)	-	1 (20%)	1 (50%)	1 (25%)	-	-	4 (16.7%)	4 (10.3%)
		Marked	-	-	-	-	1 (20%)	-	1 (25%)	-	-	2 (8.3%)	2 (5.1%)
		Subtotal	-	5 (62.5%)	4 (30.8%)	4 (80%)	2 (40%)	2 (100%)	3 (75%)	-	9 (60%)	11 (45.8%)	20 (51.3%)
		No	1 (100%)	3 (37.5%)	9 (69.2%)	1 (20%)	3 (60%)	-	1 (25%)	1 (100%)	6 (40%)	13 (54.2%)	19 (48.7%)
Histological pneumonic lesions	Yes		-	7 (87.5%)	10 (76.9%)	-	4 (80%)	1 (50%)	2 (50%)	1 (100%)	8 (53.3%)	17 (70.8%)	25 (64.1%)
	No		1 (100%)	1 (12.5%)	3 (23.1%)	5 (100%)	1 (20%)	1 (50%)	2 (50%)	-	7 (46.7%)	7 (29.2%)	14 (35.9%)

PTB, paratuberculosis; MS LNs, mesenteric lymph nodes; ICV LP, ileocecal valve lamina propria; V, vaccinated; nV, non-vaccinated.

The respiratory tract was affected in 17/39 (43.6%), mainly by bronchopneumonia (CBP) (16/39; 41%) characterized by cranioventral consolidation of the lungs, affecting at least 30% of the parenchyma, with dark red to maroon discoloration and an oedematous cut surface (Figure 1e,f). In some cases, purulent or catarrhal material oozed from the small airways. Fibrous adhesions between the parietal and visceral pleura (fibrous pleuritis) were observed in 5/39 (12.8%), along with CBP. The presence of fibrin strands or vails on the pleural surface (fibrinous pleuritis) along with CBP was described in seven (17.9%) (Figure 1g,h). No association with PTB macroscopic lesions was observed ($p = 0.792$) and no differences between the ages of the animals affected were demonstrated ($p = 0.854$).

The gastrointestinal tract exhibited gross lesions not categorized as PTB, in particular, a nonspecific diffusely thickened appearance in 10/39 animals (25.6%) and hemorrhagic enteritis in one animal (2.6%). The lesions detected had statistically significant associations with PTB gross lesions in MS LNs only in the non-vaccinated group ($p = 0.013$).

The liver exhibited a uniform pale yellow color, round edges and friable texture (steatosis) in 7/39 (17.9%) of the cases and multifocal, white, firm, well demarcated areas of capsular and/or subcapsular fibrosis in 6/39 (15.4%). One animal presented multifocal well-demarcated mineralization.

The heart was affected in 13% (5/39) of the cases by a variety of lesions including hydropericardium (2/39; 5.1%), fibrous pericarditis (2/39; 5.1%) and petechial haemorrhages on the pericardial surface (2/39; 5.1%).

Kidneys were affected in 5% (2/39) of the cases, presenting multiple small white nodules of up to 1 cm diameter throughout the cortex, consistent with nonsuppurative interstitial nephritis.

A summary of the main gross findings affecting the 39 goats included in the study can be found as Appendix A (Table A1).

3.2. Microscopical Findings

Lesions consistent with PTB were observed in 100% of the cases (granulomatous enteritis and/or granulomatous lymphadenitis of the MS LNs). In affected MS LNs (92.3%, 36/39), grade IV granulomas were described in 27/36 (75%) of the animals (Figure 2). Although, severe granuloma formation (grade IV) was not always the main PTB lesions detected in affected LNs. In fact, in 16/36 (44%) of the cases, grade I multifocal microgranulomas constituted the predominant lesion type in the examined samples (Figure 2a). All vaccinated goats (15/15, 100%) exhibited grade IV lesions in contrast with the non-vaccinated ones where 7/24 (29.2%) and 2/24 (8.3%) animals only presented grade I and grade III lesions, respectively (Table 2). The PPs could be evaluated in 33 animals, 70% (23/33) of which presented PTB-compatible lesions of mild severity (46%, 15/33) and multifocal distribution (55%, 18/33). The LP was affected in 51.3% (20/39) of the cases, mainly by mild lesions (35.9%, 14/39) with multifocal distribution (48.7%,19/39) (Figure 2b and Table 2). In the vaccinated group, 9/15 (60%) animals were affected, presenting mild ICV lesions affecting the LP. In contrast, the non-vaccinated PTB affected goats had moderate and marked lesions in 4/24 (16.7%) and 2/24 (8.3%) of the cases, respectively. No significant differences were described between the age groups of the animals studied ($p = 0.085$), as well as between the medians of the vaccinated (14 months) and non-vaccinated (17 months) goats ($p = 0.691$).

Reactive hyperplastic lesions consisted in follicular hyperplasia with larger follicles with a paler germinal center containing an increased numbers of apoptotic lymphocytes and tingible body macrophages were observed in 53% of the animals affecting different lymph nodes (Figure 2c).

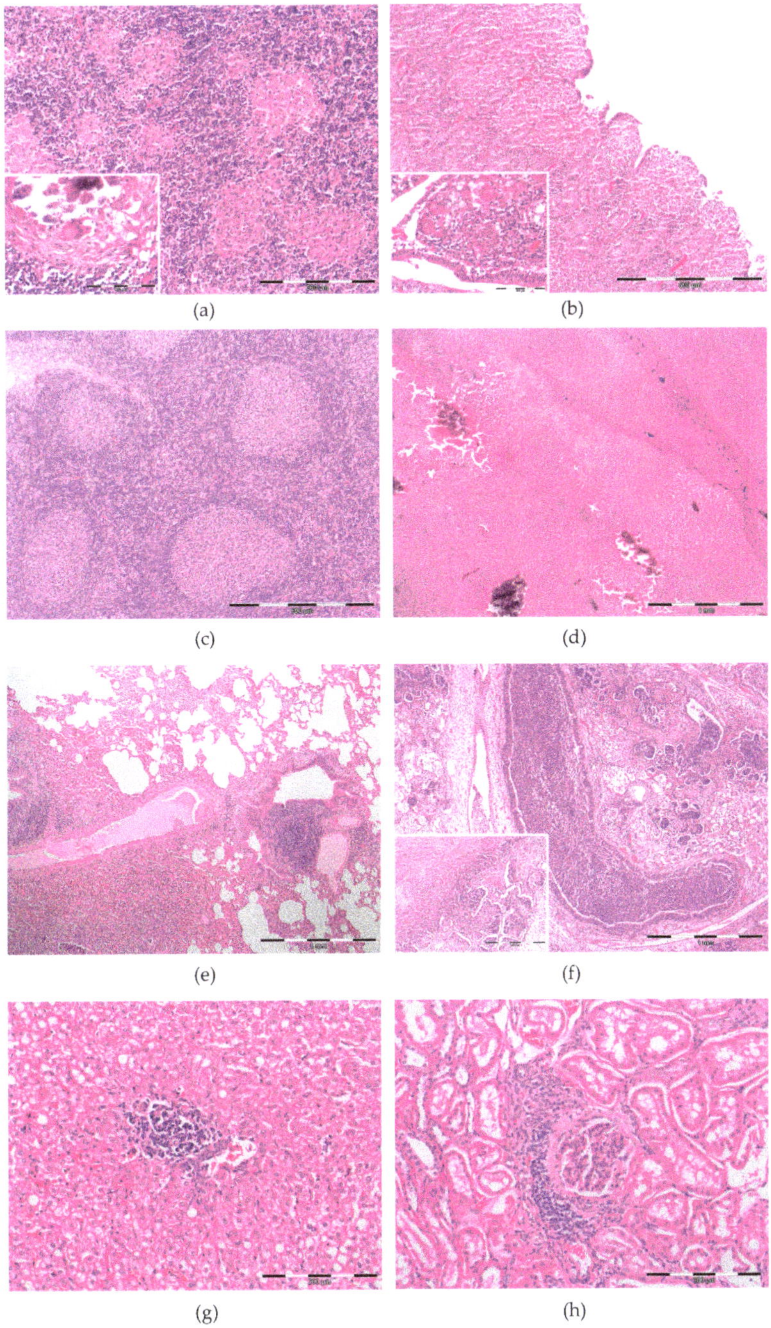

Figure 2. Main histopathological findings in 39 naturally infected goats with PTB lesions. (**a**) Mesenteric lymph node (MS LN), scale: 200 μm: multifocal aggregates of epithelioid macrophages

inset (scale: 100 μm): Langhans-type multinucleated giant cells (MGCs) in a calcified granuloma. (**b**) Jejunum (scale: 500 μm): granulomatous enteritis formed by groups of macrophages and diffuse infiltration of lymphocytes; inset (scale: 100 μm): Langhans-type MGCs with epithelioid macrophages aggregates in intestinal villi tip. (**c**) Lymph node (LN) (scale: 500 μm): reactive lymphoid hyperplasia characterized by large lymphoid follicles with prominent germinal centers. (**d**) LN (scale: 1 mm): caseous lymphadenitis/suppurative lymphadenitis (CLA/SLA) with vast areas of lytic necrosis with mineralization. (**e**) Lung (scale: 1 mm): bronchointerstitial pneumonia (BIP) characterized by proliferation of type II pneumocytes, associated lymphoid tissue and thickening of the alveolar septa by lymphoid infiltration. (**f**) Lung (scale: 1 mm): severe fibrinonecrotizing bronchopneumonia (FNBP) with degenerated neutrophils, fibrin and necrotic debris filling bronchioles and alveoli and large areas of coagulative necrosis (inset, scale: 500 μm). (**g**) Liver (scale: 200 μm): perivascular lymphohistiocytic hepatitis characterized by foci of lymphocytes and occasional macrophages and mild steatosis. (**h**) Kidney, cortex (scale: 100 μm): interstitial nephritis with lymphohistiocytic infiltration expanding the interstitium.

LNs were mainly affected by CLA and SLA (Figure 2d). CLA was characterized by variable-sized areas of lytic necrosis admixed with degenerated and viable neutrophils surrounded by epithelioid macrophages, lymphocytes and plasma cells, and a thick capsule of fibrous connective tissue. On the other hand, SLA LNs exhibited marked hyperaemia with diffuse infiltration of a moderate to large number of neutrophils, vast areas of necrosis and mineralization. The affected lymph nodes were the RPh, PE, MD and MA with lesions observed in 8/39 (20.5%), 4/39 (10.3%), 8/39(20.5%) and 4/39 (10.3%) of the cases, respectively. Abscess formation was confirmed only grossly and was not included in the histological analysis. The GLA described grossly in RPh and MA LNs were histologically classified as SLA/CLA.

Various organs also showed diverse and different lesions, mainly the respiratory tract, digestive tract, liver, cardiovascular system and kidneys.

The main lesions observed in the respiratory tract were inflammatory pneumonia processes (64.1%, 25/39) (Table 2). Necrotic-suppurative bronchopneumonia characterized by the presence of viable and degenerated neutrophils, fibrin and cell debris, filling bronchioles and alveoli, admixed with areas of necrosis and haemorrhage was detected in 12/39 animals (30.8%) (Figure 2f). Lymphohistiocytic interstitial pneumonia (IP) with increased number of lymphocytes and histiocytes thickening the alveolar septa was diagnosed in 6/39 (15.4%) of the cases. Bronchointerstitial pneumonia (BIP) was described in 3/39 animals (7.7%) and included bronchiolar necrosis, diffuse alveolar damage, thickening of the alveolar septa by lymphoid infiltration and hyperplasia of type II pneumocytes, perivascular lymphoid infiltration and hyperplasia of associated lymphoid tissue (Figure 2e). In 4/39 cases (10.3%), the lungs presented only fibrous pleuritis. In the vaccinated and non-vaccinated animals, pneumonia lesions were detected in 8/15 (53.3%) and 17/24 (70.8%), respectively (Table 2).

Differences between the age groups of vaccinated and non-vaccinated animals were detected regarding the presence of microscopic lung inflammation. Most of the vaccinated animals with histological lesions had between 12 and 24 months of age (p = 0.035) (Figure 3a). Those lesions in the non-vaccinated goats were distributed in ages from 12 months up to >120 months, with no significant differences between the number of animals in each age range (p = 0.677) (Figure 3b).

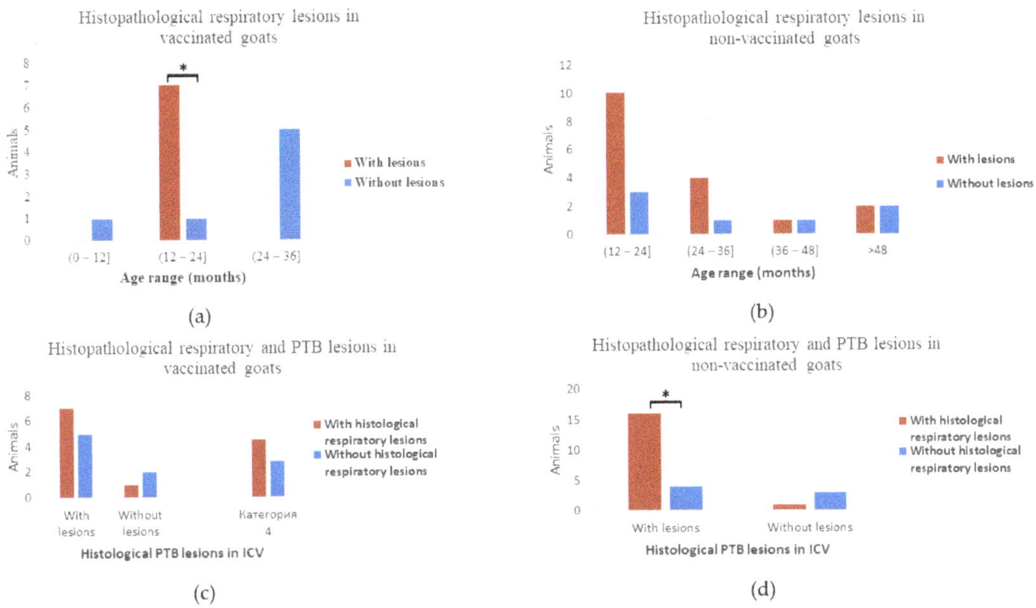

Figure 3. Associations between age groups, histopathological respiratory lesions and PTB histological lesions affecting the ileocecal valve (ICV) in vaccinated and non-vaccinated goats. (**a**) Distribution of vaccinated goats in different age groups in relation with presence of histopathological lesions of the respiratory tract. * $p = 0.035$ animals affected were mainly >12 < 24 months of age. (**b**) Distribution of non-vaccinated goats in different age groups in relation with presence of histopathological lesions of the respiratory tract. (**c**) Associations between the presence of histopathological respiratory lesions and histopathological PTB lesions in the ICV of vaccinated goats. (**d**) Associations between the presence of histopathological respiratory lesions and histopathological PTB lesions in the ICV of non-vaccinated goats. * $p = 0.027$ more non-vaccinated animals with PTB lesions affecting the ICV presented histological respiratory lesions.

In relation to the presence of histological inflammation affecting the respiratory tract (25/39), grade IV granulomas in the MS LNs were present in 16/25 (64%) animals ($p = 0.145$). Furthermore, significantly more non-vaccinated animals with MAP-induced lesions in the ICV presented lung inflammation ($p = 0.027$) (Figure 3d). No such association was described in the vaccinated group ($p = 0.438$) (Figure 3c).

Regarding the digestive system, lymphoplasmacytic (LPE) and eosinophilic enteritis (EE) were the main findings observed. LPE, characterized by a diffuse discrete-moderate number of lymphocytic mucosal infiltrate, was detected in 1/39 (2.6%) of the animals. EE characterized by eosinophils with a smaller number of globular leucocytes and plasma cells, infiltrating and expanding the intestinal mucosa was described in 2/39 (5.1%). The presence of coccidia of *Eimeria* spp. was confirmed in one animal with EE. Mild lymphohistiocytic abomasitis or gastric non-perforated ulcers were described in 4/39 (10.3%) of the animals.

There was no difference between the ages of the eight animals affected ($p = 0.430$). Seven (7) of those were non-vaccinated and only one was vaccinated ($p = 0.090$). Five (5) of the non-vaccinated affected goats (71%) exhibited grade IV granulomas affecting the MS LNs ($p = 0.170$).

The most frequent hepatic finding was lymphohistiocytic hepatitis (64.1%, 25/39), characterized by multiple foci of lymphocytes and occasional macrophages, mainly with a periportal distribution (Figure 2g). Hepatocellular steatosis with hepatocytes presenting one large or multiple variably sized discrete intracytoplasmic vacuoles, frequently displacing and compressing the nucleus to the periphery, was also observed (11/39, 28.2%).

Kidneys presented non-suppurative interstitial nephritis, mostly affecting the cortex and to a lesser extent, the medulla, with the diffuse infiltration of lymphocytes and occasional plasma cells and macrophages, interstitial fibrosis and tubular atrophy in 25.6% (10/39) of the cases (Figure 2h).

The heart was predominantly affected by the presence of apicomplexan schizont, containing large numbers of oval basophilic bradyzoites, compatible with *Sarcocystis* spp., expanding cardiomyocytes (9/39, 23.1%). Non-suppurative myocarditis characterized by multifocal lymphocytic infiltration and occasional myofiber degeneration was also observed (3/39, 7.7%). Fibrous pericarditis was confirmed in 3/39 (7.7%).

A summary of the histological lesions detected in the 39 animals studied can be found as Appendix A (Table A1).

3.3. MAP Identification (Immunohistochemistry, Ziehl–Neelsen, Bacterial Culture and qPCR)

In all cases included in the study, MAP positivity was confirmed in MS LNs and/or VIC by at least one of the techniques used. Positivity by both IHC and ZN was detected as cytoplasmic infiltrate in epithelioid macrophages and giant cells. In the case of IHC, immunoreaction was seen as brown cytoplasmic granules and as red bacilli by ZN (Figure 4). In MS LNs, results for ZN and IHC staining suggested MAP positivity in 82.1% (32/39) and 84.6% (33/39) of the cases, respectively (Table 3). The LP was positive by ZN in 53.9% (21/39) and in 59% (23/39) by IHC. The PPs were present in 33/39 of the cases and AFB were observed in 61% (20/33) by ZN and MAP-antigens were detected in 67% (22/33) by IHC. Moderate correlation between the detection capacity of ZN and IHC in MS LNs (κ = 0.537), ICV LP (κ =0.584) and ICV PP (κ = 0.609) was demonstrated. The main difference between the two techniques was observed in the intensity of the reaction which was stronger with IHC.

Table 3. Summary of Ziehl–Neelsen (ZN) and immunohistochemistry (IHC) results for MAP detection in 39 naturally infected goats.

			Age Range (Months)								Total		
			(0–12]	(12–24]		(24–36]		(36–48]	>48	Unknown			
Vaccination			V *n* = 1	V *n* = 8	nV *n* = 13	V *n* = 5	nV *n* = 5	nV *n* = 2	nV *n* = 4	V *n* = 1	V *n* = 15	nV *n* = 24	*n* = 39
MS LN	ZN	(+)	1 (100%)	6 (75%)	11 (84.4%)	4 (80%)	4 (80%)	2 (100%)	3 (75%)	1 (100%)	12 (80%)	20 (83.3%)	32 (82.1%)
		(−)	-	2 (35%)	2 (15.4%)	1 (20%)	1 (20%)	-	1 (25%)	-	3 (20%)	4 (16.7%)	7 (17.9%)
	IHC	(+)	1 (100%)	7 (87.5%)	10 (76.9%)	5 (100%)	4 (80%)	1 (50%)	4 (100%)	1 (100%)	14 (93.3%)	19 (79.2%)	33 (84.6%)
		(−)	-	1 (12.5%)	3 (23.1%)	-	1 (20%)	1 (50%)	-	-	1 (6.7%)	5 (20.8%)	6 (15.4%)
ICV LP	ZN	(+)	-	4 (50%)	7 (53.8%)	3 (60%)	3 (60%)	-	4 (100%)	-	7 (46.7%)	14 (58.3%)	21 (53.9%)
		(−)	1 (100%)	4 (50%)	6 (46.2%)	2 (40%)	2 (40%)	2 (100%)	-	1 (100%)	8 (53.3%)	10 (41.7%)	18 (46.1%)
	IHC	(+)	1 (100%)	5 (62.5%)	5 (38.5%)	3 (60%)	4 (80%)	1 (50%)	4 (100%)	-	9 (60%)	14 (58.3%)	23 (59%)
		(−)	-	3 (37.5%)	8 (61.5%)	2 (40%)	1 (20%)	1 (50%)	-	1 (100%)	6 (40%)	10 (41.7%)	16 (41%)

MS LN, mesenteric lymph node; ICV LP, ileocecal valve lamina propria; ZN, Ziehl–Neelsen; IHC, immunohistochemistry; (+) positive; (−) negative; V, vaccinated; nV, non-vaccinated.

A total of 29/39 animals were examined for MAP confirmation by qPCR (13/29) and/or bacterial culture (16/29). Four animals (4/16, 25%) tested positive by bacterial culture and 9/13 (69.2%) by qPCR. Samples from 3/29 animals were tested with both techniques, and all were identified as negative by bacterial culture and as positive by qPCR. The same three animals were positive by both ZN and IHC.

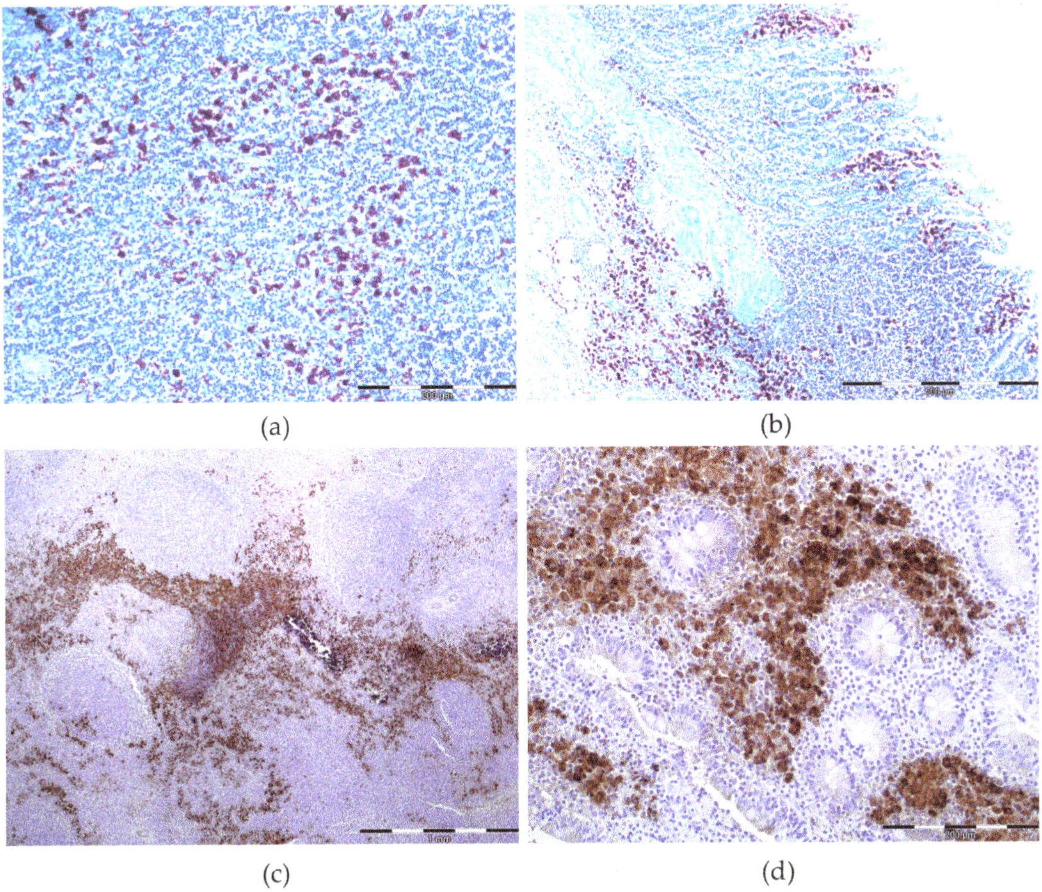

Figure 4. MAP confirmation by Ziehl–Neelsen stain (ZN) and immunohistochemistry (IHC) in mesenteric lymph nodes (MS LNs) and ileocecal valve (ICV). (a) MS LN, ZN (scale: 200 μm): numerous acid-fast bacilli (red) in the cytoplasm of macrophages and multinucleated giant cells; (b) ICV, ZN (scale: 500 μm): large number of acid-fast bacilli (red) in both lamina propria (LP) and Peyer's patches (PPs) of severely affected ICV; (c) LN, IHC (scale: 1 mm): strong immunoreaction in an affected LN showed as brown granular cytoplasmatic stain in macrophages and multinucleated giant cells; (d) ICV, IHC (scale: 200 μm): strong brown granular immunolabelling in ICV mucosa.

4. Discussion

Paratuberculosis is an important health issue affecting the goat farm industry worldwide [4]. Complete postmortem analysis is crucial for adequate diagnosis and herd status confirmation [5,6,16,29]. Nevertheless, gross PTB-compatible lesions were only observed in 62% of the animals enrolled in this study and the majority were confined to the MS LNs. Similar results have been previously reported in both experimentally [18] and naturally [16] infected goats, highlighting the importance of the histopathological examination of the MAP target organs.

Regarding the vaccination status, vaccinated animals exhibited gross PTB lesions primarily between 12 and 24 months in contrast with non-vaccinated goats where granulomatous MS LNs lesions were detected macroscopically in all age groups including older animals. Our results agree with previous studies which confirm that the anti-MAP vaccination therapeutic effect consists in the reduction of the severity of lesions and clinical signs in both early vaccinated animals (around 5 months of age) and adult goats [2,10,21,31,32].

Granuloma formation is commonly reported but not always present in MS LNs of goats with PTB in both experimental [18] and natural infection [4,6,16,17]. Thus, an adapted score, proposed by Wangoo et al., 2005 [30], for tuberculosis-induced lesions was applied in the present study. Different scores for PTB histological evaluation in goats have been described, although none of them was selected since they are majorly centered in the PPs and LP lesions and do not grade the MS LNs granuloma stages separately [16,18].

Lesions observed in the MS LNs in our study were characterized by the presence of well-formed granulomas with vast areas of calcification in 24/39 cases, but in only 16/24 were both LAM and GLA described. To our knowledge, there is no data regarding the relation between gross LMA and GLA lesions [4,6,16,17]. Thus, the correct gross inspection of the cut surface is of crucial importance for the macroscopic diagnosis of PTB lesions, even in not apparently enlarged LNs.

MS LNs presented histologically detectable granulomatous lesions in 36/39 of the cases. In 27/36 of the animals, grade IV lesions were described, although in 16/36, grade I micrograngulomas were the most frequent lesion type. Those results suggest that granuloma formation has a chronic evolution which has been confirmed in other mycobacterial infections such as cow tuberculosis [30,33,34]. Our findings also highlight that histological confirmation is needed for lesion detection in the early stage of granuloma formation.

Gross PTB lesions affecting the intestines were detected in only 7/39 of the cases. The absence of macroscopic lesions has been previously described as a common finding in goat PTB [3,4,16,17] in contrast with bovine PTB, where classic diffusely thickened corrugated intestinal mucosa folds are commonly observed [6,35]. Some authors describe ulceration of the intestinal mucosa in both clinical [36] and subclinical cases [18], but such findings were not observed in the animals studied. Histological evaluation of the intestine was carried out following the score proposed per Krüger et al., 2015 [18]. This score is based on histological lesions observed only with hematoxylin-eosin stain and is not influenced by AFB load detected with ZN as the one proposed by Corpa et al., 2000 [16], for naturally infected goats [16]. Modifications were introduced to separately evaluate the LP and the PP based on differences described in previous studies regarding the difference between the type of lesions described in those two locations [16,18]. Regarding the histological lesions described, our result agrees with previous studies which suggest that goats have limited ability to control mycobacterial infections and focal lesions are rarely seen [16,37].

MAP confirmation in formalin fixed-paraffin embedded tissues was carried out by ZN and IHC. Moderate correlation was observed between the detection capacity of the two techniques, but the intensity of the reaction was stronger in the case of IHC, agreeing with previous reports [38,39]. Low ZN staining could be explained by the fact that only the intact bacilli take up the stain and IHC anti-MAP antigens can detect dead bacterial cells with compromised cellular wall [38,39].

Etiological confirmation was also performed by qPCR/bacterial culture in 29/39 animals, 13 of which tested positive. Although bacterial culture is considered the gold standard for MAP detection, three animals which tested negative were positive by qPCR, IHC and ZN. Various studies describe qPCR as a more sensitive test for MAP detection in different species than bacterial culture, although data in goats is limited [35,40].

Both PTB and respiratory disease affect the goat farming industry worldwide, causing significant financial losses [12,17,30–33]. Nevertheless, data about the prevalence of both entities in goat herds is limited. In this study respiratory lesions were observed in 64.1% of the animals. Inflammatory lung disease was histologically detected more frequently in animals with PTB lesions in the ICV. Therefore, in naturally infected herds, there could be significant relations between the development of those two processes. Based on PTB pathogenesis and the demonstrated long subclinical stage and chronic onset [6,16,32,41], we could hypothesize that it has been the predisposal factor for lung inflammation in the present study. Additionally, vaccination against *Pasteurella* spp. was commonly issued as a control strategy in the farms studied and could have played a role, although, it is worth mentioning that lung disease in goats is multifactorial, including different causative agents and various environmental, management, herd and breed factors [24,25,42].

Pneumonia lesions had a 17.5% higher prevalence in the non-vaccinated animals in contrast with vaccinated ones. Furthermore, the majority of the vaccinated animals which exhibited histological lesions of pulmonary inflammation had between 12 and 24 months. In non-vaccinated animals, respiratory lesions were diagnosed histologically in different age groups including older animal. Additionally, significantly more non-vaccinated goats presented PTB lesions affecting the ICV, although no such difference was demonstrated regarding the MS LNs. Various studies have confirmed that vaccination against PTB has a generalized beneficial effect, reducing the overall number of losses in the herd, regardless of the cause, although there are no specific reports on reduction of respiratory inflammation [2,10,32,43].

Bronchopneumonia lesions could be associated with opportunistic bacterial pathogens including *Mannheimia haemolytica*, *Pasteurella multocida*, *Histophilus somni*, hemolytic *Streptococci*, *Helcococcus ovis* and *Mycoplasma* spp. [15–17,30]. Interstitial and bronchointerstitial pneumonia, on the other hand, have been described in a broad range of both infectious and non-infectious processes such as virus, mycoplasmosis, parasitosis, protozoal infection, toxicity and hypersensitivity reactions [19,24,25,42,44]. Since histological appearance is similar irrespective to etiology, diagnosis is usually based on clinical signs observed and lesions in other organs [15–17,30], and ancillary tests (e.g., PCR, microbial culture) are needed to identify the aetiologic agent [19,24,25,42,44].

Besides PTB, other lesions were described affecting the gastrointestinal tract including mainly lymphohistiocytic and eosinophilic enteritis which could be associated with parasitic agents such as *Ostertagia*, *Nematodirus* and *Trichostrongylus* [6,19,45].

In the present study, 15.3% more of the non-vaccinated animals presented non-PTB gastrointestinal lesions, although no statistical differences were demonstrated, probably due to the small sample size. Previous studies report a general reduction of the losses in vaccinated herds but do not analyze the different systems separately which were less affected [2,10,32,43]. Our results do suggest that anti-MAP vaccination could in fact have a protective effect against concomitant gastrointestinal pathologies.

Non-PTB lymphadenitis, on the other hand, was detected grossly in more than 50% of the animals. This reflects the invasion of infectious agents in the lymph node resulting from the drainage of products of a distant inflammation and their respective progression involving the node directly [26]. In the cases of *Corynebacterium pseudotuberculosis* compatible CLA, where LNs are the target organs, our results reported RPh, MD, and MA LNs as common lesion sites, which agrees with recent reports in goats [46,47].

The liver was mainly affected by randomly distributed multifocal lymphohistiocytic hepatitis. This unspecific inflammatory pattern is frequently reported in ruminants as secondary to systemic bacteremia caused by the agents affecting other organs [6,27]. Hepatocellular steatosis was also observed both grossly and histologically. Since all animals included presented PTB-attributed weight loss, the steatosis is most likely caused by the excessive entry of fatty acids into the liver due to their increased metabolization from the adipose tissue [6,27]. The granulomatous hepatitis recorded in only three animals could be associated with severe cases of PTB due to MAP spreading [4,6].

The cardiovascular system presented gross lesions only affecting the pericardium. The main type detected was pericarditis, which was both macroscopically and microscopically confirmed. Pericarditis is commonly associated with hematogenous microbial infections, with a high prevalence of association with pneumonic lesions [24,25,48], which were detected in the animals analyzed. Non-suppurative myocarditis was described histologically in animals and could be associated with a wide variety of systemic diseases but is rarely primary [48]. The *Sarcocystis* spp. cysts reported histologically in the studied animals are a very common finding in ruminants and an increase in the numbers of cysts with age is confirmed [48].

Regarding the urinary system, interstitial nephritis was the most common histological finding, which was rarely detected grossly, which agrees with results of previous studies in goats [19,49].

5. Conclusions

The present study describes the most common concomitant pathologies found in goat herds with natural PTB infection mainly affecting the respiratory and gastrointestinal tracts. Our results emphasize the importance of routine histological examination for disease assessment, as well as the importance of anatomic pathology as a useful tool for the diagnosis and management of PTB in goat herds, since the majority of the described findings were only detectable histologically. Furthermore, vaccination against PTB might potentially have a positive effect on the reduction of respiratory and gastrointestinal pathologies, based on the findings described and the statistically significant differences demonstrated between the vaccinated and non-vaccinated animals. Further studies on a greater scale are needed to determine and understand the multiple etiologies and factors associated with the pathologies described, as well as the effect of PTB vaccination on their prevalence.

Author Contributions: Conceptualization, Y.P.-S., Ó.Q.-C., V.P., A.C. and M.A.; methodology, E.P.S., Y.P.-S., Ó.Q.-C., V.P. and M.A.; validation, E.P.S., Y.P.-S., Ó.Q.-C., V.P. and M.A.; formal analysis, Y.P.-S., Ó.Q.-C. and M.A.; investigation, E.P.S., Y.P.-S., Ó.Q.-C., V.P., A.C. and M.A.; resources, A.E.d.l.M., M.A.R., Ó.Q.-C. and M.A.; data curation, M.d.P.Q.-M. and M.A.; writing—original draft preparation, E.P.S., Y.P.-S., Ó.Q.-C., M.J.C. and M.A.; writing—review and editing, E.P.S., Y.P.-S., Ó.Q.-C., M.d.P.Q.-M., M.J.C. and M.A.; visualization, E.P.S., Y.P.-S., Ó.Q.-C. and M.A.; supervision, A.E.d.l.M., M.A.R., M.J.C. and M.A.; project administration, E.P.S., Ó.Q.-C., Y.P.-S., A.E.d.l.M., M.A.R. and M.A.; funding acquisition, A.E.d.l.M., M.A.R. and M.A. All authors have read and agreed to the published version of the manuscript.

Funding: This research was funded by CajaCanarias Foundation and "la Caixa" Foundation, grant reference number FCC-FC-2019-03 and by the PROID2020010047 grant from the Canary Regional Ministry of Economy, Knowledge and Work through the Canarian Agency for Research, Innovation and Information Society (ACIISI), for the implementation of I+D projects in the Canary RIS-3 priority areas, co-financed through PO FEDER Canarias 2014–2020. Elena Plamenova Stefanova was the recipient of a predoctoral contract of the The Ministry of Universities, Spain (FPU20/03693) and Charter100 2021 scholarship for PhD students' mobility.

Institutional Review Board Statement: This manuscript has been exempted by the Ethical Committee for Animal Experimentation of the University of Las Palmas of Gran Canaria (OEBA-ULPGC-13/2023), which notified that the procedure falls within the exceptions; thus, according to article 2 of RD 53/2013, non-experimental clinical veterinary practices are excluded from the scope of this Royal Decree, and therefore, it does not require approval by an Ethical Committee for Animal Experimentation.

Informed Consent Statement: Not applicable.

Data Availability Statement: The data presented in this study are available on request from the corresponding author. The data are not publicly available due to privacy.

Acknowledgments: The authors thank the veterinarians from the Livestock Health Protection Groups responsible for the management of the farms for providing the animals for the study, postmortem technician Javier Ramos, the laboratory technicians of Institute of Animal Health and Food Safety (IUSA) Ana Afonso Almeda and Alba María Santana Zerpa for their technical assistance, the staff of the Instituto de Ganadería de Montaña (CSIC-ULE), Grulleros, Spain and the technician Marta Silva Samos for the knowledge exchange and laboratory training, and the VISAVET Health Surveillance, Complutense University of Madrid, Madrid, Spain for performing the bacterial culture.

Conflicts of Interest: The authors declare no conflict of interest.

Appendix A

Table A1. Summary of pathological findings in 39 naturally infected goats with PTB lesions.

System	Organ	Gross	Frequency (%)	Histology	Frequency (%)
Hemolymphatic	MS	LAM GLA	21 (53.8%) 24 (61.5%)	GLA	36 (92.3%)
	RPh	GLA CLA Abscess	2 (5.1%) 2 (5.1%) 3 (7.7%)	SLA/CLA	8 (20.5%)
	PE	SLA Abscess	6 (15.4%) 4 (10.3%)	SLA/CLA	4 (10.3%)
	MD	SLA CLA Abscess	3 (7.7%) 3 (7.7%) 2 (5.1%)	SLA/CLA	8 (20.5%)
	MA	SLA GLA CLA Abscess	1 (2.6%) 2 (5.1%) 2 (5.1%) 1 (2.6%)	SLA/CLA	4 (10.3%)
Respiratory	Lungs	CBP CBP + fibrinous pleuritis CBP + fibrous pleuritis	4 (10.3%) 7 (17.9%) 5 (12.8%)	FPBP/FNBP IP BIP Pleuritis only	12 (30.8%) 6 (15.4%) 3 (7.7%) 4 (10.3%)
Gastrointestinal	Intestine	Diffusely thickened appearance	10 (25.6%)	EE LPE	2 (5.1%) 1 (2.6%)
	Abomasum	PTB-compatible enteritis Hemorrhagic enteritis	7 (17.9%) 1 (2.6%)	GE	32(82.1%)
				Abomasitis/abomasal ulcer	4 (10.3%)
	Liver	Steatosis Capsular fibrosis Mineralizations	7 (17.9%) 6 (15.4%) 1 (2.6%)	Steatosis Lymphohistiocytic hepatitis Granulomatous hepatitis	11 (28.2%) 25 (64.1%) 3 (7.7%)
Urinary	Kidney	Interstitial nephritis	2 (5.1%)	Interstitial nephritis	10 (25.6%)
Cardiovascular	Heart	Hydropericardium Fibrous pericarditis Petechial hemorrhages	2 (5.1%) 2 (5.1%) 2 (5.1%)	Non-suppurative myocarditis Fibrous pericarditis Myocardial sarcocystosis	3 (7.7%) 3 (7.7%) 9 (23.1%)

MS, mesenteric lymph nodes; RPh, retropharyngeal lymph nodes; PE, prescapular lymph nodes; MD, mediastinal lymph nodes; MA, mammary lymph nodes; LAM, lymphadenomegaly; GLA, granulomatous lymphadenitis; SLA, suppurative lymphadenitis; CLA, caseous lymphadenitis; CBP, cranioventral bronchopneumonia; FNBP, fibrinonecrotic brochopneumonia; FPBP, fibrinopurulent bronchopneumonia; IP, interstitial pneumonia; BIP, bronchointerstitial pneumonia; GE, granulomatous enteritis; EE, eosinophilic enteritis; LPE, lymphoplasmacytic enteritis.

References

1. Karuppusamy, S.; Kirby, G.M.; Mutharia, L.; Tripathi, B.N. An Update on Mycobacterium Avium Subspecies Paratuberculosis Antigens and Their Role in the Diagnosis of Johne's Disease. *World J. Microbiol. Biotechnol.* **2019**, *35*, 120. [CrossRef] [PubMed]
2. Fernández, M.; Royo, M.; Fuertes, M.; Arteche-Villasol, N.; Ferreras, M.C.; Benavides, J.; Pérez, V. Effects of Paratuberculosis Vaccination at Different Ages in a Dairy Goat Herd: A 2-Year Follow-Up. *Animals* **2022**, *12*, 3135. [CrossRef] [PubMed]
3. Paliwal, O.P.; Rehbinder, C. Ultrastructural studies of paratuberculosis (Johne's disease) in goats. *Acta Vet. Scand.* **1981**, *22*, 180. [CrossRef] [PubMed]

4. Idris, S.M.; Eltom, K.H.; Okuni, J.B.; Ojok, L.; Elmagzoub, W.A.; El Wahed, A.A.; Eltayeb, E.; Gameel, A.A. Paratuberculosis: The Hidden Killer of Small Ruminants. *Animals* **2022**, *12*, 12. [CrossRef] [PubMed]
5. Benavides, J.; González, L.; Dagleish, M.; Pérez, V. Diagnostic Pathology in Microbial Diseases of Sheep or Goats. *Vet. Microbiol.* **2015**, *181*, 15–26. [CrossRef]
6. Uzal, F.A.; Plattner, B.L.; Hostetter, J.M. Chapter 1—Alimentary System. In *Jubb, Kennedy & Palmer's Pathology of Domestic Animals*, 6th ed.; Maxie, M.G., Ed.; W.B. Saunders: Philadelphia, PA, USA, 2016; Volume 2, pp. 1–257.e2. ISBN 978-0-7020-5318-4.
7. Garcia, A.; Shalloo, L. Invited Review: The Economic Impact and Control of Paratuberculosis in Cattle. *J. Dairy Sci.* **2015**, *98*, 5019–5039. [CrossRef]
8. Barrero-Domínguez, B.; Luque, I.; Huerta, B.; Gomez-Laguna, J.; Galán-Relaño, Á.; Gómez-Gascón, L.; Sánchez, M.; Astorga, R.J. Paratuberculosis in Dairy Goat Flocks from Southern Spain: Risk Factors Associated with Seroprevalence; Paratuberculosis in Dairy Goat Flocks from Southern Spain: Risk Factors Associated with Seroprevalence. *Vet. Rec.* **2019**, *185*, 600. [CrossRef]
9. Delgado, L.; Marín, J.F.G.; Muñ Oz, M.; Benavides, J.; Juste, R.A.; García-Pariente, C.; Fuertes, M.; González, J.; Ferreras, M.C.; Pé Rez, V. Infectious Disease Research Pathological Findings in Young and Adult Sheep Following Experimental Infection with 2 Different Doses of Mycobacterium Avium Subspecies Paratuberculosis. *Vet. Pathol.* **2013**, *50*, 857–866. [CrossRef]
10. Agulló-Ros, I.; Andrada, M.; Pérez-Sancho, M.; Roy, Á.; Bezos, J.; Bonnet, T.; Moreno, I.; Paz-Sánchez, Y.; Domínguez, M.; Gómez-Villamandos, J.C.; et al. Effect of Heat-Inactivated Mycobacterium Avium Subspecies Paratuberculosis (MAP) Vaccine on the Lesions and Immunopathology Developed in Target Tissues of Naturally MAP-Infected Goats. *Vet. Microbiol.* **2022**, *273*, 109543. [CrossRef]
11. Stau, A.; Seelig, B.; Walter, D.; Schroeder, C.; Ganter, M. Seroprevalence of Mycobacterium Avium Subsp. Paratuberculosis in Small Ruminants in Germany. *Small Rumin. Res.* **2012**, *105*, 361–365. [CrossRef]
12. Mercier, P.; Baudry, C.; Beaudeau, F.; Seegers, H.; Malher, X. Estimated Prevalence of Mycobacterium Avium Subspecies Paratuberculosis Infection in Herds of Dairy Goats in France. *Vet. Rec.* **2010**, *167*, 412–415. [CrossRef] [PubMed]
13. Hosseiniporgham, S.; Cubeddu, T.; Rocca, S.; Sechi, L.A. Identification of Mycobacterium Avium Subsp. Paratuberculosis (Map) in Sheep Milk, a Zoonotic Problem. *Microorganisms* **2020**, *8*, 1264. [CrossRef]
14. Hosseiniporgham, S.; Rebechesu, L.; Pintore, P.; Lollai, S.; Dattena, M.; Russo, S.; Ruiu, A.; Sechi, L.A. A Rapid Phage Assay for Detection of Viable Mycobacterium Avium Subsp. Paratuberculosis in Milk. *Sci. Rep.* **2022**, *12*, 475. [CrossRef]
15. De Juan Ferré, L.; Domínguez Rodríguez, L.; Suárez Fernández, G.; Mateos García, A. *Paratuberculosis Caprina: Aportaciones a Su Diagnóstico, Epidemilogía Molecular y Control*; Universidad Complutense de Madrid: Madrid, Spain, 2005; ISBN 8466928529.
16. Corpa, J.M.; Garrido, J.; García Marín, J.F.; Pérez, V. Classification of Lesions Observed in Natural Cases of Paratuberculosis in Goats. *J. Comp. Pathol.* **2000**, *122*, 255–265. [CrossRef]
17. Tafti, A.K.; Rashidi, K. The Pathology of Goat Paratuberculosis: Gross and Histopathological Lesions in the Intestines and Mesenteric Lymph Nodes. *J. Vet. Med. B* **2000**, *47*, 487–495. [CrossRef] [PubMed]
18. Krüger, C.; Köhler, H.; Liebler-Tenorio, E.M. Sequential Development of Lesions 3, 6, 9, and 12 Months After Experimental Infection of Goat Kids with Mycobacterium Avium Subsp Paratuberculosis. *Vet. Pathol.* **2015**, *52*, 276–290. [CrossRef] [PubMed]
19. Smith, M.C.; Sherman, D.M. *Goat Medicine*; John Wiley & Sons: Hoboken, NJ, USA, 2009; ISBN 0781796431.
20. Derakhshandeh, A.; Namazi, F.; Khatamsaz, E.; Eraghi, V.; Hemati, Z. Goat Paratuberculosis in Shiraz: Histopathological and Molecular Approaches. *Vet. Res. Forum* **2018**, *9*, 253–257. [CrossRef]
21. Bastida, F.; Juste, R.A. Paratuberculosis Control: A Review with a Focus on Vaccination. *J. Immune Based Ther. Vaccines* **2011**, *9*, 8–17. [CrossRef]
22. Van Metre, D.C.; Tyler, J.W.; Stehman, S.M. Diagnosis of enteric disease in small ruminants. *Vet. Clin. N. Am. Food Anim. Pract.* **2000**, *16*, 87–115. [CrossRef]
23. Tariq, K.A. A Review of the Epidemiology and Control of Gastrointestinal Nematode Infections of Small Ruminants. *Proc. Natl. Acad. Sci. India Sect. B-Biol. Sci.* **2015**, *85*, 693–703. [CrossRef]
24. Caswell, J.L.; Williams, K.J. Chapter 5—Respiratory System. In *Jubb, Kennedy & Palmer's Pathology of Domestic Animals*, 6th ed.; Maxie, M.G., Ed.; W.B. Saunders: Philadelphia, PA, USA, 2016; Volume 2, pp. 465–591.e4. ISBN 978-0-7020-5318-4.
25. Chakraborty, S.; Kumar, A.; Tiwari, R.; Rahal, A.; Malik, Y.; Dhama, K.; Pal, A.; Prasad, M. Review Article Advances in Diagnosis of Respiratory Diseases of Small Ruminants. *Veter-Med. Int.* **2014**, *2014*, 1–16. [CrossRef] [PubMed]
26. Valli, V.E.O.; Kiupel, M.; Bienzle, D.; Wood, R.D. Chapter 2—Hematopoietic System. In *Jubb, Kennedy & Palmer's Pathology of Domestic Animals*, 6th ed.; Maxie, M.G., Ed.; W.B. Saunders: Philadelphia, PA, USA, 2016; Volume 3, pp. 102–268.e1. ISBN 978-0-7020-5319-1.
27. Brown, D.L.; Van Wettere, A.J.; Cullen, J.M. Chapter 8—Hepatobiliary System and Exocrine Pancreas1. In *Pathologic Basis of Veterinary Disease*, 6th ed.; Zachary, J.F., Ed.; Mosby: London, UK, 2017; pp. 412–470.e1. ISBN 978-0-323-35775-3.
28. Cullen, J.M.; Stalker, M.J. Chapter 2—Liver and Biliary System. In *Jubb, Kennedy & Palmer's Pathology of Domestic Animals*, 6th ed.; Maxie, M.G., Ed.; W.B. Saunders: Philadelphia, PA, USA, 2016; Volume 2, pp. 258–352.e1. ISBN 978-0-7020-5318-4.
29. King, J.M.; Roth-Johnson, L.; Dodd, D.C.; Newsom, M.E. *The Necropsy Book: A Guide for Veterinary Students, Residents, Clinicians, Pathologists, and Biological Researchers*; The Internet-First University Press: Ithaca, NY, USA, 2014.
30. Wangoo, A.; Johnson, L.; Gough, J.; Ackbar, R.; Inglut, S.; Hicks, D.; Spencer, Y.; Hewinson, G.; Vordermeier, M. Advanced Granulomatous Lesions in Mycobacterium Bovis-Infected Cattle Are Associated with Increased Expression of Type I Procollagen, Γδ (WC1+) T Cells and CD 68+ Cells. *J. Comp. Pathol.* **2005**, *133*, 223–234. [CrossRef]

31. Juste, R.A.; Perez, V. Control of Paratuberculosis in Sheep and Goats. *Vet. Clin. N. Am. Food Anim. Pract.* **2011**, *27*, 127–138. [CrossRef]
32. Corpa, J.M.; Pérez, V.; García Marín, J.F.; Sánchez, M.A. Control of Paratuberculosis (Johne's Disease) in Goats by Vaccination of Adult Animals. *Vet. Rec.* **2000**, *146*, 195–196. [CrossRef] [PubMed]
33. Domingo, M.; Vidal, E.; Marco, A. Pathology of Bovine Tuberculosis. *Res. Vet. Sci.* **2014**, *97*, S20–S29. [CrossRef] [PubMed]
34. Cassidy, J.P.; Bryson, D.G.; Pollock, J.M.; Evans, R.T.; Forster, F.; Neill, S.D. Early Lesion Formation in Cattle Experimentally Infected with Mycobacterium Hovis. *J. Comp. Pathol.* **1998**, *119*, 27–44. [CrossRef]
35. Gilardoni, L.R.; Paolicchi, F.A.; Mundo, S.L. Diagnosis of Bovine Paratuberculosis 201 Bovine Paratuberculosis: A Review of the Advantages and Disadvantages of Different Diagnostic Tests. *Rev. Argent. Microbiol.* **2012**, *44*, 201–215. [PubMed]
36. Lybeck, K.R.; Storset, A.K.; Djønne, B.; Valheim, M.; Olsen, I. Faecal Shedding Detected Earlier than Immune Responses in Goats Naturally Infected with Mycobacterium Avium Subsp. Paratuberculosis. *Res. Vet. Sci.* **2011**, *91*, 32–39. [CrossRef]
37. Gutiérrez, M.; Samper, S.; Gavigan, J.; Francisco Garci, J.; Mari, A.; Marti, C. Differentiation by Molecular Typing of Mycobacterium Bovis Strains Causing Tuberculosis in Cattle and Goats. *J. Clin. Microbiol.* **1995**, *33*, 2953–2956. [CrossRef]
38. Thoresen, O.F.; Falk, K.; Evensen, Ø. Comparison of Immunohistochemistry, Acid-Fast Staining, and Cultivation for Detection of Mycobacterium Paratuberculosis in Goats. *J. Vet. Diagn. Investig.* **1994**, *6*, 195–199. [CrossRef]
39. Kheirandish, R.; Tafti, A.K.; Hosseini, A. Classification of Lesions and Comparison of Immunohistochemical and Acid Fast Staining in Diagnosis of Naturally Occurring Paratuberculosis in Goats. *Small Rumin. Res.* **2009**, *87*, 81–85. [CrossRef]
40. Bögli-Stuber, K.; Kohler, C.; Seitert, G.; Glanemann, B.; Antognoli, M.C.; Salman, M.D.; Wittenbrink, M.M.; Wittwer, M.; Wassenaar, T.; Jemmi, T.; et al. Detection of Mycobacterium Avium Subspecies Paratuberculosis in Swiss Dairy Cattle by Real-Time PCR and Culture: A Comparison of the Two Assays. *J. Appl. Microbiol.* **2005**, *99*, 587–597. [CrossRef] [PubMed]
41. Sweeney, R.W. Pathogenesis of Paratuberculosis. *Vet. Clin. N. Am. Food Anim. Pract.* **2011**, *27*, 537–546. [CrossRef] [PubMed]
42. Matthews, J.G. Chapter 15—Respiratory System. In *Diseases of the Goat*; John Wiley & Sons: Hoboken, NJ, USA, 2016; pp. 258–274. ISBN 9781119073543.
43. Mercier, P.; Brémaud, I.; Gautier, M.P. Vaccination of Kids under One Month of Age with a Killed Vaccine and Reduction in the Frequency of Faecal Shedding of Mycobacterium Avium Subspecies Paratuberculosis. *Small Rumin. Res.* **2014**, *121*, 425–433. [CrossRef]
44. Carvallo, F.R.; Stevenson, V.B. Interstitial Pneumonia and Diffuse Alveolar Damage in Domestic Animals. *Vet. Pathol.* **2022**, *59*, 586–601. [CrossRef]
45. Matthews, J.G. Chapter 14—Diarrhoea. In *Diseases of the Goat*; John Wiley & Sons: Hoboken, NJ, USA, 2016; pp. 204–231. ISBN 9781119073543.
46. Debien, E.; Hélie, P.; Buczinski, S.; Leboeuf, A.; Bélanger, D.; Drolet, R. Article Proportional Mortality: A Study of 152 Goats Submitted for Necropsy from 13 Goat Herds in Quebec, with a Special Focus on Caseous Lymphadenitis. *Can. Vet. J.* **2013**, *54*, 581.
47. Pointon, A.; Hamilton, D.; Kiermeier, A. Comparison of Postmortem Inspection Procedures for Detecting Caseous Lymphadenitis of Australian Sheep and Goats. *Vet. Rec.* **2019**, *185*, 54. [CrossRef]
48. Robinson, W.F.; Robinson, N.A. Chapter 1—Cardiovascular System. In *Jubb, Kennedy & Palmer's Pathology of Domestic Animals*, 6th ed.; Maxie, M.G., Ed.; W.B. Saunders: Philadelphia, PA, USA, 2016; Volume 3, pp. 1–101.e1. ISBN 978-0-7020-5319-1.
49. Cianciolo, R.E.; Mohr, F.C. Chapter 4—Urinary System. In *Jubb, Kennedy & Palmer's Pathology of Domestic Animals*, 6th ed.; Maxie, M.G., Ed.; W.B. Saunders: Philadelphia, PA, USA, 2016; Volume 2, pp. 376–464.e1. ISBN 978-0-7020-5318-4.

animals

Article

Mycobacterium avium subsp. *Paratuberculosis* in Different Environmental Samples from a Dairy Goat Barn—Implications for Sampling Strategies for Paratuberculosis Diagnostic and Prevention

Chris Pickrodt [1,2], Karsten Donat [2,3], Udo Moog [3] and Heike Köhler [1,*]

1 Institute of Molecular Pathogenesis, Friedrich-Loeffler-Institut, Federal Research Institute for Animal Health, Naumburger Straße 96a, 07743 Jena, Germany; chris.pickrodt@fli.de
2 Clinic for Obstretics, Gynaecology and Andrology for Large and Small Animal Science with Veterinary Ambulance, Justus-Liebig-University Gießen, Frankfurter Straße 106, 35392 Gießen, Germany; kdonat@thtsk.de
3 Thuringian Animal Diseases Fund, Victor-Goerttler-Straße 4, 07745 Jena, Germany; umoog@thtsk.de
* Correspondence: heike.koehler@fli.de; Tel.: +49-3641-804-2240

Simple Summary: Paratuberculosis is a chronic inflammatory disease leading to emaciation and production losses in ruminants. Important aspects of disease control are the detection of infected herds and environmental areas where contact between animals and the causative pathogen *Mycobacterium avium* subsp. *paratuberculosis* (MAP) is possible. Bedding, dust, feed, and water samples were collected from the barn of a paratuberculosis-infected dairy goat herd and analyzed for culturable MAP and MAP DNA. Cultivation was successful in 28 bedding and one dust sample, while MAP DNA was detected in 117 out of 256 samples from all materials. Samples collected from areas of high animal traffic, such as the milking parlor, were more likely to reveal positive results than adult and young goat areas. Positive culture results were also obtained from kidding pens, constituting this barn area as a possible infection site. Overall, environmental sampling may be suitable for the detection of MAP in goat herds and the identification of potential critical areas for pathogen transmission. These results should be taken into account for paratuberculosis control programs in goat herds to improve their efficiency and, thus, animal welfare.

Abstract: Environmental samples are often used to classify the paratuberculosis status of cattle herds. The disease is caused by *Mycobacterium avium* subsp. *paratuberculosis* (MAP), predominantly through oral ingestion during infancy. In this explorative study, the presence of MAP was determined in the barn environment of a paratuberculosis-infected vaccinated dairy goat herd. A total of 256 bedding, dust, feed, and water samples were collected at eight time points and examined using culture and qPCR. Detection rates of both methods were compared, and factors determining MAP confirmation were identified. MAP was cultured from 28 bedding and one dust sample, while MAP DNA was detected in all materials (117/256). Samples from high animal traffic areas and those collected during the indoor season were more likely to yield positive culture and qPCR results. Cultivation of MAP from kidding pens indicated this area as a possible infection site. Dust proved to be the most suitable material for detecting MAP DNA, as bedding was for MAP culture. Environmental sampling was demonstrated to be an effective way to detect MAP in a dairy goat herd. qPCR results could confirm herd infection, while culture results provided insight into crucial areas for MAP transmission. These findings should be considered when designing farm-specific paratuberculosis control plans.

Keywords: small ruminants; Johne's disease; MAP; environmental sampling; bedding; dust; feed; water; transmission; control program

1. Introduction

Paratuberculosis is a chronic inflammatory intestinal disease of major economic importance in domestic ruminants. It is caused by an infection with *Mycobacterium avium* subsp. *paratuberculosis* (MAP) which usually occurs in the first days of life. The slow disease progression results in a long incubation period until symptoms such as diarrhea, decreasing milk yield, and emaciation up to death appear. Possible routes of transmission include oral ingestion of the pathogen, inhalation, or in-utero infection [1–6].

MAP is shed with the feces of infected animals and can remain infectious in the environment for almost one year, even outside a suitable host [1,7]. Detection of MAP in samples taken from the immediate environment of the animals can be exploited in two directions. First, when applied in the monitoring of paratuberculosis, this approach can be used for the identification of MAP-positive herds, ensuring a high probability of truly positive results. This is an important aspect of many control programs [8–10]. Environmental sampling for MAP detection, whether molecular biological or cultural, is considered to be cost-effective and highly specific [10–12]. Secondly, this diagnostic approach can identify areas with a high bacterial load that may serve as infection sources for susceptible animals in the herd. This can be valuable information in the framework of veterinary risk assessment within a herd aiming to improve hygiene and management.

Most studies concerning MAP detection in environmental samples have been conducted in cattle farms. Due to the partly considerable differences in the husbandry conditions of cattle and small ruminants, it is not possible to transfer the results obtained to other animal species without further verification. Only one previous study focused on grazing sheep flocks, including some combined herds of sheep and goats [13]. Systematic environmental sampling in dairy goat barns has not yet been conducted.

Different materials have been analyzed for the presence of MAP. As the pathogen is shed with feces, most studies conducted focused on manure samples. However, MAP has also been found in dust, water, feed, and soil samples from in and outside the barn [9–17]. A comparison between different matrices collected in the same barn to determine which is best suited for MAP detection in goat barns has been lacking. Barn areas with high animal traffic, such as alleyways and the milking parlor, revealed the highest incidence of MAP-positive samples in several studies in cattle, whereas calving pens and youngstock areas were less affected [9–12,18]. In this regard, both culture and qPCR appear to be appropriate detection methods and comparable in terms of their performance [14–16]. However, herds with a low prevalence may be misclassified as uninfected due to negative environmental samples caused by intermittent pathogen shedding or dilution as well as dissemination effects [17]. For this reason, repeated collection of multiple environmental samples is recommended to increase sensitivity [9,19,20].

The approach of drawing implications for disease management from environmental sampling results has been rarely investigated in cattle [21] but so far not in goats. The main objective during youngstock rearing should always be to prevent exposure of kids, lambs, and calves to MAP as long as possible. Although the minimal infectious dose has not yet been determined, the later exposure occurs, and the less MAP is taken up, the lower the likelihood for successful infection and or fast clinical expression may be [1,22,23]. High-risk areas for pathogen transmission should be identified, results critically evaluated, and measures for MAP reduction and contact minimization implemented in every individual farm-level paratuberculosis control plan.

Here, we present the results of an explorative study conducted on one dairy goat farm. It focused on MAP detection in environmental samples from a goat barn, as previously shown for cattle and sheep farms. The objective was to evaluate the suitability of different sample materials, sites, seasons, and analysis methods (qPCR and culture) for identifying a paratuberculosis-infected goat herd. In addition, the study aimed at determining crucial barn areas for exposure of kids to MAP to enable their integration into management strategies for effective disease control.

2. Materials and Methods

2.1. Animal Welfare and Legislation

The study was approved by the Animal Health and Welfare Unit of the Thuringian State Office for Consumer Protection (file reference: 2684-04-04-BFI-20-103) and carried out in accordance with European and national laws. Every effort was made to minimize suffering.

2.2. Study Herd

The study was conducted in an organic dairy goat farm in Thuringia, Germany, between 2020 and 2022. The herd was affected by paratuberculosis. MAP was first confirmed culturally in 2018 in different tissues of a clinically affected goat after diagnostic necropsy. Because of numerous clinical cases, all adult goats and the annual female youngstock were subsequently vaccinated with an inactivated vaccine (Gudair, CZ Vaccines, O Porriño, Spain), while goats showing severe clinical signs of paratuberculosis were culled.

Young and adult goats were kept in loose housing on deep litter in different open pens within the same building. The rotary milking parlor was located in a separate part of the building attached to the animal husbandry area. The parlor was entered from a passage from the waiting area adjacent to a part of the lactating goat pen. All lactating goats were driven into the waiting pen for milking twice a day.

Kids were raised at the farm for replacement. They were separated from the doe immediately after birth. About 20 age-grouped kids each were kept in temporary detached pens until reaching a body weight of about 15 kg. At about five months of age, the juvenile goats were moved from the group pen in the kid-rearing area to the juvenile goat pen on the opposite side of the building, and the kid pens were deconstructed afterwards. The kid-rearing area was cleaned and disinfected using white hydrated and slaked lime (calcium hydroxide) before new kid pens were built during the next kidding season.

Rotational grazing was performed during the grazing season between May and November. Young goats were turned out to graze after their first kidding. In 2020 and 2021, dried-off goats were grouped in separate pens approximately two months before kidding, whereas no dry goat/kidding pens were set up in 2022. The deep litter was removed from the pens before the end of each grazing season, followed by cleaning and disinfection using slake lime. In between, new bedding material was added every day as necessary to ensure clean udder skins and coats.

2.3. Study Design

Environmental samples were collected at eight environmental sampling events (ES) between 2020 and 2022. ES 1, 2, and 3 were conducted within a six-month interval, while the sampling frequency was enhanced to approximately every three months from ES 4 onwards (Figure 1).

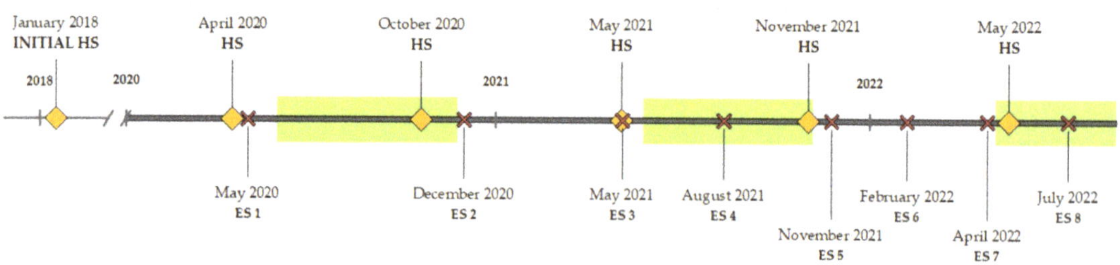

Figure 1. Sampling schedule for fecal and environmental samples between 2018 and 2022. HS = herd sampling (♦), ES = environmental sampling (×). Grazing seasons are marked as green boxes.

Herd sampling to determine fecal shedding of MAP was first conducted in February 2018 and then biannually between 2020 and 2022 (Figure 1).

2.4. Environmental Sample Collection

The following nine sampling sites were defined: entrances to the barn, waiting area in front of the milking parlor, milking parlor, exit of the milking parlor, lactating goat pen (divided into four quadrants: Q1–Q4), dry goat/kidding pen(s), buck pen, juvenile goat pen, and goat kid pen(s). If more than one dry goat/kidding or goat kid pen existed at the same time, each pen was sampled individually. Four different materials (bedding, dust, water, feed) were collected from each site if available. Due to herd management, it was not possible to sample all locations or collect all different sampling materials at each ES. A detailed breakdown of all collected environmental samples ($n = 256$) at the respective ES and sampling sites is provided in Appendix A (Table A1).

Bedding material from approximately five spots within a whole pen or pen quadrant was combined into one composite sample. Deviating from this procedure, four individual samples were collected from kidding pens. Feed samples were collected out of racks or from the ground when forage was offered in front of the respective pen. Each feed and bedding sample was collected manually using a fresh glove. The samples contained between 50 and 250 g of forage or bedding with fecal material. The material was placed in a plastic bag, transferred to the laboratory, and stored at −20 °C. Dust samples were scraped from the surface of the barn facilities in a 50 mL tube using a scalpel. Water from drinking troughs was collected using a 1 L vial scooped through the trough. Dust and water samples were transferred to the laboratory and stored at room temperature until further processing.

2.5. Sample Preparation and Analysis

2.5.1. Fecal Samples

Fecal samples from every goat older than one year were examined by bacterial culture according to the official manual of diagnostic procedures published by the Friedrich-Loeffler-Institute [24]. In short, 3 g of feces were decontaminated over 48 h at room temperature using 30 mL of 0.75% hexadecylpyridinium chloride monohydrate solution (Sigma Aldrich, Taufkirchen, Germany). The supernatants were discarded, and 200 µL of the remainder were transferred on each of three slopes of Herrold's Egg Yolk Agar with Mycobactin J and Amphotericin, Nalidixic acid and Vancomycin (HEYM, Becton Dickinson, Sparks, MD, USA). Cultures were incubated at 37 °C for up to six months. Examination of bacterial growth was conducted every second week, starting after 42 days. The presence of MAP was confirmed by IS*900* PCR [25] of characteristic colonies.

2.5.2. Bedding and Feed Samples

Samples were thawed overnight and processed according to Whittington et al. [7], modified as follows: 50 g of each sample was placed in a lockable plastic container. Then, 500 mL of sterile distilled water was added to cover the material completely. After intensively shaking by hand, containers were agitated in a shaking incubator overnight at 37 °C and 50 rpm. Stepwise, the liquid was poured off and collected in 50 mL vials, followed by centrifugation at 4200 rpm for 20 min. Two aliquots of 3 g each from the obtained pellet were portioned into two vials for further cultural and molecular biological examination. If the total pellet weight was below 6 g, the pellet was divided into two equal parts, and volumes of further additives were adjusted proportionally.

Based on the procedure applied to the fecal samples, the first aliquot of this pellet was processed for bacterial cultivation in the same way as described above using three slopes of HEYM (Becton Dickinson, Sparks, MD, USA). Examination of bacterial growth was conducted over six months, and the presence of MAP was confirmed by IS*900* PCR [25] of characteristic colonies. Based on the approach of Köhler et al. [26], a colony score (CS) from 0 to 5 was used to estimate MAP growth semi-quantitatively (0 = 0; 1 = 1–10; 2 = 11–20; 3 = 21–50; 4 = 51–100 individual colonies per slope; 5 = bacterial lawn). The week of appearance (WA) was also documented. For numerical estimation of the MAP concentration in the sample, a growth index (GI) for each slope was calculated using the formula:

$$GI = CS \times 100/WA \tag{1}$$

followed by determination of the mean of all inoculated slopes of the sample.

For qPCR analysis, 20 mL of sterile distilled water was added to the second aliquot of the pellet. After vigorous vortexing, the sample was left in an upright position for 20 min at room temperature for sedimentation. Then, 10 mL of the supernatant was centrifiltrated at 3000× *g* for 5 min using an ADIAFILTER (Adiagene, Bio-X Diagnostics S.A., Rochefort, Belgium) to concentrate bacteria and remove PCR-inhibiting substances. The resulting pellet was suspended in 500 µL sterile distilled water. Then, 300 mg zirconia/glass beads (Carl Roth GmbH + Co. KG, Karlsruhe, Germany) were added and the sample was placed in a Mixer Mill MM 400 (Retsch GmbH, Haan, Germany) for 10 min at 30 Hz to mechanically disrupt contained bacteria and centrifugated at 15,000× *g* for 5 min afterwards. DNA extraction was performed from the supernatant using a QIAamp DNA Mini Kit (Qiagen, Hilden, Germany) according to the manufacturer's instructions. The DNA extracts were analyzed in duplicate with the ADIAVET PARATB REAL TIME PCR kit (Adiagene, Bio-X Diagnostics S.A., Rochefort, Belgium), a detection method based on the amplification of the insertion element IS*900* of MAP, following instructions provided by the manufacturer. Data were analyzed using QuantStudio Design & Analysis Software, v1.5.1 (Life Technologies, Carlsbad, CA, USA). Cycle threshold (C_t) values of the duplicates were determined. Values without a specific qPCR signal were assigned to 45.0. The mean C_t values of the duplicates were calculated and used for sample classification. Samples with mean C_t values $\leq$ 40.0 were considered positive, and values > 40.0 were considered negative.

2.5.3. Dust Samples

A total of 20 mL of sterile distilled water was added to each sample and the tube vigorously vortexed. After overnight incubation at room temperature, the liquid was divided into two aliquots, and each aliquot was centrifiltrated at 3000× *g* for 5 min using an ADIAFILTER (Adiagene, Bio-X Diagnostics S.A., Rochefort, Belgium). Supernatants were discarded, and the pellets were further used, one for cultural and one for qPCR analysis as described above, with slight modifications. The pellet used for bacterial cultivation was decontaminated with only 5 mL of 0.75% hexadecylpyridinium chloride monohydrate solution (Sigma Aldrich, Taufkirchen, Germany).

2.5.4. Trough Water Samples

The whole sample volume of 1 L was stepwise centrifugated in tubes of 50 mL at 4200 rpm for 20 min. The resulting pellet was suspended in 2 mL of sterile distilled water, from which 1 mL was further processed for bacterial cultivation according to the method described for the dust samples.

From samples obtained during ES 1, 2 and 3, 500 µL of the remaining 1 mL specimen were processed for qPCR analysis as described for the bedding and feed samples. From ES 4 onwards, 9 mL sterile distilled water was added to the remaining 1 mL specimen, and further DNA extraction was performed using the ADIAPURE PARATB MILK kit (Adiagene, Bio-X Diagnostics S.A., Rochefort, Belgium) according to the manufacturer's instructions to extract MAP DNA by immunomagnetic separation. Subsequently, the extracts were analyzed in duplicate with the ADIAVET PARATB REAL TIME PCR kit (Adiagene, Bio-X Diagnostics S.A., Rochefort, Belgium) under the same conditions as described above.

2.6. Data and Statistical Analysis

The Kappa coefficient was calculated for dichotomized MAP detection results (positive or negative) in environmental samples using culture and qPCR to measure the agreement between both detection methods. The result was classified as described by Landis and Koch [27].

Spearman's rank correlation coefficient was calculated to analyze the relationship between the GI (semi-quantitative culture result) and C_t values from qPCR sample analysis for all analyzed samples as well as each individual material if at least one positive culture result was obtained.

Characteristics listed in Table 1 were recorded for each environmental sample. The sampling sites were assigned to the following sampling locations: adult goat area (lactating goat pen, dry goat/kidding pen, buck pen), high animal traffic area (waiting area in front of the milking parlor, milking parlor, exit of the milking parlor), and youngstock area (juvenile goat pen, goat kid pen). Entrances to the barn were allocated to high animal traffic area during the grazing season and to adult goat area during the indoor season.

Table 1. Overview of recorded characteristics and conditions of environmental samples. (MAP = *Mycobacterium avium* subsp. *paratuberculosis*, HS = herd sampling).

Characteristic	Conditions
Material	Bedding Dust Feed Water
Location	Adult goat area High animal traffic area Youngstock area
Season	Grazing Indoor
Percentage of MAP shedders at nearest HS	Number culture-positive fecal samples/number analyzed fecal samples × 100

Logistic regression models were used to determine the likelihood of a positive environmental culture and qPCR result related to the sample characteristics. The dichotomized result of MAP detection by the respective method was set as the dependent variable. Material, location, and season as categorial, as well as the percentage of MAP shedding goats at the nearest herd examination as a numeric independent variable, were entered into the qPCR model. Because there were no cases in certain characteristic condition groups, the independent variable 'material' was excluded from the culture model. For the same reason, samples from the youngstock area were not included. The fit of the models was assessed using the −2 log-likelihood value.

In general, the statistical significance level was set at $p = 0.05$. Data evaluation and statistical analysis were performed using Microsoft Office Excel version 2019 (Microsoft Corporation, Redmond, WA, USA) and MedCalc version 14.8.1; (MedCalc Software Ltd., Ostend, Belgium). Figures were created using MedCalc and Microsoft Visio version 2019 (Microsoft Corporation, Redmond, WA, USA).

3. Results

3.1. Herd Examination Using Fecal Culture

Herd examination using fecal culture was conducted six times between 2018 and 2022. Only a small number of the inoculated slants showed fungal growth. However, this did not result in any sample that could not be assessed. The proportion of positive fecal samples dropped from 30.0% (January 2018) to 1.7% (May 2022) over the period of four years, but fecal shedding was present at all time points during the study period (Table 2).

Table 2. Number and culture result of fecal samples collected during six examinations between 2018 and 2022 from goats older than one year.

Herd Examination	Number of Collected Samples	Positive Fecal Culture	
		n	%
2018 January	307	92	30.0
2020 April	286	27	9.4
2020 October	309	16	5.2
2021 May	414	13	3.1
2021 November	382	9	2.4
2022 May	407	7	1.7

3.2. Analysis Results of Environmental Samples

MAP was cultivated out of 28 (25.2%) bedding samples and one (1.8%) dust sample, whereas all water and feed samples were negative for viable MAP. This adds up to cultural MAP detection in 10.5% (29/256) of all environmental samples. Calculated GI ranged from 3.85 to 50.00 (see also Table 4). Fungal growth was partially observed in some dust samples and a few bedding samples, especially towards the end of the cultivation period. Positive culture results were obtained at 7 of 8 sampling time points. In August 2021 (ES4), cultivation of MAP was not possible from any environmental sample (Table A2). Culture-positive samples were obtained from various sites of the goat housing area but not from the youngstock area. Bacterial cultivation succeeded several times from bedding samples from the kidding pen and lactating goat pen as well as the milking area (waiting area, parlor, and exit). This area showed the most culture-positive bedding samples (47.6%), while the milking parlor was the sampling site with the highest proportion of MAP-positive samples (6/8; 75.0%). These sites, grouped together with the barn entrance during the grazing season as areas of high animal traffic, yielded 44.4% (12/27) of the culture-positive bedding samples. Bedding samples from other areas of the lactating goat pen were culturally positive in 38.1% of cases, and thus, this was the sampling site with the second most MAP detections. Kidding pens of adult goats were ranked third with 34.8% (8/23) (Figure 2). The culture-positive dust sample was collected from a kidding pen.

Positive qPCR results were obtained from 37 (64.9%) dust, 55 (49.6%) bedding, 12 (34.3%) water, and 13 (24.5%) feed samples, resulting in MAP DNA detection in 45.7% (117/256) of all environmental samples. The lowest measured C_t value was 27.57, obtained from a water sample. Minimum values of the other matrices were 28.58 (bedding), 31.29 (dust), and 36.11 (feed). The mean C_t value over all positive samples from all matrices was 35.62 (see also Table 4). During every ES, MAP DNA was detected in dust and bedding samples, whereas this was not the case for water and feed specimens (Table A2). In contrast to MAP culture, positive qPCR results were obtained from all areas of the stable, including the area of youngstock rearing and the buck pen (Figure 3).

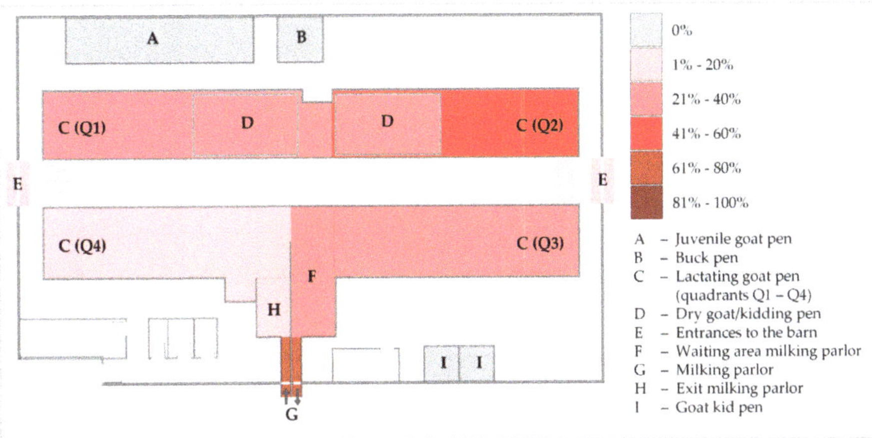

Figure 2. Heatmap indicating the percentage of culture-positive bedding samples for *Mycobacterium avium* subsp. *paratuberculosis* from the different sampling sites in the dairy goat barn over the study period. Dry goat/kidding pens (D) were set up temporarily within quadrants Q1 and Q2 during the kidding season in 2020 and 2021.

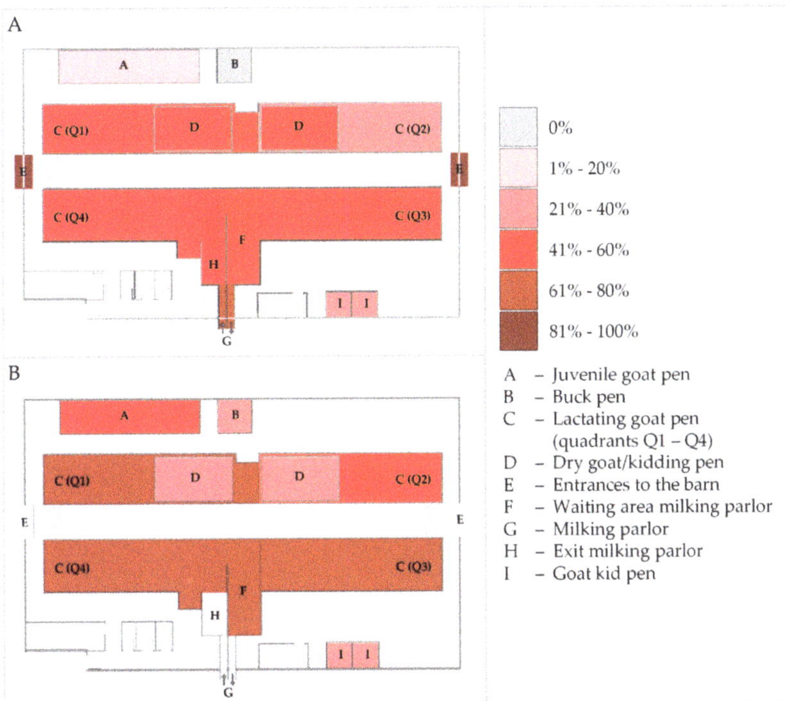

Figure 3. Heatmap indicating the percentage of qPCR-positive bedding (**A**) and dust (**B**) samples for *Mycobacterium avium* subsp. *paratuberculosis* from the different sampling sites in the dairy goat barn over the study period. Dry goat/kidding pens (D) were set up temporarily within quadrants Q1 and Q2 during the kidding season in 2020 and 2021. No dust samples were collected from the barn exits (E), the milking parlor (G), and its exit (H).

3.3. Statistical Analysis

3.3.1. Cohen's Kappa

A slight agreement (κ = 0.20, 95% CI 0.11–0.28) [27] was observed between the MAP detection results (positive or negative) from environmental sample analysis by cultivation and qPCR (Table 3).

Table 3. Contingency table of *Mycobacterium avium* subsp. *paratuberculosis* detection results by culture and qPCR from 256 analyzed environmental samples used for Kappa test analysis.

qPCR Result	Culture Result		Σ
	Negative	Positive	
Negative	135 (52.7%)	4 (1.6%)	139
Positive	92 (35.9%)	25 (9.8%)	117
Σ	227	29	256

3.3.2. Spearman's Rank Correlation

GI (semi-quantitative culture results) for cultivation on HEYM and qPCR C_t values of all environmental samples were significantly negatively correlated, with a corresponding Spearman's rank correlation coefficient of $\rho = -0.401$ (95% CI −0.499–−0.293). Bedding samples revealed the strongest correlation ($\rho = -0.529$), followed by dust samples ($\rho = -0.204$). No correlation was calculated for feed and water samples due to complete negative culture results (Figure 4, Table 4).

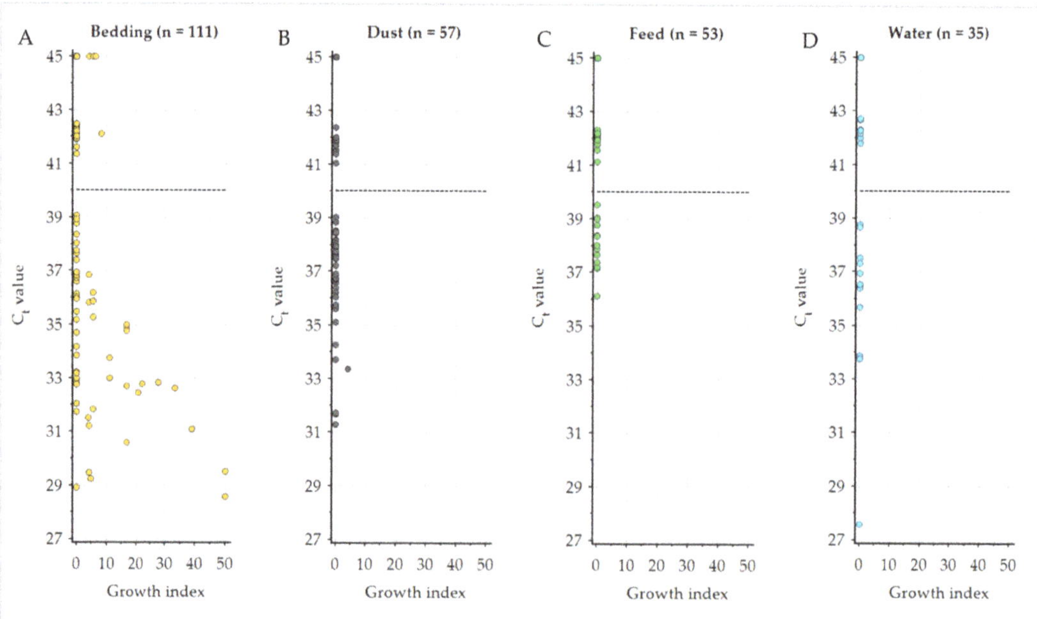

Figure 4. Distribution of the C_t values of qPCR analysis relative to the calculated growth index (semi-quantitative culture results) of *Mycobacterium avium* subsp. *paratuberculosis* in the different sampling materials (**A**): bedding, (**B**): dust, (**C**): feed, (**D**): water displayed with the qPCR classification cut-off (40.0; dashed line).

Table 4. Spearman's rank correlation for growth index (GI) and C_t values of environmental samples tested for *Mycobacterium avium* subsp. *paratuberculosis* using bacterial culture and qPCR.

Material	ρ	p-Value	Mean GI of Culture Positive Samples (Minimum, Maximum)	Mean C_t of qPCR Positive Samples (Minimum, Maximum)
All	−0.401	<0.001 *	14.64 (3.85, 50.00)	35.62 (27.57, 39.52)
Bedding	−0.529	<0.001 *	15.01 (3.85, 50.00)	34.41 (28.58, 39.08)
Dust	−0.204	0.128	4.17 (4.17, 4.17)	36.52 (31.29, 39.02)
Feed	—	—	—	38.03 (36.11, 39.52)
Water	—	—	—	35.79 (27.57, 38.78)

ρ = Spearman's rank coefficient. *: $p \leq 0.05$.

3.3.3. Logistic Regression

The logistic regression model for positive qPCR results revealed no significant effect of the percentage of MAP-shedding goats on the likelihood of MAP DNA detection (p = 0.07). The odds for a positive qPCR result were higher in dust than in any other material, with an odds ratio (OR) of 2.49 compared with bedding as reference and for samples collected during the indoor season with an OR for grazing season of 0.49. Samples from feed had a lower OR (0.385) than bedding. Further, samples collected from areas of high animal traffic were associated with higher odds for MAP DNA detection (OR: 3.03) compared to those from areas where adult goats are kept, while no difference was observed between the latter and samples from youngstock areas (p = 0.095) (Table 5).

Table 5. Outcome of the logistic regression model to estimate the relationship of the qPCR results of *Mycobacterium avium* subsp. *paratuberculosis* (MAP) to different environmental sample characteristics.

Characteristic		Positive qPCR Result		Negative qPCR Result		β [a]	S.E. [b] β	p-Value	OR [c]	95% CI [d] (OR)
		n	%	n	%					
Material	Bedding	55	49.55	56	50.45	reference	—	—	—	—
	Dust	37	64.91	20	35.09	0.913	0.363	0.012 *	2.49	1.22–5.08
	Feed	13	24.53	40	75.47	−0.965	0.399	0.016 *	0.38	0.17–0.83
	Water	12	34.29	23	65.71	−0.359	0.431	0.405	0.70	0.30–1.63
Location	Adult goat area	70	44.87	86	55.13	reference	—	—	—	—
	High animal traffic area	23	71.88	9	28.12	1.109	0.475	0.020 *	3.03	1.20–7.68
	Youngstock area	24	35.29	44	64.71	−0.563	0.338	0.095	0.57	0.29–1.10
Season	Grazing	25	40.32	37	59.68	−0.710	0.342	0.038 *	0.49	0.25–0.96
	Indoor	92	47.42	102	52.58	reference	—	—	—	—
Percentage of MAP shedders at nearest herd sampling						−0.105	0.058	0.069	0.90	0.80–1.01

[a] β: logistic regression coefficient, [b] S.E.: standard error, [c] OR: odds ratio, [d] 95% CI: lower and upper limits of the 95% confidence interval. *: $p \leq 0.05$

Another logistic regression was performed for positive culture results. The sampling location was modified by only distinguishing between high animal traffic areas and adult goat areas because MAP was not cultivated out of any sample collected from the youngstock area. The season and the percentage of MAP-shedding goats in the herd at the nearest herd examination were used in the same way as in the model described above. Of the three variables entered into the regression model, the location and season contributed significantly to the prediction of a positive culture result (p = 0.029; p = 0.034), while the percentage of MAP-shedding goats showed no significant effect. The odds for MAP cultivation were higher in samples collected from high animal traffic (OR: 3.34) compared to those from other adult goat areas. In addition, sampling during grazing season revealed a lower likelihood (OR: 0.24) for positive culture results (Table 6).

Table 6. Outcome of the logistic regression model to estimate the relationship of the culture results for *Mycobacterium avium* subsp. *paratuberculosis* (MAP) to different bedding sample characteristics.

Characteristic		Positive Culture Result		Negative Culture Result		β [a]	S.E. [b] β	*p*-Value	OR [c]	95% CI [d] (OR)
		n	%	*n*	%					
Location	Adult goat area	16	27.12	43	72.88	reference	—	—	—	—
	High animal traffic area	12	44.44	15	55.56	1.025	0.552	0.029 *	3.34	1.13–9.84
Season	Grazing	4	18.18	18	81.82	−1.443	0.681	0.034 *	0.24	0.06–0.90
	Indoor	24	37.50	40	62.50	reference	—	—	—	—
Percentage of MAP shedders at nearest herd sampling						−0.105	0.058	0.069	0.90	0.80–1.01

[a] β: logistic regression coefficient, [b] S.E.: standard error, [c] OR: odds ratio, [d] 95% CI: lower and upper limits of the 95% confidence interval. *: $p \leq 0.05$

4. Discussion

This study was the first to evaluate the presence of MAP in various environmental materials collected from a dairy goat barn. The results allowed us to derive recommendations (1) for suitable sampling sites for the identification of infected goat herds using environmental sampling and (2) for measures to reduce MAP exposure of goat kids by classification of potential high-risk areas in the barn.

Overall, the detection of MAP in environmental samples was possible by both, qPCR and culture. This led to the conclusion that environmental samples are a suitable matrix for paratuberculosis confirmation in dairy goat herds. However, certain limitations must be considered.

MAP was only cultivated from bedding and dust, whereas samples from all materials were MAP DNA-positive. Dust samples had a significantly higher probability for qPCR-positive results than the other sampling matrices, while 96.6% of the culture-positive environmental samples were obtained from bedding. This indicates that the environmental sampling material should be selected depending on the intended analysis method.

The agreement between both detection methods was assessed as slight (κ = 0.20). This is mainly caused by 93 samples (35.9%) that were qPCR-positive but culture-negative. Fecal shedding of MAP by infected animals leads to the spread of the pathogen in the environment and, thereby, to multiple potential infection sources [5,13]. The first contact material, if present, would be bedding. Drying of the feces in combination with air circulation can lead to dissemination and sedimentary deposition. This sediment, which may or may not contain MAP, can settle on the bedding itself, on feed and water surfaces, or accumulate as dust on diverse objects in the barn [28]. The main factor for higher odds of MAP DNA detection in the dust samples is probably accumulation. Fresh forage is presented twice a day, the water in all troughs is changed daily, and new bedding material is provided when necessary. In contrast, the barn is only cleaned once a year in total, which provides a long period of time for dust to accumulate on the surfaces of barn facilities. In addition, Eisenberg et al. [14] showed that even after a cattle farm was completely destocked and cleaned with a high-pressure cleaner, MAP DNA could be found in analyzed dust samples. MAP can survive outside a suitable host for up to one year if appropriate environmental conditions are present [7]. However, in addition to viable, qPCR methods can also detect non-viable pathogens. Moreover, the existence of dormant or viable but non-cultivable states is confirmed for MAP. Dormancy describes the ability of bacteria to survive as a non-spore-forming cell without replication, mainly induced by unfavorable environmental conditions [7]. Bacteria in this reversible state may have also been detected by qPCR but caused false negative culture results.

Sample handling might also play a role in the rate of detection of culturable MAP. Freezing environmental manure samples before cultural analysis leads to lower numbers of cultivated colony-forming units per sample compared to direct sample processing [15]. In addition, decontamination during the cultivation process may adversely affect MAP viability. For samples containing small numbers of MAP which may be irregularly distributed, pellet partitioning during sample preparation can lead to false negative results of one or both analytical methods. This is demonstrated by four bedding samples from which MAP could be cultivated while being negative in qPCR. The impact of the environmental sample preparation methods used in this study on the analytical sensitivity of cultural and molecular biological MAP detection was not determined. Principle procedures were based on techniques successfully used in former studies [7,14]. A decrease of 10^2 viable organisms was identified for soil samples spiked with feces from paratuberculosis-infected animals [13], leading to the assumption that low numbers of culturable MAP in the environmental samples may not have been detected, resulting in false negative sample classification. Taking all the mentioned factors into account, an underestimation of the true contamination of the barn environment with viable MAP is likely.

Nevertheless, a negative correlation between the C_t value and the GI was observed. A comparable relationship was reported for environmental manure samples from dairy cow herds in the US [15]. The Spearman's rank correlation coefficient reported by Aly et al. [15] for frozen environmental samples ($\rho = -0.611$) was higher than the one calculated in the current study for all sampling materials ($\rho = -0.401$). However, when only bedding samples, which would be equivalent to the manure samples, were considered, the negative correlation was even stronger ($\rho = -0.529$), demonstrating the influence of the sample material.

Since there are no former studies concerning MAP detection in dairy goat barns, comparisons can only be made with studies conducted in cattle farms. Most of these studies analyzed mixed manure samples and, thus, areas frequented daily by large numbers of adult animals, such as alleyways and the milking parlor. In each study, these areas were identified as having the highest probability for culturable MAP [9–11,17,18]. Similar results were obtained in the goat housing. Here, the milking area, and especially the milking parlor, revealed the most culture-positive bedding samples. Logistic regression results underscore this finding by calculating an OR of 3.34 for the probability of bedding samples from areas of high animal traffic being positive for MAP culture compared to samples from other adult goat areas. However, viable MAP was detected in more than one-third of the bedding samples from the kidding pens of adult goats. Opposing results were reported from cattle farms where MAP was cultivated only occasionally from calving pens (14.3–17%) which were considered a sampling site with lower contamination [9,10,17,18]. The most likely explanation for this finding is the difference in the time span cows and goats normally spend in these pens. Dairy cows are usually dried off and separated from lactating cows approximately six weeks before calving. While separate dry cow and calving pens exist on most cattle farms, goats in the study herd stayed in the same pen until after kidding. In addition, in most dairy cow herds, calving pens are cleaned daily or at short time intervals, whereas in the goat kidding pen, only fresh bedding was added throughout the whole kidding season. This can lead to an accumulation of MAP even if just one shedding doe is present in the pen.

Because the probability of pathogen shedding rises with the age of the animal [5], the lack of detection of culturable MAP from samples from the kid and juvenile goat areas is plausible. Raizman et al. [18] were also unable to cultivate MAP from samples from pre-weaned calf pens, whereas 3% of samples from post-weaned calf areas were MAP-positive. This may be due to early pathogen shedding or, more likely in herds with high paratuberculosis prevalence, an introduction from the adult cow area through common devices, footwear, etc., as also mentioned by Field et al. [21], who detected MAP DNA in

21% of boot swab samples from pre-weaned calf areas. The construction modalities of the goat kid pens at the study farm, which do not require entering under normal circumstances, most likely avoid this route of pathogen entry.

However, MAP DNA was detected at every sampling site in the goat barn. This includes areas that are accessed by adult goats as well as pens where kids and juvenile goats are kept. Other studies reported comparable results for cattle husbandries [12,14]. Environmental samples from areas of high animal traffic had a higher probability of being qPCR positive (OR: 3.03) than those from other adult goat areas. No significant difference was observed between the young and adult goat area, which is in clear contrast to the culture results, where MAP was not cultivated from any sample collected in the youngstock area. Dust was the matrix with the highest proportion of positive MAP DNA results. Because the milking herd and kids are kept in one common air space, airborne distribution of MAP, viable or not, from adult to young goat areas is likely, especially because of the proximity between the milking parlor and the goat kid pens in the studied barn. Higher average C_t values from all materials, in particular bedding, from young compared to adult goat areas support the assumption of directed dissemination between these barn compartments. Eisenberg et al. [14] confirmed airborne distribution of MAP by pathogen detection high above the animal level where fecal contamination was unfeasible. In contrast to samples from the ground of cubicles, there was no difference concerning the likelihood of positive dust samples collected above animal level between stocked and non-stocked barn areas.

Considering the results of qPCR and cultural analysis, the milking parlor is the most recommended sampling site for both analysis methods. The farm management and husbandry conditions of the studied goat herd were very similar to those that can be found in other commercial dairy goat farms in Germany as well as other countries with a commercial dairy industry in goats. However, since samples from only one herd and barn were examined in this explorative study, the current results must be confirmed in a larger comprehensive study, including several dairy goat farms.

Two of the eight ES (ES 4 and 8) were conducted during grazing season when only the juvenile goats were in the barn throughout the whole day. The remaining dairy herd was driven inside twice a day for milking. MAP DNA was detected at every ES, while MAP could be cultivated at 7 of 8 ES. No cultivation out of any sample was possible in August 2021 (ES 4). The logistic regression model for both analysis methods identified the season (indoor vs. grazing) as a significant variable concerning the likelihood of a positive analysis result. In both cases, samples collected during grazing season had a lower probability for MAP detection (qPCR: OR = 0.49; culture: OR = 0.24). Other studies in cattle farms determined no seasonal effect [11] or reported a higher percentage of positive samples during spring and summer in comparison to autumn and winter [9]; however, some frequently positive locations were not sampled at all times of the year in the second study. The studies provide no information concerning pasture use at the farms. The data from the current study indicate that the time animals spend inside the barn is important. Prolonged stays raise the likelihood of defecation and, thereby, the risk of MAP spread by shedding animals. Furthermore, intense animal movement may contribute to dust formation and airborne distribution in the housing [12,28]. Thus, MAP detection in environmental samples will be more successful during time periods when the majority of the herd stays inside the barn continuously.

Eisenberg et al. [14] cultivated MAP from several dust and water samples from a cattle farm. These data contrast with the present study. Firstly, fecal contamination of the trough water was possible in the cattle barn, whereas this was most likely not the case in the studied goat barn. More important, in contrast to the current study, only paratuberculosis-positive cattle which had previously shed MAP were introduced into the barn. Biannual testing of the goat herd showed that during the study period, only between 9.4% and 1.7% of the animals were active shedders of MAP. As intermittent shedding is known to occur during the subclinical disease stage [3] and the herd samplings were not carried out together with the ES, these numbers only allow an approximate estimation of the precise situation at the particular ES.

Environmental manure or bedding samples are considered to be a form of pooled fecal sample due to the potential composition of feces from several animals. Thus, a dilution effect may lead to false negative results in herds with low disease prevalence [11]. Therefore, different studies tried to evaluate the correlation between MAP-positive environmental samples and disease prevalence. Donat et al. [10] estimated 87% sensitivity for the cultivation of manure samples from herds with a fecal prevalence > 2.0%. Eisenberg et al. [12] detected culturable MAP in settled-dust samples from cattle herds with low (one enzyme-linked immunosorbent-assay positive animal) but more frequently in herds with higher (two or more enzyme-linked immunosorbent assay positive animals) paratuberculosis prevalence. In the present study, logistic regression revealed no association between the number of MAP-shedding goats at the nearest herd sampling and the likelihood of MAP detection in environmental samples by culture or qPCR. A possible explanation for continuously positive qPCR results is that, despite the decreasing number of MAP shedders, the pathogen's DNA persists in the environment for a long time and, as mentioned earlier, is not completely eliminated by normal cleaning and disinfection routines. This assumption is supported by Eisenberg et al. [12], who detected MAP DNA in more than 50% of low- and high-prevalence cattle herds. The amount of MAP shed with the feces can vary with regard to the disease stage and the individual animal [3]. It is conceivable that one high-shedding doe introduces the same number of bacteria into the environment as several light shedders, leading to consistent cultural detection of MAP. Therefore, areas that are passed by each animal are favorable for MAP detection, even in herds with only a few active shedders.

The herd examinations indicate a decreasing number of MAP-shedding goats within the herd during the course of the study. However, the data on MAP detection in environmental samples imply that a small number of active MAP shedders in the herd can sufficiently contaminate the barn environment and, thus, maintain potential infection risks.

In 2018, two years before the first ES was conducted, vaccination was included in herd management as a measure to control paratuberculosis. Despite vaccination, the barn environment was still considerably contaminated with MAP. It has to be assumed that new infections occurred in each birth cohort. The findings of this study point out that vaccination as a single measure in disease control does not lead to a complete prevention of environmental contamination with MAP and, thereby, enables the disease to maintain in the herd.

In addition to demonstrating that MAP can be found in the environment of a paratuberculosis-infected dairy goat herd, the results of the present study also allow the identification of potential risk areas for MAP transmission at the farm level. MAP was detected in dust samples in the kid-rearing area, but only by molecular biological methods. As viable pathogens are needed to cause an infection, this route of exposure is probably less critical for kids. Nevertheless, a clear structural separation of adult and young animals would be preferable in general for a further reduction of airborne MAP dissemination between these groups. However, the successful cultivation of MAP from a dust sample from the kidding area is a cause for concern. Even if one out of 57 samples is still far less than what was detected in studies in cattle farms [12,14], the sampling site is the problematic aspect. In addition to the dust sample, MAP could be cultivated from several bedding samples in this area. If kids are not immediately separated from the does after parturition, as realized using the snatching method [5,29], the likelihood of MAP intake increases over time. To minimize the environmental MAP burden of this sensitive barn area, it is recommended to set up separate kidding pens for known MAP-infected animals, to shorten the cleaning and disinfection intervals, and not to use kids from infected does for replacement.

Bedding samples from the kidding and lactating goat pens revealed a high bacterial load. This is a possible risk for contamination of the udder skin, enabling MAP intake when kids are allowed to suckle from the does. Even though a study to detect MAP on the udder skin of the goats from this herd revealed only low pathogen prevalence (10.3% qPCR-positive udder swabs) [30], the detection limits of the methods may underestimate the issue. This is another argument for the recommendation of immediate separation of the kids from the does.

The high MAP contamination of the milking parlor only plays a minor role in disease transmission since this area is only accessed by adult animals for a short time. It is assumed that a relatively high dose would be required to infect these animals [5]. In addition, environmental MAP contamination of the milk during the milking process was not detected [30]. However, the present results highlight this location as a suitable sampling site to determine the paratuberculosis-herd status using environmental samples.

Each analysis method has its specific advantages and disadvantages. When decisions about the appropriate method have to be made, the discussed key points, in combination with the objective of the investigation, should be taken into account. qPCR is an appropriate method for the identification of infected herds. Confirmation of MAP in the environment is a strong initial indicator of the presence of infected animals. MAP viability does not influence the analysis results, which increases the probability of pathogen detection. In this study, MAP DNA was detected in 117 out of 256 environmental samples, and thus, if only one sample had been taken, the MAP infection of the goat herd would have been identified in almost half of the instances. The likelihood can be increased if recommendations concerning sampling material, location, and season are considered during sample collection. Cultivation of environmental samples revealed fewer positive results (11.3%). However, MAP could be cultivated from 60.0% of bedding samples collected from high animal traffic areas during the indoor season. Therefore, this analysis method can also be suitable for paratuberculosis status determination and for confirmation of positive qPCR results. The lower limits of the within-herd prevalence of MAP-shedding animals resulting in a positive test outcome using environmental samples still have to be determined. A similarly important application of bacterial culture, however, is the identification of barn areas of high contamination with culturable MAP which serve as possible infection sites.

5. Conclusions

Environmental sampling can be an appropriate method for the identification of MAP-infected dairy goat herds and barn areas with a high bacterial load. To increase the probability of MAP detection, the collection of bedding or dust samples from areas with high animal traffic, especially from the milking parlor, is recommended. The applied analysis method should be chosen depending on the aim of the examination. qPCR is suitable for making a general statement about whether paratuberculosis is present in the herd or whether the direct animal environment is contaminated with MAP. Instead, cultural analysis can confirm that animals are actually shedding MAP, and identify crucial areas where the youngstock is at risk of being exposed to the pathogen and from which MAP can be spread in the barn, respectively. Detection of culturable MAP should be considered more important than solely DNA detection because of their potential role in paratuberculosis transmission. Incorporating these findings into the development of individual paratuberculosis control plans for husbandries of goats may promote efficient and effective disease control. However, to ensure the external validity of the obtained results, further research in a larger number of goat herds with different husbandry conditions and prevalence levels is needed.

Author Contributions: Conceptualization, C.P., K.D., U.M. and H.K.; methodology, C.P. and H.K.; formal analysis, C.P. and H.K.; investigation, C.P. and H.K.; resources, C.P. and H.K.; data curation, C.P. and H.K.; writing—original draft preparation, C.P. and H.K.; writing—review and editing, C.P., K.D., U.M. and H.K.; visualization, C.P.; supervision, K.D. and H.K.; project administration, K.D. and H.K.; funding acquisition, K.D. and H.K. All authors have read and agreed to the published version of the manuscript.

Funding: This research was funded by the European Agricultural Fund for Rural Development (EAFRD) via the Thüringer Aufbaubank, grant number 2019 LFE 0017.

Institutional Review Board Statement: The animal study protocol was approved by the Thuringian State Office for Consumer Protection, which is the competent authority for research ethics approval in Thuringia (file reference: 2684-04-04-BFI-20-103, date of approval 1 July 2020).

Informed Consent Statement: Not applicable.

Data Availability Statement: The dataset analyzed during the current study is available from the corresponding author upon reasonable request.

Acknowledgments: We wish to acknowledge the dairy goat farmer involved in this study for his cooperation and friendly support.

Conflicts of Interest: The authors declare no conflict of interest.

Appendix A

Table A1. Number of environmental samples of different materials from nine sampling sites in a dairy goat barn collected at eight environmental sampling events (ES) (total $n = 256$). B = bedding ($n = 111$), D = dust ($n = 57$), F = feed ($n = 53$), W = water ($n = 35$).

Sampling Site	ES 1				ES 2				ES 3				ES 4				ES 5				ES 6				ES 7				ES 8			
	B	D	F	W	B	D	F	W	B	D	F	W	B	D	F	W	B	D	F	W	B	D	F	W	B	D	F	W	B	D	F	W
Buck pen	—	—	—	—	—	—	—	—	—	—	—	—	1	1	1	1	1	1	1	-	1	1	1	1	1	1	1	1	1	1	—	—
Dry goat/kidding pen(s)	8	2	—	—	7	2	—	—	12	3	2	—	—	—	—	—	—	—	—	—	—	—	—	—	—	—	—	—	—	—	—	—
Entrances to the barn	2	—	—	—	2	—	—	—	2	—	—	—	2	—	—	—	2	—	—	—	2	—	—	—	2	—	—	—	2	—	—	—
Goat kid pen(s)	9	7	1	7	—	1	—	—	—	1	—	—	—	—	—	—	—	—	—	—	—	—	—	—	3	3	2	1	1	1	1	1
Juvenile goat pen	—	—	—	—	3	1	1	1	—	—	—	—	1	1	1	-	1	1	1	1	1	1	1	1	1	1	1	1	1	1	—	—
Lactating goat pen	—	—	—	—	1	1	10	—	—	—	—	—	4	4	6	1	4	4	6	4	4	4	6	4	4	4	6	4	4	4	6	4
Milking rotary	1	—	—	—	1	—	—	—	1	—	—	—	1	—	—	—	1	—	—	—	1	—	—	—	1	—	—	—	1	—	—	—
Milking rotary waiting area	1	—	—	—	1	—	—	—	1	—	—	—	1	1	—	—	1	1	—	—	1	1	—	—	1	1	—	—	1	1	—	—
Milking rotary exit	—	—	—	—	—	—	—	—	—	—	—	—	1	—	—	—	1	—	—	—	1	—	—	—	1	—	—	—	1	—	—	—

Table A2. Number of positive samples of bacterial culture and qPCR for the detection of *Mycobacterium avium* subsp. *paratuberculosis* in environmental samples per environmental sampling event (ES) (total $n = 256$).

Analysis Result	Sampling Material	Environmental Sampling Event								Σ
		ES 1	ES 2	ES 3	ES 4	ES 5	ES 6	ES 7	ES 8	
Culture positive	Bedding	5/21	3/15	5/16	0/11	5/11	6/11	2/14	2/12	28/111
	Dust	1/9	0/5	0/4	0/7	0/7	0/7	0/10	0/8	1/57
	Feed	0/1	0/11	N.s.	0/8	0/8	0/8	0/10	0/7	0/53
	Water	0/3	0/1	0/2	0/2	0/5	0/6	0/7	0/5	0/35
qPCR positive	Bedding	8/21	6/15	7/16	2/11	9/11	10/11	6/14	7/12	55/111
	Dust	7/9	2/5	1/4	6/7	3/7	5/7	7/10	6/8	37/57
	Feed	0/1	3/11	N.s.	0/8	1/8	3/8	6/10	0/7	13/53
	Water	1/7	0/1	0/2	1/2	0/5	3/6	6/7	1/5	12/35

N.s. = no sample of this material was collected.

References

1. Sweeney, R.W. Transmission of paratuberculosis. *Vet. Clin. N. Am. Food Anim. Pract.* **1996**, *12*, 305–312. [CrossRef]
2. Whittington, R.J.; Windsor, P.A. In utero infection of cattle with *Mycobacterium avium* subsp. *paratuberculosis*: A critical review and meta-analysis. *Vet. J.* **2009**, *179*, 60–69. [CrossRef]
3. Windsor, P.A. Paratuberculosis in sheep and goats. *Vet. Microbiol.* **2015**, *181*, 161–169. [CrossRef]
4. Eisenberg, S.W.F.; Koets, A.P.; Nielen, M.; Heederik, D.; Mortier, R.; de Buck, J.; Orsel, K. Intestinal infection following aerosol challenge of calves with *Mycobacterium avium* subspecies *paratuberculosis*. *Vet. Res.* **2011**, *42*, 117. [CrossRef]

5. Windsor, P.A.; Whittington, R.J. Evidence for age susceptibility of cattle to Johne's disease. *Vet. J.* **2010**, *184*, 37–44. [CrossRef]
6. Smith, M.C.; Sherman, D.M.; van Metre, D.C. *Goat Medicine*, 3rd ed.; Wiley-Blackwell: Hoboken, NJ, USA, 2022; pp. 476–485.
7. Whittington, R.J.; Marshall, D.J.; Nicholls, P.J.; Marsh, I.B.; Reddacliff, L.A. Survival and dormancy of *Mycobacterium avium* subsp. *paratuberculosis* in the environment. *Appl. Environ. Microbiol.* **2004**, *70*, 2989–3004. [CrossRef]
8. Whittington, R.; Donat, K.; Weber, M.F.; Kelton, D.; Nielsen, S.S.; Eisenberg, S.; Arrigoni, N.; Juste, R.; Sáez, J.L.; Dhand, N.; et al. Control of paratuberculosis: Who, why and how. A review of 48 countries. *BMC Vet. Res.* **2019**, *15*, 198. [CrossRef]
9. Wolf, R.; Barkema, H.W.; Buck, J.d.; Orsel, K. Sampling location, herd size, and season influence *Mycobacterium avium* ssp. *paratuberculosis* environmental culture results. *J. Dairy Sci.* **2015**, *98*, 275–287. [CrossRef] [PubMed]
10. Donat, K.; Schau, U.; Soschinka, A. Identifizierung von mit *Mycobacterium avium* ssp. *paratuberculosis* (MAP) infizierten Milchviehbeständen mithilfe von Umgebungskotproben. *Berl. Münch. Tierärztl. Wochenschr.* **2011**, *124*, 360–367. [CrossRef] [PubMed]
11. Lombard, J.E.; Wagner, B.A.; Smith, R.L.; McCluskey, B.J.; Harris, B.N.; Payeur, J.B.; Garry, F.B.; Salman, M.D. Evaluation of Environmental Sampling and Culture to Determine *Mycobacterium avium* subspecies *paratuberculosis* Distribution and Herd Infection Status on US Dairy Operations. *J. Dairy Sci.* **2006**, *89*, 4163–4171. [CrossRef] [PubMed]
12. Eisenberg, S.W.F.; Koets, A.P.; Hoeboer, J.; Bouman, M.; Heederik, D.; Nielen, M. Presence of *Mycobacterium avium* subsp. *paratuberculosis* in environmental samples collected on commercial Dutch dairy farms. *Appl. Environ. Microbiol.* **2010**, *76*, 6310–6312. [CrossRef] [PubMed]
13. Whittington, R.J.; Marsh, I.B.; Taylor, P.J.; Marshall, D.J.; Taragel, C.; Reddacliff, L.A. Isolation of *Mycobacterium avium* subsp *paratuberculosis* from environmental samples collected from farms before and after destocking sheep with paratuberculosis. *Aust. Vet. J.* **2003**, *81*, 559–563. [CrossRef] [PubMed]
14. Eisenberg, S.W.F.; Nielen, M.; Santema, W.; Houwers, D.J.; Heederik, D.; Koets, A.P. Detection of spatial and temporal spread of *Mycobacterium avium* subsp. *paratuberculosis* in the environment of a cattle farm through bio-aerosols. *Vet. Microbiol.* **2010**, *143*, 284–292. [CrossRef] [PubMed]
15. Aly, S.S.; Mangold, B.L.; Whitlock, R.H.; Sweeney, R.W.; Anderson, R.J.; Jiang, J.; Schukken, Y.H.; Hovingh, E.; Wolfgang, D.; van Kessel, J.A.S.; et al. Correlation between Herrold egg yolk medium culture and real-time quantitative polymerase chain reaction results for *Mycobacterium avium* subspecies *paratuberculosis* in pooled fecal and environmental samples. *J. Vet. Diagn. Investig.* **2010**, *22*, 677–683. [CrossRef]
16. Hahn, N.; Failing, K.; Eisenberg, T.; Schlez, K.; Zschöck, P.-M.; Donat, K.; Einax, E.; Köhler, H. Evaluation of different diagnostic methods for the detection of *Mycobacterium avium* subsp. *paratuberculosis* in boot swabs and liquid manure samples. *BMC Vet. Res.* **2017**, *13*, 259. [CrossRef]
17. Pillars, R.B.; Grooms, D.L.; Kaneene, J.B. Longitudinal study of the distribution of *Mycobacterium avium* subsp. *paratuberculosis* in the environment of dairy herds in the Michigan Johne's disease control demonstration herd project. *Can. Vet. J.* **2009**, *50*, 1039–1046.
18. Raizman, E.A.; Wells, S.J.; Godden, S.M.; Bey, R.F.; Oakes, M.J.; Bentley, D.C.; Olsen, K.E. The distribution of *Mycobacterium avium* ssp. *paratuberculosis* in the environment surrounding Minnesota dairy farms. *J. Dairy Sci.* **2004**, *87*, 2959–2966. [CrossRef]
19. Wichert, A.; Kasbohm, E.; Einax, E.; Wehrend, A.; Donat, K. Detection of Low MAP Shedder Prevalence in Large Free-Stall Dairy Herds by Repeated Testing of Environmental Samples and Pooled Milk Samples. *Animals* **2022**, *12*, 1343. [CrossRef]
20. Zoche-Golob, V.; Donat, K.; Barkema, H.W.; de Buck, J.; Kastelic, J.; Wolf, R. Predicting sensitivity of repeated environmental sampling for *Mycobacterium avium* subsp. *paratuberculosis* in dairy herds using a Bayesian latent class model. *Vet. J.* **2021**, *275*, 105728. [CrossRef]
21. Field, N.L.; Mee, J.F.; McAloon, C.G. Evaluation of Environmental Sampling for Detection of *Mycobacterium avium* subspecies *paratuberculosis* in the Pre-Weaned Calf Area and Calving Area of Infected Dairy Farms Enrolled in a Voluntary Johne's Disease Control Programme. *Animals* **2023**, *13*, 669. [CrossRef]
22. Mortier, R.A.R.; Barkema, H.W.; Orsel, K.; Wolf, R.; de Buck, J. Shedding patterns of dairy calves experimentally infected with *Mycobacterium avium* subspecies *paratuberculosis*. *Vet. Res.* **2014**, *45*, 71. [CrossRef] [PubMed]
23. Groenendaal, H.; Nielen, M.; Jalvingh, A.W.; Horst, S.H.; Galligan, D.T.; Hesselink, J.W. A simulation of Johne's disease control. *Prev. Vet. Med.* **2002**, *54*, 225–245. [CrossRef] [PubMed]
24. Friedrich-Loeffler-Institut. Paratuberkulose: Amtliche Methode und Falldefinition. Available online: https://www.openagrar.de/receive/openagrar_mods_00058039 (accessed on 5 May 2020).
25. Englund, S.; Ballagi-Pordány, A.; Bölske, G.; Johansson, K.-E. Single PCR and nested PCR with a mimic molecule for detection of *Mycobacterium avium* subsp. paratuberculosis. *Diagn. Microbiol. Infect. Dis.* **1999**, *33*, 163–171. [CrossRef] [PubMed]
26. Köhler, H.; Soschinka, A.; Meyer, M.; Kather, A.; Reinhold, P.; Liebler-Tenorio, E. Characterization of a caprine model for the subclinical initial phase of *Mycobacterium avium* subsp. *paratuberculosis* infection. *BMC Vet. Res.* **2015**, *11*, 74. [CrossRef]
27. Landis, J.R.; Koch, G.G. The Measurement of Observer Agreement for Categorical Data. *Biometrics* **1977**, *33*, 159. [CrossRef]
28. Carpenter, G.A. Dust in livestock buildings—Review of some aspects. *J. Agric. Eng. Res.* **1986**, *33*, 227–241. [CrossRef]

29. Nagel-Alne, G.E.; Asheim, L.J.; Hardaker, J.B.; Sølverød, L.; Lindheim, D.; Valle, P.S. The Norwegian Healthier Goats programme—A financial cost-benefit analysis. *Prev. Vet. Med.* **2014**, *114*, 96–105. [CrossRef]
30. Pickrodt, C.; Donat, K.; Moog, U.; Köhler, H. Analysis of Colostrum and Udder Skin Swabs from a Dairy Goat Herd in Germany regarding the Occurrence of *Mycobacterium avium* Subsp. *paratuberculosis*. *Animals* **2022**, *12*, 1779. [CrossRef]

animals

MDPI

Article

High-Impact Risk Factors for *Mycobacterium avium* ssp. *paratuberculosis* in Dairy Herds in Germany

Mette Krieger [1,*], Susanne Eisenberg [2], Karsten Donat [3] and Amely Campe [1]

1 Department of Biometry, Epidemiology and Information Processing, WHO Collaborating Centre for Research and Training for Health at the Human-Animal-Environment Interface, University of Veterinary Medicine Hannover, 30559 Hanover, Germany; amely.campe@tiho-hannover.de
2 Lower Saxonian Animal Diseases Fund, 30169 Hanover, Germany; susanne.eisenberg@ndstsk.de
3 Animal Health Service, Thuringian Animal Diseases Fund, 07745 Jena, Germany; kdonat@thtsk.de
* Correspondence: mette.krieger@tiho-hannover.de

Simple Summary: Previous studies have investigated the role of different factors in paratuberculosis introduction in dairy farms or dairy cattle. Because paratuberculosis between-herd prevalence was shown to differ substantially between three regions of Germany, this study evaluated different management factors for their impact on the paratuberculosis status of a farm. The most obvious impact was found for an increase of 100 or more cows per herd, followed by the purchase of cattle with unknown paratuberculosis status and limitations in calf feeding management within the barn. These aspects should be prioritized in paratuberculosis control beneath a tailored control approach for individual farms and regions.

Abstract: In a cross-sectional study, it was identified that three regions in Germany differed with respect to their herd-level prevalence for paratuberculosis in dairy cattle. In the study presented here, the same farms were analyzed to identify those components of biosecurity and farm management with the highest impact on *Mycobacterium avium* ssp. *paratuberculosis* (MAP) introduction and establishment in a farm. Hence, the data analyzes included 183, 170 and 104 herds from the study regions north, east and south, respectively. A herd was considered MAP-positive if at least one fecal environmental sample was positive. Twenty-six different possible risk factors from five different components of biosecurity and farm management were analyzed. We show that the average management of calf feeding increased the odds for a MAP-positive farm by 5.22 times (95% confidence interval (CI) = 1.25–21.83). With every 100-cow increase in farm size, the risk for a farm to test MAP-positive increased by 1.94 times (CI = 1.15–3.27), 1.14 times (CI = 1.02–1.27) and 5.53 times (CI = 0.44–68.97) in the north, east and south study regions, respectively. Furthermore, the purchase of cattle with an unknown MAP status increased the risk for a farm testing MAP-positive by 2.86-fold (CI = 1.45–5.67). Our results demonstrate that herd size, unknown MAP status of the purchased cattle and different aspects of calf feeding play an important role in the MAP status of a farm and should be in focus in regions with different MAP between-herd prevalence. Additionally, farm individual risk patterns should be identified during (veterinary) biosecurity consultancy.

Keywords: *Mycobacterium avium* ssp. *paratuberculosis*; risk factor; dairy herd; odds ratio; Johne's disease

1. Introduction

In previous studies, several biosecurity and farm management factors were identified as risk factors for paratuberculosis infection in a herd. Herd size [1–3] and animal purchase [3–5] play an important role in the risk of a herd testing *Mycobacterium avium* ssp. *paratuberculosis* (MAP)-positive. Also other risk factors regarding outdoor hygiene were identified as risk factors for paratuberculosis, for example, keeping heifers on contaminated pastures increased MAP excretion within infected herds [4], and keeping animals other than cattle on the farm could be a source of infection [6,7]. Besides many management factors which are recommended for paratuberculosis control, hygiene of equipment and animals is also considered a useful tool on a farm [8]. Furthermore, several studies have investigated the role of calf management for MAP transmission within a herd. The contact of calves with the feces of adults was shown to be the most important factor for transmission within a herd [9], and suckling of the dam's udder by calves increased the risk for a positive MAP status of the herd [3,10]. Furthermore, it was shown that the use of calving pens to house sick or lame cows [11] and group housing of preweaned calves [12,13] increased the risk of MAP transmission and infection. The evaluation of calf feeding revealed that MAP can also be shed into colostrum and milk [14,15], and fecal contamination of milk is possible [16].

Herd-level diagnosis of the MAP status by fecal culture or polymerase chain reaction (PCR) of fecal environmental samples has been described as an appropriate tool [17,18]. In addition, combining PCR and fecal culture of boot swabs and liquid manure samples [19] or of environmental fecal samples from different locations [20] results in an accurate detection of paratuberculosis-positive herds.

The MAP status in herds enrolled in this study was determined using a standardized sampling scheme for environmental samples, liquid manure and sampling socks [21]. Different values of between-herd prevalence for three different regions of Germany (apparent prevalence: study region north = 12.0%, east = 40.6% and south = 2.9%; corrected true prevalence: study region north = 14.8%, east = 50.1%, south = 3.6%) were identified. Each region reflects different agricultural structures, with a high animal density in region north, large dairy herds with a low farm density in region east and a high farm density with small herds in region south [22].

The objective of this study was to identify those components of biosecurity and farm management that have the highest impact on MAP introduction and establishment in a farm. Furthermore, noticeable risk factors are to be described in order to derive a tailored control approach for the investigated farms and regions.

2. Materials and Methods

2.1. Study Design

The herds were located in Germany, and the analysis was conducted according to a random stratified sampling scheme for each region. The data originated from a cross-sectional study conducted between 2016 and 2020 [21,23]. In the PraeRi study (Animal health, hygiene and biosecurity in German dairy cow operations—a prevalence study [23]), the sample size was set at 250 farms per region (power 80%, significance level 5%), calculated using NCSS PASS Version 13.0.8 [24]. The dairy farms in the study region north and east were randomly selected from the German database of animal identification and registration, stratified by region and herd size [23]. In the study region south, the dairy farms were randomly selected from dairy farms organized in the dairy herd improvement organization for Bavarian dairy farms (Milchprüfring Bayern e.V., Wolnzach, Germany), stratified by region and herd size as well [23]. Our study was part of the PraeRi study, and only those farms that agreed to be sampled for paratuberculosis were analyzed. The farms were visited by study veterinarians to evaluate farm characteristics such as animal health, biosafety, feeding and animal housing on a farm.

2.2. Analyzed Data

Twenty-six variables were analyzed as possible risk factors for paratuberculosis infection at the herd level. Some of these variables were measured at the animal level (e.g., breed and hygienic score of cows and calves), and other variables (e.g., regarding feeding, animal housing and biosafety) were evaluated with an interview and data entry forms at the herd level or at a stable compartment level [23]. Variables that were measured at the animal level were aggregated on herd-level values, i.e., percentage of breed and hygienic scores for a farm.

As we aimed to demonstrate from which component of biosecurity and farm management (BFM) the highest risk for MAP infection and transmission may originate, we assigned the variables of interest to different components of BFM.

General farm management practices were sorted as internal biosecurity and external biosecurity, and calf management practices were sorted as calving, calf housing and calf feeding with the assigned variables of interest (Figure 1).

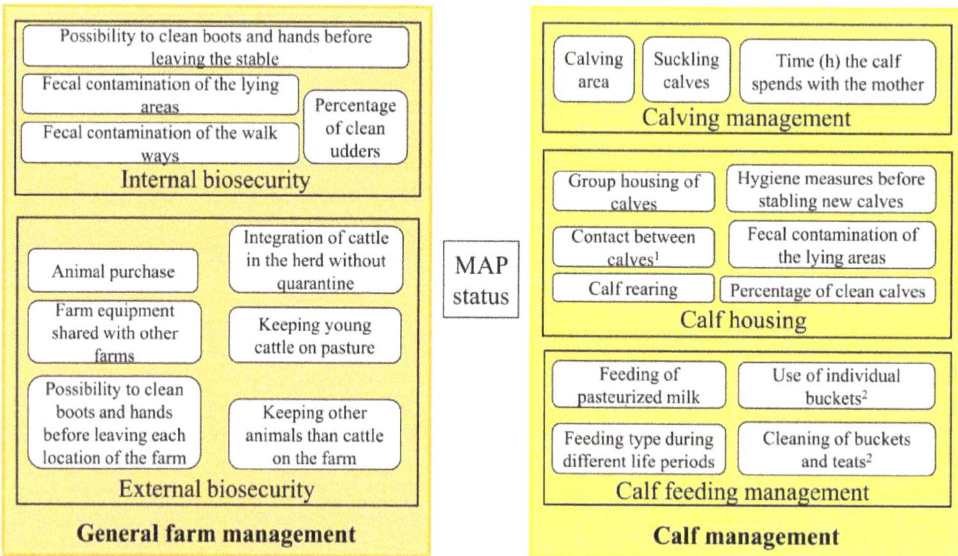

Figure 1. Components of biosecurity and farm management (BFM) for paratuberculosis on a farm with the assigned variables of interest; [1] = contact between calves up to two weeks if housing in individual calf igloo; [2] = in the case of bucket drinking/automatic feeding.

The variables were scored as "good" (reference category), "average" and "insufficient" regarding the level of hygienic quality (i.e., MAP prevention potential). The quantitative variables, except herd size, were scored likewise using their 33% and 66% quantiles as cut-offs for each region. The modus of the quality scores of all respective variables was then used to assign an overall hygienic quality score to each component of BFM (i.e., "good", "average" and "insufficient").

The housing type ($\geq$80% of cows were housed in the according housing type) and herd size (i.e., current number of dry and lactating cows) were analyzed as possible confounders in multivariable regression models. Herd size was included as a quantitative variable.

Paratuberculosis samples were collected between July 2017 and July 2019. Detailed information regarding environmental sampling and laboratory analysis has been published previously [21]. The MAP herd status was determined by PCR and fecal culture. Farms with at least one positive environmental sample, determined by PCR and/or fecal culture, were classified as MAP-positive.

2.3. Statistical Analysis

The data were analyzed with SAS® software, Version 9.4 (SAS Institute Inc., Cary, NC, USA). The statistical unit was the herd. A descriptive analysis was performed stratified by region and MAP status. To evaluate the association and the strength of the association between categorical variables, Fisher's exact test (cut off: $\leq$0.05) and Cramer's V test (cut off: >0.5 for strong association) were used. To evaluate the association between categorical and quantitative variables, Kruskal–Wallis test (cut off: $\leq$0.05) and Wilcoxon two-sample test (cut off: $\leq$0.05) were used. Spearman's rank correlation coefficient (cut off: >0.7) was used to evaluate the strength of the association between quantitative variables. If significant associations were identified and the descriptive analysis underlined the association between variables, only one variable was included in the multifactorial model. We used univariable and multivariable logistic regression models to assess the association between risk factors (independent variable/s) and MAP status (dependent variable). For each of the three regions, one separate model was fit. Univariable logistic regression analysis was performed for each variable. In study region south, herd prevalence and sample size were too small for model fitting including categorical variables. Therefore, we performed only descriptive analyses for categorical variables in study region south. Farms with missing data for an independent variable were not included in the univariable analyses.

Multivariable logistic regression was only performed for components of BFM. One multivariable analysis was performed with general farm management and calf management. The other multivariable analysis was performed with internal biosecurity, external biosecurity, calving, calf housing and calf feeding as a stepwise backward selection. Here, external biosecurity and calving were not allowed for selection due to their contextual importance for the entrance and establishment of MAP in a farm. Although housing type and herd size were to be analyzed as possible confounders, only herd size was finally included in the models. This was due to the high association between herd size and housing type. Herd size was included as a confounder in all models, considering parameter estimates and model fit criteria to identify the best fitted model. The models with herd size showed no effect on the parameter estimates of the other included variables, and lower values of the model fit criteria were reached.

To estimate the proportion of MAP-positive farms in exposed farms and in the population, the attributable fraction exposed (AFe) and the population attributable fraction (AFp) were also calculated.

3. Results

A total of 660 dairy farms (study region north: 199 farms, study region east: 201 farms, study region south: 260 farms) were invited to participate in the study, out of which 457 farms (69.24%) participated and could be analyzed, with 183 farms (91.96%) in the study region north, 170 (84.58%) in the study region east and 104 (40%) in the study region south.

3.1. Herd Size and Housing Type

The mean herd sizes in tie stalls were 33 cows (study region north; n = 7 farms), 28 cows (study region east; n = 3 farms) and 22 cows (study region south; n = 33 farms). The mean herd sizes in free stalls were 104 cows (study region north; n = 154 farms), 331 cows (study region east; n = 129 farms) and 54 cows (study region south; n = 68 farms). The herd size was associated with the housing type ($p < 0.01$ in Wilcoxon two-sample test for the study regions north, east and south). With every 100-cow increase in farm size, the risk for a farm to test MAP-positive increased by 1.94 times (95% confidence interval (CI) = 1.15–3.27, $p = 0.01$), 1.14 times (CI = 1.02–1.27, $p = 0.02$) and 5.53 times (CI = 0.44–68.97, $p = 0.18$) in the north, east and south study regions, respectively (Figure 2; Table 1).

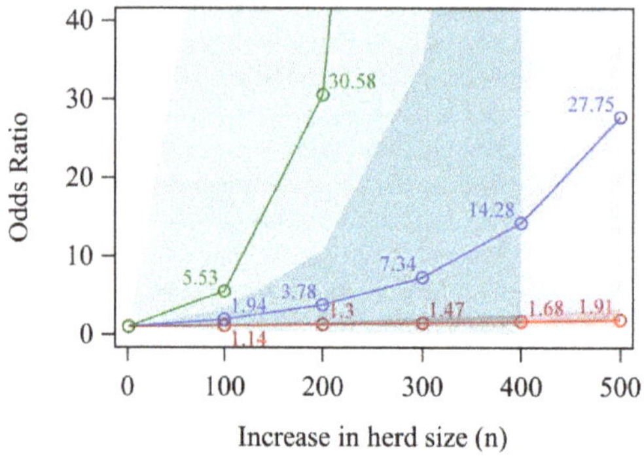

Figure 2. Odds ratio from univariable analysis for quantitative herd size on the MAP status of a farm. Study region north = blue line; study region east = red line; study region south = green line; colored shaded areas = approximate confidence intervals (95% C.I.'s); MAP = *Mycobacterium avium* spp. *paratuberculosis*; *n* = number of cows.

Table 1. Descriptive and single-factorial analyses of quantitative variables considered to be risk factors for MAP-positive dairy herds (*Mycobacterium avium* ssp. *paratuberculosis* was isolated from at least one environmental sample).

Variable	MAP-Positive Herds				MAP-Negative Herds				OR	LCL-UCL	*p*
	n	Mean	Median	CV	*n*	Mean	Median	CV			
Herd characteristics											
Herd size	N: 22 E: 69 S: 3	N:133.64 E:371.14 S: 66	N: 112 E: 250 S: 72	N: 56.84 E: 90.56 S: 37.21	N: 161 E: 101 S: 101	N: 90.40 E:241.74 S: 42.65	N: 73 E: 144 S: 37	N: 72.49 E:136.22 S: 64.28	N: 1.01 E: 1.00 S: 1.02	N: 1.00–1.01 E: 1.00–1.00 S: 0.99–1.04	N: 0.01 E: 0.02 S: 0.18
General farm management											
Internal biosecurity: Percentage of clean udders [1]	N: 22 E: 69 S: 3	N: 33.82 E: 49.74 S: 62.33	N: 27 E: 51 S: 63	N: 60.32 E: 37.76 S: 22.48	N: 161 E: 101 S: 99	N: 44.42 E: 55.53 S: 38.90	N: 45 E: 55 S: 38	N: 47 E: 35.63 S: 53.48	N: 0.98 E: 0.99 S: 1.05	N: 0.95–1.0 E: 0.97–1.0 S: 1.0–1.1	N: 0.03 E: 0.06 S: 0.08
Calf management											
Calving: Time (h) calf spends with the mother	N: 12 E: 26 S: 1	N: 23.83 E: 14.12 S: 6	N: 12 E: 8.5 S: 6	N:138.27 E:174.74 S: n.p.	N: 90 E: 57 S: 31	N: 55.96 E: 23.53 S: 11.74	N: 12 E: 12 S: 5	N:610.14 E: 150.2 S: 117.27	N: 1.0 E: 0.98 S: 0.95	N: 0.99–1.00 E: 0.96–1.01 S: 0.75–1.21	N: 0.78 E: 0.25 S: 0.69
Calf housing: Percentage of clean calves [2]	N: 20 E: 69 S: 3	N: 54.8 E: 49.78 S: 62	N: 50.5 E: 50 S: 71	N: 51.52 E: 49.08 S: 41.47	N: 153 E: 97 S: 97	N: 65.7 E: 62.21 S: 65.63	N: 67 E: 67 S: 67	N: 39.34 E: 44.66 S: 45.22	N: 0.99 E: 0.98 S: 1.0	N: 0.97–1.00 E: 0.97–0.99 S: 0.96–1.04	N: 0.09 E: <0.01 S: 0.84

N = Study region north, E = Study region east, S = Study region south, CV = coefficient of variation, OR = Odds Ratio, LCL = Lower Confidence Limit, UCL = Upper Confidence Limit, n.p. = calculation not possible, [1] clean udders = udders that were not covered with any dirt, [2] clean calves = calves with no or minor dirt located on the abdomen, body side and back side.

3.2. General Farm Management

3.2.1. Internal Biosecurity

In multivariable analysis of biosecurity components, farms with insufficient internal biosecurity in the study region east had a 1.72-fold higher chance of becoming MAP-positive compared to farms with good internal biosecurity ($p = 0.05$; Table 2). We found that 52% of MAP cases in the eastern population were due to bad internal biosecurity and could have been prevented, if internal biosecurity had been better (Table 3).

Table 2. Multivariable analyses for different biosecurity components with quantitative herd size.

Component of Biosecurity	Category	OR	LCL	UCL	p
	First multivariable analysis				
Herd size		N. 1.01 E. 1.00	N:1.00 E: 1.00	N: 1.01 E: 1.00	N: 0.01 E: 0.01
General farm management (GFM)	Good [a]				N: 0.42 [b] E: 0.17 [b]
	Average	N: 3.68 E: <0.001	N: 0.33 E: <0.001	N: 41.65 E: >999.999	N: 0.3 E: 0.97
	Insufficient	N: 1.62 E: 1.99	N: 0.62 E: 0.98	N: 4.19 E: 4.07	N: 0.33 E: 0.06
Calf management (CM)	Good [a]				N: 0.91 [b] E: 0.49 [b]
	Average	N: 0.85 E: 1.66	N: 0.21 E: 0.58	N: 3.46 E: 4.73	N: 0.83 E: 0.34
	Insufficient	N: 0.8 E: 1.41	N: 0.29 E: 0.7	N: 2.23 E: 2.86	N: 0.67 E: 0.33
	Second multivariable analysis				
Herd size		N: 1.01 E: 1.00	N: 1.00 E: 1.00	N. 1.01 E: 1.00	N: 0.02 E: 0.05
GFM: Internal biosecurity	Good [a]				E: 0.05 [b]
	Average	E: 0.46	E: 0.1	E: 2.09	E: 0.31
	Insufficient	E: 1.72	E: 0.52	E: 5.64	E: 0.37
GFM: External biosecurity	Good [a]				N: 0.63 [b] E: 0.42 [b]
	Average	N: 0.85 E: 1.2	N: 0.26 E: 0.54	N: 2.78 E: 2.65	N: 0.78 E: 0.66
	Insufficient	N: 1.74 E: 0.25	N: 0.47 E: 0.03	N: 6.47 E: 2.37	N: 0.41 E: 0.23
CM: Calving	Good [a]				N: 0.95 [b] E: 0.35 [b]
	Average	N: 0.75 E: 0.85	N: 0.12 E: 0.21	N: 4.86 E: 3.44	N: 0.76 E: 0.81
	Insufficient	N: 0.78 E: 0.52	N: 0.13 E: 0.13	N: 4.79 E: 2.15	N: 0.78 E: 0.36
CM: Calf feeding	Good [a]				N: 0.03 [b]
	Average	N: 5.22	N: 1.25	N: 21.83	N: 0.02
	Insufficient	N: 1.6	N: 0.4	N: 6.34	N: 0.5

[a] = Reference category, N = Study region north, E = Study region east, OR = Odds Ratio, LCL = Lower Confidence Limit, UCL = Upper Confidence Limit, [b] = global p-value.

Table 3. Descriptive and univariable analyses of different biosecurity components considered to be a risk for *Mycobacterium avium* ssp. *paratuberculosis* (MAP)-positive dairy herds (*Mycobacterium avium* ssp. *paratuberculosis* was isolated from at least one environmental sample).

Component of Biosecurity (*n* = Herds with Information Available)	Category	MAP-Positive Herds *n* (%)	MAP-Negative Herds *n* (%)	OR	LCL-UCL	*p*	AFe	AFp
General farm management (GFM) (N: *n* = 183, E: *n* = 170, S: *n* = 104)	Good [a]	N: 10 (45.45) E: 46 (66.67) S: 3 (100)	N: 89 (55.28) E: 78 (77.23) S: 79 (78.22)			N: 0.56 [b] E: 0.27 [b]		
	Average	N: 1 (4.55) E: 0 (0.0) S: 0 (0.0)	N: 3 (1.86) E: 1 (0.99) S: 1 (0.99)	N: 2.97 E: <0.001	N: 0.28–31.29 E: <0.001–> 999.999	N: 0.44 E: 0.97	N: 0.60 E: n.p.	N: 0.05 E: −0.01
	Insufficient	N: 11 (50.00) E: 23 (33.33) S: 0 (0.0)	N: 69 (42.86) E: 22 (21.78) S: 21 (20.79)	N: 1.42 E: 1.77	N: 0.57–3.53 E: 0.89–3.53	N: 0.78 E: 0.97	N: 0.27 E: 0.27	N: 0.14 E: 0.14
GFM: Internal biosecurity (N: *n* = 183, E: *n* = 170, S: *n* = 104)	Good [a]	N: 2 (9.09) E: 5 (7.25) S: 1 (33.33)	N: 14 (8.70) E: 10 (9.90) S: 21 (20.79)			N: 0.43 [b] E: 0.11 [b]		
	Average	N: 1 (4.55) E: 6 (8.70) S: 1 (33.33)	N: 25 (15.53) E: 20 (19.80) S: 11 (10.89)	N: 0.28 E: 0.6	N: 0.02–3.37 E: 0.15–2.46	N: 0.23 E: 0.17	N: −2.25 E: −0.44	N: −0.75 E: −0.24
	Insufficient	N: 19 (86.36) E: 58 (84.06) S: 1 (33.33)	N:122 (75.78) E: 71 (70.30) S: 69 (68.32)	N: 1.09 E: 1.63	N: 0.23–5.18 E: 0.53–5.05	N: 0.29 E: 0.06	N: 0.07 E: 0.57	N:0.07 E: 0.52
GFM: External biosecurity (N: *n* = 183, E: *n* = 170, S: *n* = 104)	Good [a]	N: 13 (59.09) E: 52 (75.36) S: 3 (100)	N:100 (62.11) E: 73 (72.28) S: 73 (72.28)			N: 0.75 [b] E: 0.41 [b]		
	Average	N: 5 (22.73) E: 16 (23.19) S: 0 (0.0)	N: 41 (25.47) E: 22 (21.78) S: 9 (8.91)	N: 0.94 E: 1.02	N: 0.31–2.8 E: 0.49–2.13	N: 0.62 E: 0.24	N: −0.06 E: 0.01	N: −0.02 E: 0.00
	Insufficient	N: 4 (18.18) E:1 (1.45) S: 0 (0.0)	N: 20 (12.42) E: 6 (5.94) S: 19 (18.81)	N: 1.54 E: 0.23	N: 0.46–5.21 E: 0.03–2.00	N: 0.45 E: 0.18	N: 0.31 E: −1.91	N: 0.07 E: −0.04
Calf management (CM) (N: *n* = 183, E: *n* = 170, S: *n* = 104)	Good [a]	N: 11 (50.00) E: 32 (46.38) S: 3 (100.00)	N: 73 (45.34) E: 52 (51.49) S: 51 (50.50)			N: 0.92 [b] E: 0.74 [b]		
	Average	N: 3 (13.64) E: 9 (13.04) S: 0 (0.0)	N: 23 (14.29) E: 10 (9.90) S: 11 (10.89)	N: 0.87 E: 1.46	N: 0.22–3.37 E: 0.54–3.99	N: 0.95 E: 0.54	N: −0.13 E: 0.20	N: −0.03 E: 0.04
	Insufficient	N: 8 (36.36) E: 28 (40.58) S: 0 (0.0)	N: 65 (40.37) E: 39 (38.61) S: 39 (38.61)	N: 0.82 E: 1.17	N: 0.31–2.16 E: 0.61–2.25	N: 0.8 E: 0.92	N: −0.19 E: 0.09	N: −0.08 E: 0.04

Table 3. *Cont.*

Component of Biosecurity (n = Herds with Information Available)	Category	MAP-Positive Herds n (%)	MAP-Negative Herds n (%)	OR	LCL-UCL	p	AFe	AFp
	Good [a]	N: 2 (9.09) E: 6 (8.70) S: 0 (0.0)	N: 10 (6.21) E: 5 (4.95) S: 8 (7.92)			N: 0.86 [b] E: 0.15 [b]		
CM: Calving (N: n = 183, E: n = 170, S: n = 104)	Average	N: 9 (40.91) E: 33 (47.83) S: 3 (100)	N: 64 (39.75) E: 37 (36.63) S: 63 (62.38)	N: 0.7 E: 0.74	N: 0.13–3.74 E: 0.21–2.66	N: 0.82 E: 0.74	N: −0.35 E: −0.16	N: −0.29 E: −0.13
	Insufficient	N: 11 (50.00) E: 30 (43.48) S: 0 (0.0)	N: 87 (54.04) E: 59 (58.42) S: 30 (29.70)	N: 0.63 E: 0.42	N: 0.12–3.27 E: 0.12–1.5	N: 0.6 E: 0.07	N: −2.34 E: 0.15	N: −1.15 E: 0.14
	Good [a]	N: 16 (72.73) E: 40 (57.97) S: 2 (66.67)	N:101 (62.73) E: 71 (70.30) S: 73 (72.28)			N: 0.66 [b] E: 0.23 [b]		
CM: Calf housing (N: n = 183, E: n = 170, S: n = 104)	Average	N: 4 (18.18) E: 22 (31.88) S: 1 (33.33)	N: 39 (24.22) E: 21 (20.79) S: 25 (24.75)	N: 0.65 E: 1.86	N: 0.2–2.06 E: 0.91–3.79	N: 0.78 E: 0.26	N: −0.47 E: 0.30	N: −0.09 E: 0.10
	Insufficient	N: 2 (9.09) E: 7 (10.14) S: 0 (0.0)	N: 21 (13.04) E: 9 (8.91) S: 3 (2.97)	N: 0.6 E: 1.38	N: 0.13–2.81 E: 0.48–3.99	N: 0.72 E: 0.98	N: −0.57 E: 0.18	N: −0.06 E: 0.03
	Good [a]	N: 3 (13.64) E: 10 (14.49) S: 0 (0.0)	N: 48 (29.81) E: 24 (23.76) S: 26 (25.74)			N: 0.01 [b] E: 0.34 [b]		
CM: Calf feeding (N: n = 183, E: n = 170, S: n = 104)	Average	N: 9 (40.91) E: 22 (31.88) S: 1 (33.33)	N: 23 (14.29) E: 28 (27.72) S: 8 (7.92)	N: 6.26 E: 1.89	N: 1.55–25.34 E: 0.75–4.76	N: <0.01 E: 0.35	N: 0.79 E: 0.33	N: 0.59 E: 0.23
	Insufficient	N: 10 (45.45) E: 37 (53.62) S: 2 (66.67)	N: 90 (55.90) E: 49 (48.51) S: 67 (66.34)	N: 1.78 E: 1.81	N: 0.47–6.77 E: 0.77–4.25	N: 0.48 E: 0.39	N: 0.41 E: 0.32	N: 0.32 E: 0.25

[a] = Reference category, n = Herds with information available, N = Study region north, E = Study region east, S = Study region south, OR = Odds Ratio, LCL = Lower Confidence Limit, UCL = Upper Confidence Limit, Afe = Attributable fraction exposed, Afp = Population attributable fraction, [b] = global p-value.

3.2.2. External Biosecurity

In univariable analysis, animal purchase without inquiring about the MAP status increased the chance of a farm testing MAP-positive by 2.86 times in the study region east (CI = 1.45–5.67, p = 0.02) compared to no purchase, and 44% of the MAP cases in MAP-positive farms could have been prevented, if a farm had not purchased cattle (see Table S1). This effect could not be detected in the other regions. In contrast, in the other regions, animal purchase with inquiring about the MAP status increased the chance for a MAP-positive farm compared to no purchase by 1.97 times (CI = 0.62–6.26; p = 0.15; region north) or by 1.28 times (CI = 0.43–3.81; p = 0.6; region east) (see Table S1).

3.3. *Calf Management*

3.3.1. Calving

If the calf stayed with the mother after birth, the time (h) of contact was evaluated. The suckling time of the calves differed between regions (33% quantile = 6 h in study region north, 8 h in study region east, 1 h in study region south; 66% quantile = 24 h in study region north, 12 h in study region east and south). The suckling of calves reduced the risk for being tested MAP-positive in the eastern region by 0.57 times (CI = 0.31–1.07; $p = 0.08$; see Table S1) compared to farms where the calves were not allowed to stay with their dam.

3.3.2. Calf Housing

Within univariable analysis of calf rearing, farms in the study region east that reared calves from other farms had a 7.86-fold higher chance (CI = 0.9–68.82; $p = 0.16$) for a MAP-positive status than farms that reared only their own calves. In addition, 53% of the MAP cases in MAP-positive farms were due to foreign calves and could have been prevented, if calf rearing had only been accomplished from the own herd (see Table S1). Farms in the study regions north and east with no calf rearing (i.e., rearing of own or purchased calves until weaning or the age of six months) had a 3.69-fold higher chance (CI = 0.63–21.5; $p = 0.09$) in the study region north and a 1.57-fold higher chance (CI = 0.1–25.58; $p = 0.7$) in the study region east for a MAP-positive status compared to farms that only reared their own calves (see Table S1). We found that 64% of the MAP cases in MAP-positive farms in the northern region could have been prevented, if calf rearing had have better organized (see Table S1).

3.3.3. Calf Feeding

In multivariable analysis, farms in the study region north with average calf feeding had a 5.22-fold higher chance (CI = 1.23–21.83) for a MAP-positive status ($p = 0.02$) than farms with a good calf feeding management. Farms in the study region north with insufficient calf feeding had a 1.6-fold higher chance (CI = 0.4–6.34) for an MAP-positive status compared to farms with good calf feeding ($p = 0.6$; Table 2). In univariable analysis, the results were similar. It appeared that 59% of the MAP cases in the northern population were due to average calf feeding and could have been prevented, if calf feeding had been better (Table 3).

The cleaning of buckets less often than after every feeding increased the risk for a farm to test MAP-positive. Buckets that were not cleaned at all increased the risk for a farm to test MAP-positive by 3.4 times (CI = 0.5–23.03; $p = 0.63$ in study region north) and by 1.86 times (CI = 0.62–5.62; $p = 0.78$ in study region east; see Table S1). Buckets that were cleaned only once a day increased the risk for a MAP-positive farm by 4.58 times (CI = 1.09–19.17; $p = 0.15$ in study region north) and by 3.51 times (CI = 1.37–9.0; $p = 0.06$ in study region east). In addition, 74% of the MAP cases in MAP-positive farms could have been prevented, if the buckets had been cleaned more often than once a day (study region north; see Table S1).

4. Discussion

This study aimed to evaluate the effect of different components of biosecurity and farm management on the MAP herd status in the three regions of Germany that are characterized by different MAP between-herd prevalence rates [21] and different agricultural structures. Furthermore, we wanted to illustrate high-impact risk factors for MAP. Due to different agricultural characteristics in Germany [22], the analysis was stratified into three regions. Hence, we wanted to investigate possible infection pathways in herds of these specific regions to offer tailored advice for disease control. Because of the different MAP between-herd prevalence [21] and agricultural structures [22], the results of this study can be transferred to other regions of the world with similar between-herd prevalence and agricultural structures.

Recommendations to control MAP in a herd generally consist of a combination of measures to reduce the risk of infection in a certain age group [8,25,26]. Therefore, it can be assumed that single risk factors play only a minor role in the development of paratuberculosis on a farm. The improvement of several management measures at the same time will potentiate the effect compared to one single improved step. Furthermore, from a population perspective, it might be difficult to identify a certain set of risk factors responsible for the most MAP-infected herds within a region, which is what we experienced during the study presented here. Therefore, we investigated different components of BFM and could identify calf feeding as an important farm management component which should be considered, especially, in MAP-positive farms.

4.1. Herd Size and Housing Type

We demonstrated that larger herds are more likely to test MAP-positive (Figure 2), which is in accordance with previous studies [1–3]. The farms in study regions north and south generally seemed to have a higher risk of testing positive compared to those in study region east, given a certain herd size. However, at the time of sampling, the herds in regions north and south were on average smaller than those in farms in study region east. If farms in study region north or south increase their herd size to similar values as in study region east, a drastic increase in the risk of becoming MAP-positive should be anticipated. The reasons for an increased risk of testing MAP-positive with an increase in the herd size, especially in study regions north and south, can only be hypothesized. One reason might be different hygiene or other farm management factors which are introduced when the herd size increases. It is possible that farms with larger herds purchase more cattle or that these herds include more cattle with unknown paratuberculosis status [3]. Furthermore, at the same time, large herds have both more animals shedding the organism and more animals at risk for infection than small herds, even if the within-herd prevalence is similar. Furthermore, the possibility for identifying a MAP-positive animal increases with an increase in the herd size if more animals are tested in larger herds [3]. Vilar et al. [1] discussed that contact between young and adult animals and, therefore, with MAP-contaminated feces occurs more frequently in larger herds because of missing paddocks to separate age groups from each other. In larger herds, the risk for transmission could increase because of multiple cows calving together, with imperfect calving hygiene and weak hygiene in calf rearing [27]. However, these factors were not investigated here. On the other hand, we have to consider that the smaller sample size and the low number of MAP-positive farms in study region south might have led to more imprecise estimates compared to study regions north and east [28]. However, regardless of precision, herd size seems to play an important role in predicting the MAP status of a herd in Germany as well as in other countries. Therefore, the ongoing agricultural changes characterized by a reduction in the number of farms, on the one hand, and an increase in herd size, on the other hand, should motivate the improvement of MAP control in the future.

As in previous studies [29,30], herd size was associated with housing type in our study. Tie stalls are less common than free stalls but are more commonly found in farms with smaller herds than in those with large herds. In general, most of the cows in the study region north are kept in free stalls (84.15%) compared to those (65.38%) in the study region south [21]. Moreover, herds in free stalls have a higher MAP between-herd prevalence than herds kept in tie stalls [29]. This may be due to reduced contact between cows in tie stalls compared to those in free stalls [29] or to different fecal contamination patterns in these different housing types [30]. However, a higher risk of testing MAP-positive for free-stall herds compared to tie-stall herds could not be detected in this study.

4.2. General Farm Management

4.2.1. Internal Biosecurity

The hygiene of housing, equipment and cows is important for paratuberculosis control [8]. By including the hygiene of boots and hands, lying areas and walkways, as well as the hygiene of the udders in our analysis, hygienic aspects of equipment and animals were considered. None of these aspects of internal biosecurity appeared to have a considerable impact on the MAP status in the study presented here. This might be explained by the fact that these indisputably useful measures alone do not protect susceptible animals such as calves and might therefore be of less importance when compared with all components of BFM.

4.2.2. External Biosecurity and Calf Housing

One of the general recommendations to prevent the introduction of paratuberculosis into a herd is to close the herd by rearing only its own calves and youngstock for replacement [31]. As this was practiced by most of the farms in all regions, our results support this recommendation. As already shown elsewhere [3,4], our results support the possibility that the purchase of (adult) cattle with an unknown MAP status poses a risk for MAP infection, explaining part of the higher MAP between-herd prevalence in the study region east. Furthermore, this is also important for the introduction of calves from other calves into a herd. This effect was most obvious in study region east, where the odds for a MAP-positive status were increased by 7.86 times (CI: 0.9–68.82; see Table S1) when own calves and calves from other farms were reared in the respective farms. Therefore, farmers should inquire about the MAP status of purchased animals before introducing them into their herd, as recommended for paratuberculosis control [8,25,26]. In addition, also animal purchase after inquiring increased the risk of a herd to test MAP-positive; therefore, animal purchase should be avoided whenever possible.

4.3. Calf Management

4.3.1. Calving

Surprisingly, our data suggest that suckling calves for a certain time after birth seems to decrease the MAP risk for a herd. It is known that calves can take up infectious fecal material from the dam's udder [10]. Ansari-Lari, Haghkhah, Bahramy and Baheran [10] determined an odds ratio of 6.4 for a herd to test MAP-positive, if the udders of periparturient cows were contaminated with feces compared with no contamination. Besides the fecal contamination of colostrum, MAP has been shown to be shed directly into colostrum in clinical stages of the disease [14]; therefore, colostrum might play a role in MAP transmission between dams and their offspring [32]. It is common agricultural practice in some German herds to have the calf stay with the cow until the next milking and feeding time for reasons such as efficiency and personnel resources. Hence, any changes from this might indicate reverse causation in the way that calves are kept from a contact with the cow for hygienic reasons in a MAP control attempt. We assume that the farmers were already aware of the disease and had an ongoing or initiated control strategy with hygienic or other management measures before the farm visit. This effect of reverse causation [33] is well known in cross-sectional observational studies, where the status of a disease and the risk factors are investigated at the same point in time. However, we can only assume that reverse causation is the reason for these confusing results. Hence, further research should be conducted using personal interviews to elicit the true reasons for these results.

4.3.2. Calf Feeding

In the study we presented here, "calf feeding" was the only component of BFM with a statistically significant impact on the MAP status, indicating its importance for future biosecurity and farm management consultancy. A closer investigation of the respective variables indicated that the neglected cleaning of the drinking buckets for different age groups could be a possible source of infection with MAP.

4.4. Limits and Strengths

Because only fecal environmental samples were taken, the MAP status might have been estimated as a false negative in some farms, resulting in a biased odds ratio. Therefore, taking individual samples could have led to more precise results for the MAP status. Univariable and multivariable analyses revealed interesting results. However, they could not always identify risk factors with statistical significance due to the comparably small sample size for logistic regression. As these risk factor analyses were only enabled through cooperation with another large project on dairy cattle health in Germany [23] and the study participation was voluntary, we had to cope with the achieved sample size of 457 farms in total which had to be analyzed per region due to different agricultural structures [22]. Only

40% (n = 104 out of 260 PraeRi participants) of the addressed farms in the study region south participated, indicating difficulties in recruiting participants for this part of the study. Therefore, a selection bias must be considered here. This could have influenced the results for the study region south. However, the majority of the addressed northern and eastern farms participated in this study, providing more meaningful results. Moreover, all farms were randomly selected, which is a strength of the study.

5. Conclusions

We identified several risk factors as known from previous studies from all around the world having a high impact for MAP in German dairy herds. The risk factors did not seem to contribute evenly to MAP infection in all farms, indicating a multifactorial background. However, calf feeding seems to have universal validity in MAP control. Furthermore, the purchase of cattle with an unknown MAP status, the purchase of calves for replacement and hygiene and usage habits of the drinking buckets seemed to have a considerable impact and should be taken into account in veterinary consultancy in Germany as well as in other regions of the world that are structurally comparable with Germany. Nevertheless, a tailored control approach is indicated for individual farms and regions.

Supplementary Materials: The following supporting information can be downloaded at: https://www.mdpi.com/article/10.3390/ani13121889/s1, Table S1: Descriptive and single-factorial analyses of collected categorical variables considered to be risk factors for MAP-positive dairy herds (*Mycobacterium avium* ssp. *paratuberculosis* had been isolated from at least one environmental sample).

Author Contributions: Conceptualization, M.K., S.E. and A.C.; methodology, S.E. and A.C.; software, M.K.; validation, S.E. and A.C.; formal analysis, M.K.; data curation, M.K.; writing—original draft preparation, M.K.; writing—review and editing, S.E., K.D. and A.C.; visualization, M.K.; supervision, A.C.; project administration, K.D. and A.C. All authors have read and agreed to the published version of the manuscript.

Funding: This Open Access publication was funded by the Deutsche Forschungsgemeinschaft (DFG, German Research Foundation)-491094227 "Open Access Publication Funding" and the University of Veterinary Medicine Hannover, Foundation. The original studies were funded by the German Animal Diseases Fund (Thuringia, Brandenburg, Lower Saxony, Mecklenburg-West Pomerania), the Animal Health Service of Bavaria, the Ministry of Energy, Agriculture, Environment, Nature and Digitalization (MELUND) of Schleswig-Holstein and the Federal Ministry of Food and Agriculture (BMEL) through the Federal Office for Agriculture and Food (BLE; grant number 2814HS006).

Institutional Review Board Statement: Participation in the original study was voluntary and anonymous. The study animals were treated in consideration of the Animal Welfare Act (09.12.2010) of Germany. Scientific research including animals was registered at the State Office for Consumer Protection and Food Safety of Lower Saxony (LAVES: reference No. 13A308), the Ministry for Energy Turnaround, Agriculture, Environment, and Rural Areas of the State Schleswig-Holstein (MELUR: reference No. V 312-7224.123-34) and the State Office for Nature, Environment and Consumer Protection of Northrhine-Westfalia.

Informed Consent Statement: Participating farmers of the PraeRi project were invited to participate in the MAP prevalence study (PraeMAP) by written consent.

Data Availability Statement: The data were made available through internal cooperation. Therefore, any data transfer to interested persons is not allowed without an additional formal contract. Data are available to qualified researchers who sign a contract with the University of Veterinary Medicine Hannover. This contract will include guarantees to the obligation to maintain data confidentiality in accordance with the provisions of the German data protection law. A data access committee will be established on demand. This committee will consist of the authors as well as of members of the University of Veterinary Medicine Hannover. Interested cooperative partners, who are able to sign a contract as described above, may contact Amely Campe, Department of Biometry, Epidemiology and Information Processing University of Veterinary Medicine, Hannover Bünteweg 2, 30559 Hannover, Email: amely.campe@tiho-hannover.de.

Acknowledgments: The work of all study veterinarians and PraeRi scientists is kindly acknowledged. The authors would like to thank all farmers who participated in the PraeRi project. The authors would also like to thank the Thuringian Animal Health Service and Bavarian Animal Health Service for provision of the paratuberculosis results.

Conflicts of Interest: The authors declare no conflict of interest.

References

1. Vilar, A.L.T.; Santos, C.S.A.B.; Pimenta, C.L.R.M.; Freitas, T.D.; Brasil, A.W.L.; Clementino, I.J.; Alves, C.J.; Bezerra, C.S.; Riet-Correa, F.; Oliveira, T.S.; et al. Herd-level prevalence and associated risk factors for *Mycobacterium avium* subsp. *paratuberculosis* in cattle in the state of Paraíba, Northeastern Brazil. *Prev. Vet. Med.* **2015**, *121*, 49–55. [CrossRef] [PubMed]
2. Kennedy, A.E.; Byrne, N.; Garcia, A.B.; Mahony, J.O.; Sayers, R.G. Analysis of Johne's disease ELISA status and associated performance parameters in Irish dairy cows. *BMC Vet. Res.* **2016**, *12*, 1–12. [CrossRef]
3. Puerto-Parada, M.; Arango-Sabogal, J.C.; Paré, J.; Doré, E.; Côté, G.; Wellemans, V.; Buczinski, S.; Roy, J.P.; Labrecque, O.; Fecteau, G. Risk factors associated with *Mycobacterium avium* subsp. *paratuberculosis* herd status in Québec dairy herds. *Prev. Vet. Med.* **2018**, *152*, 74–80. [CrossRef] [PubMed]
4. Künzler, R.; Torgerson, P.; Keller, S.; Wittenbrink, M.; Stephan, R.; Knubben-Schweizer, G.; Berchtold, B.; Meylan, M. Observed management practices in relation to the risk of infection with paratuberculosis and to the spread of *Mycobacterium avium* subsp. *paratuberculosis* in Swiss dairy and beef herds. *BMC Vet. Res.* **2014**, *10*, 132. [CrossRef]
5. Rangel, S.J.; Paré, J.; Doré, E.; Arango, J.C.; Côté, G.; Buczinski, S.; Labrecque, O.; Fairbrother, J.H.; Roy, J.P.; Wellemans, V.; et al. A systematic review of risk factors associated with the introduction of *Mycobacterium avium* spp. *paratuberculosis* (MAP) into dairy herds. *Can. Vet. J.* **2015**, *56*, 169–177.
6. Chiodini, R.J.; Van Kruiningen, H.J.; Merkal, R.S. Ruminant paratuberculosis (Johne's disease): The current status and future prospects. *Cornell. Vet.* **1984**, *74*, 218–262.
7. Cetinkaya, B.; Erdogan, H.M.; Morgan, K.L. Relationships between the presence of Johne's disease and farm and management factors in dairy cattle in England. *Prev. Vet. Med.* **1997**, *32*, 253–266. [CrossRef] [PubMed]
8. Sweeney, R.W.; Collins, M.T.; Koets, A.P.; McGuirk, S.M.; Roussel, A.J. Paratuberculosis (Johne's Disease) in cattle and other susceptible species. *J. Vet. Intern. Med.* **2012**, *26*, 1239–1250. [CrossRef] [PubMed]
9. Doré, E.; Paré, J.; Côté, G.; Buczinski, S.; Labrecque, O.; Roy, J.P.; Fecteau, G. Risk factors associated with transmission of *Mycobacterium avium* subsp. *paratuberculosis* to calves within dairy herd: A systematic review. *J. Vet. Intern. Med.* **2012**, *26*, 32–45. [CrossRef]
10. Ansari-Lari, M.; Haghkhah, M.; Bahramy, A.; Baheran, A.M.N. Risk factors for *Mycobacterium avium* subspecies *paratuberculosis* in Fars province (Southern Iran) dairy herds. *Trop. Anim. Health Prod.* **2009**, *41*, 553–557. [CrossRef]
11. McAloon, C.G.; Doherty, M.L.; Whyte, P.; More, S.J.; O'Grady, L.; Citer, L.; Green, M.J. Relative importance of herd-level risk factors for probability of infection with paratuberculosis in Irish dairy herds. *J. Dairy Sci.* **2017**, *100*, 9245–9257. [CrossRef]
12. Tiwari, A.; VanLeeuwen, J.A.; Keefe, G.P.; Haddad, J.P.; Scott, H.M.; Whiting, T. Risk factors associated with *Mycobacterium avium* subspecies *paratuberculosis* seropositivity in Canadian dairy cows and herds. *Prev. Vet. Med.* **2009**, *88*, 32–41. [CrossRef] [PubMed]
13. Wells, S.J.; Wagner, B.A. Herd-level risk factors for infection with Mycobacterium paratuberculosis in US dairies and association between familiarity of the herd manager with the disease or prior diagnosis of the disease in that herd and use of preventive measures. *J. Am. Vet. Med. Assoc.* **2000**, *216*, 1450–1457. [CrossRef] [PubMed]
14. Streeter, R.N.; Hoffsis, G.F.; Bech-Nielsen, S.; Shulaw, W.P.; Rings, D.M. Isolation of Mycobacterium paratuberculosis from colostrum and milk of subclinically infected cows. *Am. J. Vet. Res.* **1995**, *56*, 1322–1324. [PubMed]
15. Sweeney, R.W.; Whitlock, R.H.; Rosenberger, A.E. Mycobacterium paratuberculosis cultured from milk and supramammary lymph nodes of infected asymtomatic cows. *J. Clin. Microbiol.* **1992**, *30*, 166–171. [CrossRef]
16. McAloon, C.G.; Doherty, M.L.; Donlon, J.; Lorenz, I.; Meade, J.; O'Grady, L.; Whyte, P. Microbiological contamination of colostrum on Irish dairy farms. *Vet. Rec.* **2016**, *178*, 474. [CrossRef]
17. Lombard, J.E.; Wagner, B.A.; Smith, R.L.; McCluskey, B.J.; Harris, B.N.; Payeur, J.B.; Garry, F.B.; Salman, M.D. Evaluation of environmental sampling and culture to determine *Mycobacterium avium* subspecies *paratuberculosis* distribution and herd infection status on US dairy operations. *J. Dairy Sci.* **2006**, *89*, 4163–4171. [CrossRef]
18. Hahn, N.; Failing, K.; Eisenberg, T.; Schlez, K.; Zschöck, P.-M.; Donat, K.; Einax, E.; Köhler, H. Evaluation of different diagnostic methods for the detection of *Mycobacterium avium* subsp. *paratuberculosis* in boot swabs and liquid manure samples. *BMC Vet. Res.* **2017**, *13*, 259. [CrossRef]
19. Donat, K.; Hahn, N.; Eisenberg, T.; Schlez, K.; Köhler, H.; Wolter, W.; Rohde, M.; Pützschel, R.; Rösler, U.; Failing, K.; et al. Within-herd prevalence thresholds for the detection of *Mycobacterium avium* subspecies *paratuberculosis*-positive dairy herds using boot swabs and liquid manure samples. *Epidemiol. Infect.* **2016**, *144*, 413–424. [CrossRef]
20. Donat, K.; Kube, J.; Dressel, J.; Einax, E.; Pfeffer, M.; Failing, K. Detection of *Mycobacterium avium* subspecies *paratuberculosis* in environmental samples by faecal culture and real-time PCR in relation to apparent within-herd prevalence as determined by individual faecal culture. *Epidemiol. Infect.* **2015**, *143*, 975–985. [CrossRef]

21. Eisenberg, S.; Krieger, M.; Campe, A.; Lorenz, I.; Einax, E.; Donat, K. Herd prevalence estimation of *Mycobacterium avium* subspecies *paratuberculosis* burden in the three main dairy production regions of Germany (PraeMAP). *Animals* **2022**, *12*, 447. [CrossRef] [PubMed]
22. Merle, R.; Busse, M.; Rechter, G.; Meer, U. Regionalisation of Germany by data of agricultural structures. *Berl. Munch. Tierarztl. Wochenschr.* **2012**, *125*, 52–59. [CrossRef] [PubMed]
23. PraeRi. Animal Health, Hygiene and Biosecurity in German Dairy Cow Operations—A Prevalence Study (PraeRi). Available online: https://ibei.tiho-hannover.de/praeri/pages/69#_AB (accessed on 26 October 2021).
24. Dachrodt, L.; Arndt, H.; Bartel, A.; Kellermann, L.M.; Tautenhahn, A.; Volkmann, M.; Birnstiel, K.; Duc, P.D.; Hentzsch, A.; Jensen, K.C.; et al. Prevalence of disorders in preweaned dairy calves from 731 dairies in Germany: A cross-sectional study. *J. Dairy Sci.* **2021**, *104*, 9037–9051. [CrossRef]
25. USDA-APHIS. Uniform Program Standards for the Voluntary Bovine Johne's Disease Control Program. Available online: https://www.aphis.usda.gov/animal_health/animal_diseases/johnes/downloads/johnes-ups.pdf (accessed on 23 November 2021).
26. BAnz. Bekanntmachung von Empfehlungen für Hygienische Anforderungen an das Halten von Wiederkäuern in the Version of July 7, 2014. 2014. Available online: https://www.bundesanzeiger.de/pub/de/amtliche-veroeffentlichung?10 (accessed on 11 October 2022).
27. Ridge, S.E.; Heuer, C.; Cogger, N.; Heck, A.; Moor, S.; Baker, I.M.; Vaughan, S. Herd management practices and the transmission of Johne's disease within infected dairy herds in Victoria, Australia. *Prev. Vet. Med.* **2010**, *95*, 186–197. [CrossRef] [PubMed]
28. Dohoo, I.; Martin, W.; Stryhn, H. Sampling. In *Veterinary Epidemiologic Research*, 2nd ed.; McPike, S.M., Ed.; VER Inc.: Charlottetown, PE, Canada, 2009; p. 46.
29. Corbett, C.S.; Naqvi, S.A.; Bauman, C.A.; Buck, J.D.; Orsel, K.; Uehlinger, F.; Kelton, D.F.; Barkema, H.W. Prevalence of Mycobacterium avium ssp. paratuberculosis infections in Canadian dairy herds. *J. Dairy Sci.* **2018**, *101*, 11218–11228. [CrossRef]
30. Noll, I.; Eisenberg, T.; Failing, K.; Rohde, M.; Schlez, K.; Wolter, W.; Fawzy, A.; Zschöck, M. Herd prevalence of Mycobacterium avium ssp. paratuberculosis (MAP) in Hessian dairy herds. *Berl. Munch. Tierarztl. Wochenschr.* **2017**, *130*, 13–20.
31. Garry, F. Control of paratuberculosis in dairy herds. *Vet. Clin. N. Am. Food Anim. Pract.* **2011**, *27*, 599–607. [CrossRef]
32. Sweeney, R.W. Transmission of paratuberculosis. *Vet. Clin. N. Am. Food Anim. Pract.* **1996**, *12*, 305–312. [CrossRef]
33. Dohoo, I.; Martin, W.; Stryhn, H. Introduction to observational studies. In *Veterinary Epidemiologic Research*, 2nd ed.; McPike, S.M., Ed.; VER Inc.: Charlottetown, PE, Canada, 2009; pp. 161–162.

animals

MDPI

Article

Evaluation of Environmental Sampling for Detection of *Mycobacterium avium* subspecies *paratuberculosis* in the Pre-Weaned Calf Area and Calving Area of Infected Dairy Farms Enrolled in a Voluntary Johne's Disease Control Programme

Niamh L. Field [1,2,*], John F. Mee [1] and Conor G. McAloon [2]

1 Animal and Bioscience Research Department, Teagasc, Moorepark Research Centre, P61 P302 Fermoy, Ireland
2 UCD School of Veterinary Medicine, University College Dublin, D04 W6F6 Dublin, Ireland
* Correspondence: niamh.field@teagasc.ie

Simple Summary: Transmission of Mycobacterium avium subspecies paratuberculosis (MAP) from infected adult cattle to susceptible calves occurs mainly through contamination of the calf environment, i.e., the calving pen and the pre-weaned calf rearing area, with adult faeces. Control programmes for Johne's disease utilise a risk assessment and management plan (RAMP) to evaluate the risk of transmission of infection on a specific farm, however the assessment is limited by its subjective, point-in-time nature. The objective of this study was to evaluate a novel environmental sampling protocol for measurement of MAP contamination of the calf environment on infected farms. Across 28 infected farms, 46% had detectable MAP in the calf environment, with 21% of farms having detectable MAP in the pre-weaned calf area. There was no significant association found between the RAMP scores for each farm and the result of environmental testing for MAP, and there was a moderate correlation found between RAMP scores and the level of MAP contamination as quantified by PCR. We conclude that environmental sampling is a potentially useful tool to objectively measure transmission risk in the calf environment on farms, as a complement to the annual RAMP.

Abstract: One of the pillars of Johne's disease control is to break the cycle of transmission from infectious adult animals to young susceptible animals. Many control programmes utilise a risk assessment and management plan (RAMP) to identify specific risks for transmission of infection on individual farms and then recommend tailored biosecurity measures to mitigate the risk. It is important that infected farms in particular, practice effective biocontainment in the calving area and pre-weaned calf area. The objectives of this study were: (1) to determine what proportion of MAP-infected farms (PCR-positive) enrolled in a control programme had detectable MAP in their calf environment; (2) to compare RAMP scores and apparent within-herd prevalence (aWHP) of PCR-positive and PCR-negative farms; (3) to evaluate the correlation between RAMP scores, aWHP and levels of contamination based on PCR test Ct value. A novel environmental sampling protocol combining manure samples and boot swab samples was conducted in the calving area and pre-weaned calf area on 28 dairy farms with PCR-confirmed MAP infection. All samples were tested with PCR. Logistic regression modelling was used to evaluate the association between RAMP score and aWHP and PCR outcome. Overall, 46% of farms had positive PCR samples in either the calving area, pre-weaned calf area or both. The calving area was positive in 36% of farms and the pre-weaned calf area was positive in 21% of farms. There was a moderate, negative correlation ($r = -0.45$; $p = 0.08$) observed between RAMP scores and Ct value at the sample-level that slightly missed the required significance level. No significant association was identified between RAMP scores or aWHP and PCR test outcome (positive or negative). It was concluded that environmental sampling of the calving area and pre-weaned calf area has potential applications in the assessment of transmission risk on infected farms and could be used to monitor the efficacy of biosecurity measures over time. Further research conducted on a larger scale is required to support these results.

Keywords: paratuberculosis; environmental; PCR detection; transmission; risk

1. Introduction

Johne's disease is a bacterial infection of cattle and other ruminants caused by *Mycobacterium avium* subspecies *paratuberculosis* (MAP). MAP is an obligate intracellular pathogen incapable of environmental proliferation; however, the organism is sufficiently robust to survive in the environment for prolonged periods [1]. MAP is primarily transmitted orally through ingestion of faeces, colostrum or milk from an infectious animal [2]. Environmental contamination of housing and pasture with faeces from infected animals constitutes a risk for transmission of infection to susceptible animals. Cattle are usually infected as calves and develop clinical signs as adults, such as reduced productivity, progressive diarrhoea, and weight loss [2]. MAP is of particular significance in dairy herds due to impacts on animal health, associated production losses [3–5], and controversial associations with Crohn's disease in humans [6]. Even without evidence of a causal link, the association between MAP and Crohn's disease is likely sufficient to affect consumer confidence in dairy products.

Control programmes for Johne's disease usually involve a combination of herd testing to detect infection and biosecurity measures to reduce the spread of disease within herds and between herds [7,8]. Test and cull strategies have been shown to be ineffective when used in isolation to control the disease in infected herds [9,10], due to the poor sensitivity of diagnostic tests in individual animals. Therefore, the emphasis must be on breaking the cycle of transmission of MAP from potentially infectious animals to susceptible animals. Faeces is the most important source of transmission [11], both through faecal contamination of the calf environment and through faecal contamination of colostrum and milk fed to calves [12]. Direct shedding of MAP in milk and colostrum is of lesser significance, occurring mainly in clinically affected animals, and at low-levels in a small proportion of subclinical animals [13,14].

Many control programmes internationally use variations of a risk assessment and management plan (RAMP) in herds to reduce the risk of transmission of MAP on infected farms [7,8,15]. The risk assessment (RA) portion of the RAMP aims to score the risk of transmission of MAP within the herd, while the management plan formulated on completion of the RA recommends risk mitigation measures to address farm-specific risks identified during the RA that may lead to the exposure of susceptible calves to potentially infectious faeces, milk and colostrum. A combination of test-and-cull and biocontainment measures have been shown to be most effective to reduce MAP infections [9]. These measures include removing the calf from the dam immediately after birth, strict hygiene of calving pens, feeding of milk replacer instead of whole milk and preventing contamination of calf rearing areas with adult manure. Limitations to the RAMP include a reliance on farmer recollections and reporting of biocontainment actions, as well as the "point-in-time" nature of the observations of the veterinarian conducting the RA. Additionally, compliance with the management plan has been found to be poor [16,17] and implementation is inadequately assessed with indirect monitoring structures. Testing of adult animals and comparison of consecutive risk assessments remain the only way of measuring and comparing progress over time [18,19].

Environmental sampling may be a potentially useful technique to detect MAP in the pre-weaned calf area and calving area of farms. Boot swabs have been evaluated as a practical and efficient way of obtaining environmental samples for MAP testing in adult cow housing on farms and have shown sensitivity that is comparable to taking standard grab samples of manure [20,21]. Boot swabs have also been used successfully in poultry flocks to test bedding material in housing for the presence of *Salmonella* [22,23], with boot swab kits commercially available for screening flocks. Calving pens often have areas of manure concentration near the feeding area, suitable for grab samples, as well as bedded areas more suited to boot swabs. Boot swabs are likely most suitable for sampling calf pens which typically have dry bedding such as straw or peat. A blended testing strategy combining standard environmental manure sampling and boot swab sampling may be effective and practical for objectively monitoring the efficacy of biocontainment measures recommended during a RAMP. Our primary hypothesis was that detection of MAP in the calf environment is associated with transmission risk (as measured during the RAMP) and apparent within-herd prevalence (aWHP) of infection. Our secondary hypothesis was that the level of contamination in a sample (as measured by the Ct value of the PCR test) is correlated with aWHP and RAMP score. The objectives of this study were (1) to determine what proportion of MAP-infected farms enrolled in a control programme had detectable MAP in their calf/calving environment, (2) to compare RAMP scores and apparent within-herd prevalence (aWHP) of PCR-positive and PCR-negative farms, (3) to evaluate the correlation between RAMP scores, aWHP and levels of contamination based on PCR test Ct value.

2. Materials and Methods

2.1. Herd Selection

In a previous study (unpublished) conducted in 2019, 122 dairy herds were selected from the Irish Johne's Control Programme (IJCP) database using a stratified random selection process. Briefly, the stratification criteria used for the previous study was based on the MAP PCR status of the herds to allow a minimum proportion of confirmed-infected herds to be selected. The IJCP is a voluntary control programme open to both dairy and beef herds that aims to identify MAP-infected herds and control the spread of Johne's disease both between herds and within herds. Herds registered in the programme conduct annual herd testing for MAP using individual serum or milk ELISA in addition to annual RAMPs conducted by private veterinary practitioners. In total, 1658 dairy herds were registered in the IJCP at that time. From this sample of 122 herds, 39 herds with confirmed infection based on individual faecal PCR were identified. The herds all had MAP infection confirmed within the last five years. A total of 35 herds were randomly selected from this cohort using the Excel random number generator, as the maximum number of herds it would be possible to visit for sampling during the spring calving period. Of these, 28 farmers agreed to participate in sample collection during the spring period of 2022. These were all spring calving herds, located mostly in the south and east of Ireland. Herd size ranged from 41-393 cows with a mean herd size of 166 at the time of sampling.

2.2. RAMP Scores

Herds registered in the IJCP undergo a RAMP annually [18]. It is completed by an approved veterinary practitioner (AVP) who has undergone specific training in Johne's disease control and management. The RAMP comprises a non-scored bioexclusion section, and four scored sections: pre-weaned heifers risk assessment (scored out of 80), heifers over six months old risk assessment (scored out of 33), cows risk assessment (scored out of 34) and calving area risk assessment (scored out of 80) (template available on request). The higher the score assigned for an area, the greater the perceived risk of transmission of MAP.

The AVP uploads the scores to an online database hosted by the Irish Cattle Breeding Federation (ICBF), from which, with authorisation from the farmer, the first author downloaded information for each of the study herds.

2.3. Apparent Within-Herd Prevalence

The apparent within-herd prevalence (aWHP) for each herd was calculated based on the most recent annual whole-herd ELISA test recorded on the IJCP database. Herds registered in the IJCP must have all animals over two years old tested with MAP ELISA using either serum or milk samples. aWHP was calculated as the percentage of total animals tested by ELISA that had a positive or inconclusive result, as this is the procedure followed in the IJCP.

2.4. Sampling

All 28 herds were visited once between February and March 2022 during the spring calving period. The boot swabs used were sourced from Trafalgar Scientific Ltd., 190 Waterside road, Hamilton Industrial Estate, Leicester, LE5 1QZ, and were produced specifically for *Salmonella* sampling of poultry housing. Each boot swab pair consisted of two plastic overboots and two absorbent boot swabs pre moistened with maximum recovery diluent to aid recovery of organisms. On each farm three boot swab samples were obtained, one pair from the calving area and two pairs from the pre-weaned calf area. The procedure used for boot swab sampling was adapted from Eisenberg et al. (2013). Fresh overboots were used between the two locations. There was some variation between farms in the type of calving pens (individual, group, combination) and calf pens (size, floor type, number of pens) but at each location the same sampling protocol was used. Fresh overboots and boot swabs were put on at the entrance to the pen/shed, any alleyway/walkway in front of the pen was walked first, followed by entry into the pen, and then the pen was thoroughly walked in a zigzag pattern to maximise the surface area sampled. In the case of more than two separate calf sheds, the total area was roughly divided in two parts (two boot swab pairs) and boot swabs and overboots removed to walk across to the next shed before putting them back on to continue sampling. The same protocol was used for more than one calving pen. After sampling, the boot swabs were double-bagged in plastic zip-lock bags with samples from the same location pooled together to form one sample. All boot swab samples were placed in a −20 °C freezer within 48 h.

A manure sample was also obtained where possible off the floor of the calving pen, using the procedure for sample collection described by Wolf et al. (2014) [24]. Briefly, the pooled sample consisted of at least four "grabs" of mixed manure from separate sites in the pen, typically at the feeding area or around water troughs. The manure was mixed thoroughly in a mixing bag before transferring 20 g into a sample pot, which was kept refrigerated (4 °C) until it was tested in the laboratory.

2.5. Boot Swab Sample Processing

The boot swabs were defrosted just prior to processing in the laboratory. The protocol for processing was based on the method described by Eisenberg et al. (2013) with modifications made to accommodate test kit requirements, including standardising dilution rates of boot swab samples with varying weights of faecal material. Once defrosted, each boot swab was placed in a paddle blender bag and allowed to settle for 20 min at room temperature. The weight of boot swabs was subtracted from the total weight to give a faecal weight for each sample. The maximum volume of diluent that could be added to any sample was 150 mL (for sample handling reasons) and samples had to be diluted at a ratio of 6.67 mL to 1 g of faeces, (as required by the kit and in order to ensure results were comparable); therefore, the maximum faecal weight that could be processed as a single sample was 22.5 g.

Samples with a faecal weight of greater than 22.5 g had to be divided and processed as separate sub-samples. Demineralised water was added to each sub-sample in a weight of faeces to volume of water ratio of approximately 1: 6.67. Demineralised water was massaged into the boot swab sub-sample by agitating the bag, then allowing it to rest for 5 min. Each sub-sample was paddle-blended for 3 min. The recovered liquid was retrieved from the boot swab sub-sample and transferred to a sterile collection container. Particular attention was given to squeezing out the material to allow maximum recovery. The liquid sample was allowed to rest for 15 min to facilitate sample separation. Ten ml of supernatant was loaded into the ADIAfilter (Bio-X Diagnotics). The sample was centrifuged for 5 min at 3000× *g* and supernatant discarded. A volume of 500 µL of demineralised water was added to the recovered sample pellet, vortexed for 15 s then transferred to a lysis beads tube. The sample was disrupted in a bead-beater three times for 45 s each time with a 5-min rest period between each beading phase. After mechanical disruption the sample was centrifuged for 5-min at 15,000× g and 220 μL of the resultant supernatant phase was removed and was ready for the MAP DNA extraction protocol.

2.6. DNA Extraction and Amplification

The supernatant from the boot swab sub-samples and the manure sample from the calving pen floor underwent a multi-step DNA extraction process suitable for MAP. Briefly, samples were suspended in buffer then cleaned and concentrated via the use of the ADI-Afilter (Bio-X Diagnotics) and centrifugation for 5 min at 3000× g to create a bacterial pellet. Physical disruption of the bacterial pellet was achieved via beating with glass beads which results in the release of DNA. A commercial DNA extraction kit QIAmp® DNA Mini Kit (Qiagen) was then used to extract, clean and concentrate the DNA for use in the PCR reaction. The ADIAVET™ Paratb RealtimePCR kit (Bio-X Diagnotics) which detects the target gene IS900 was used for the detection of MAP DNA. In order to ensure the validity of results, four controls were included in each PCR run. A negative control was used to demonstrate that contamination of reagents or test wells did not occur during the process. An external positive control was added to each sample prior to DNA extraction to confirm that the DNA extraction process worked. A single kit positive control was included in each run to confirm that DNA amplification occurred. A validation control, composed of extracted DNA from a previously positive faecal sample, was included in every test run to challenge the cut-off of the method. The sample was considered negative if no Ct value was expressed or if greater than 45 Ct. The sample was considered inconclusive if Ct >36 as weak positive samples were not always reproducible after this value. The sample was considered positive if Ct $\leq$ 36.

2.7. Data Analysis

The results of testing are reported at sub-sample-level, sample-level (pre-weaned calf area, calving area) and farm-level, initially including negative, inconclusive and positive results but subsequently results were dichotomised so that inconclusive results were treated as negative. In order to obtain sample-level PCR results i.e., one pre-weaned calf area result and one calving area result per farm, sub-sample results for each location were pooled, including the floor samples from the calving area, such that a location was deemed to be PCR-positive if at least one sub-sample from the location was positive on PCR (Ct $\leq$ 36). For farm-level analysis, a farm was defined as PCR-positive if there was at least one positive PCR result for either calf pens or calving pens.

Logistic regression models were built with four different outcomes, categorized as follows: (1) farm-level PCR outcome; (2) pre-weaned calf area PCR outcome; (3) calving area PCR outcome; and (4) sample-level PCR outcome (all PCR results from both locations included). RAMP scores (overall score, pre-weaned calf area score or calving area score, as appropriate) and aWHP were the independent variables of interest. Model 4 had two observations per farm, so farm was included as a random effect. Variables were visually assessed for normality of distribution using histograms prior to inclusion in the models. Inconclusive results were defined as negative for the purpose of modelling.

Spearman's correlation test was used to evaluate associations at sample-level between location-specific RAMP scores and the lowest Ct value at the location, and aWHP and the lowest Ct values, in PCR-positive farms. High correlations were deemed as coefficients between 0.5–1, medium correlations as coefficients between 0.3–0.49, low correlations as coefficients below 0.30. With varying numbers of samples tested between farms for each location, the most biologically accurate method of comparing contamination quantitatively was to select the lowest Ct value recorded for each location on each farm.

3. Results

3.1. Apparent Within-Herd Prevalence

In total, 26 herds had serology results available to allow calculation of aWHP (Table 1). There were 14 farmers that used predominantly blood samples and 12 used predominantly milk samples for their herd test. The most recent blood/milk test results for study herds ranged between one and six months before or after the time of boot swab sampling. The aWHP for the herds ranged from 0–21%, with a median of 3.5%. The mean and median aWHP for PCR-positive and PCR-negative farms was 5.1% and 5%, and 4.8% and 3%, respectively, with no significant difference identified between groups ($p = 0.48$).

Table 1. Summary of PCR results, RAMP scores and apparent within-herd prevalence (aWHP) for 28 study farms.

Farm	Pre-Weaned Calf Area PCR Result	Pre-Weaned Calf Area Ct Value *	Pre-Weaned Calf Area RAMP SCORE /80	Calving Area PCR Result	Calving Area Ct Value *	Calving Area RAMP Score /80	Total FarmRAMP Score /227	aWHP (%)
1	Positive	35.76	41	Negative	0	50	131	8
2	Negative	0	44	Negative	0	41	125	21
3	Positive	31.48	32	Positive	31.56	35	104	8
4	Negative	0	29	Inconclusive	38.27	29	119	5
5	Negative	0	23	Negative	0	23	68	1
6	Inconclusive	40.51	17	Positive	31.1	26	74	2
7	Negative	0	20	Negative	0	29	83	3
8	Inconclusive	37.15	14	Positive	33.43	32	83	6
9	Negative	0	11	Negative	0	38	80	1
10	Inconclusive	39.41	26	Positive	33.17	26	104	1
11	Inconclusive	37.65	32	Positive	29.95	47	134	15
12	Negative	0	23	Positive	35.08	35	83	2
13	Positive	32.77	53	Positive	29.52	41	155	4
14	Negative	0	20	Negative	0	23	68	2
15	Negative	0	32	Inconclusive	39.34	32	98	8
16	Positive	34.34	20	Positive	32.68	41	107	2
17	Negative	0	-	Negative	0	-	-	-
18	Negative	0	29	Negative	0	17	71	3
19	Positive	35.66	11	Inconclusive	36.15	38	80	5
20	Negative	0	35	Negative	0	38	125	0
21	Negative	0	20	Negative	0	23	59	4
22	Negative	0	8	Positive	33.78	29	59	6
23	Inconclusive	36.69	23	Negative	0	35	92	10
24	Negative	0	23	Negative	38.09	44	98	3
25	Positive	35.55	32	Inconclusive	41.1	23	98	5
26	Negative	0	35	Inconclusive	40.43	56	155	2
27	Negative	0	23	Negative	0	38	101	-
28	Inconclusive	41.05	11	Positive	35.88	20	47	2

* Ct value reported for each location refers to the lowest Ct value recorded for an individual sample obtained at the location.

3.2. Environmental Samples

Calving area floor samples were obtained for 23 out of 28 herds. Eight floor samples were positive, four were inconclusive and eleven were negative. Five farms had no manure available in the pens to sample. In total, 84 boot swab pairs were obtained from all 28 farms (three pairs per farm). In total, 104 diluted faecal sub-samples for PCR extraction and amplification were obtained during processing of the boot swabs (2–7 per farm). Overall, 57 of these sub-samples were negative, 26 inconclusive and 21 positive. Table 2 shows the distribution of positive, negative and inconclusive results when boot swab sub-samples and calving area floor samples were pooled to give one result per location on each farm. 10/28 (36%) and 6/28 (21%) of farms had positive PCR results in the calving area and pre-weaned calf area, respectively. A total of 13/28 (46%) of farms had at least one positive sample from either location. There were 4/28 (14%) of farms that had at least one inconclusive sample from either location, and no positive samples.

Table 2. Distribution of pooled PCR results for pre-weaned calf pens and calving pens.

	Positive	Inconclusive	Negative	Total
Calf pens	6	6	16	28
Calving pens	10	5	13	28
Total	16	11	29	56

3.3. RAMP Scores

The most recent RAMP scores were available for 27 out of 28 herds and are described in Table 1. Overall RAMP scores ranged from 47–155 (out of total score of 227). The lower the score, the lower the perceived risk of transmission of MAP. The scores for the pre-weaned calf area ranged from 8 to 53, and for the calving area from 17–56 (each out of total 80). The mean overall RAMP score in PCR-positive and PCR-negative farms was 96.8 and 95.9, respectively, with no significant difference between mean scores ($p = 0.93$). The distribution of scores between both groups is illustrated in Figure 1.

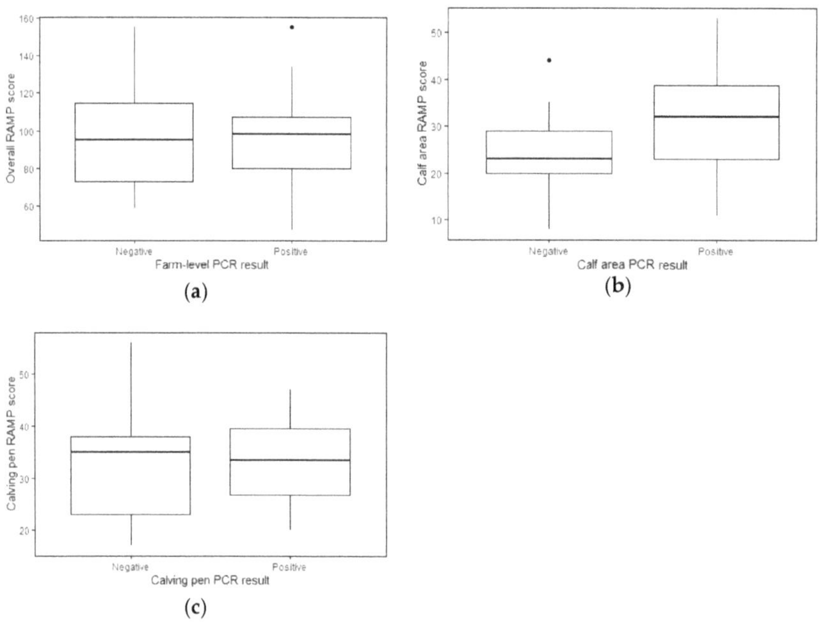

Figure 1. Boxplots illustrating the distribution of RAMP scores for the overall farm (**a**), pre-weaned calf area (**b**) and the calving area (**c**). The black dots in the graphs represent outliers in the data.

The mean RAMP scores for calf pens across farms that were PCR-positive for calf pens and PCR-negative for calf pens were 31.5 and 23.7, respectively, with no significant difference between mean scores ($p = 0.12$). The distribution of scores between both groups is illustrated in Figure 1.

The mean RAMP scores for calving pens across farms that were PCR-positive and PCR-negative for calving pens was 33.2 and 33.9, respectively, with no significant difference between means ($p = 0.85$). The distribution of scores between both groups is illustrated in Figure 1.

3.4. Logistic Regression Models

A log transformation was applied to the aWHP data prior to modelling due to positive skewness. As one farm had an aWHP of 0, it was necessary to add a constant value (1) to aWHP prior to transformation, hence the variable included in the models was log(aWHP+1). RAMP score and aWHP were not significantly associated with PCR outcome in any of the models. The model results are summarised in Table 3.

Table 3. Summary of outcomes of logistic regression models with different dependent variables: (1) Farm-level PCR result; (2) Pre-weaned calf area PCR result; (3) Calving area PCR result; (4) Sample-level PCR result (includes both pre-weaned calf area and calving pen samples).

		(1) Farm-Level		
	Estimate	Odds Ratio	95% CI	*P*-Value
(Intercept)	−0.60395	0.54665	0.024, 11.11	0.692975
log(aWHP + 1)	0.422476	1.525734	0.477, 5.508	0.483722
Overall RAMP	−0.00051	0.999494	0.971, 1.029	0.971782
		(2)Pre-weaned calf area only		
	Estimate	Odds Ratio	95% CI	*P*-Value
(Intercept)	−3.42849	0.032436	0.001, 0.584	0.034072
Calf pen RAMP	0.06061	1.062484	0.97, 1.185	0.211207
log(aWHP + 1)	0.348346	1.416723	0.36, 6.733	0.626936
		(3) Calving area only		
	Estimate	Odds Ratio	95% CI	*P*-Value
(Intercept)	−0.34837	0.705836	0.029, 16.388	0.825372
Calving pen RAMP	−0.00639	0.993632	0.907, 1.083	0.883681
log(aWHP + 1)	0.059468	1.061271	0.314, 3.697	0.922076
		(4) Sample level		
	Estimate	Odds Ratio	95% CI	*P*-Value
(Intercept)	−2.07968	0.12497	0.004, 1.512	0.118357
Overall RAMP	0.008885	1.008925	0.984, 1.038	0.450377
log(aWHP + 1)	0.226056	1.253646	0.447, 4.114	0.641828

3.5. Correlation Tests

The association between the lowest Ct value for a location and the RAMP score for that location was assessed by selecting all PCR-positive sample-level results ($n = 16$) from all farms. There was moderate negative correlation found using Spearman's rank correlation coefficient (−0.45; (p-value = 0.08)) illustrated in Figure 2.

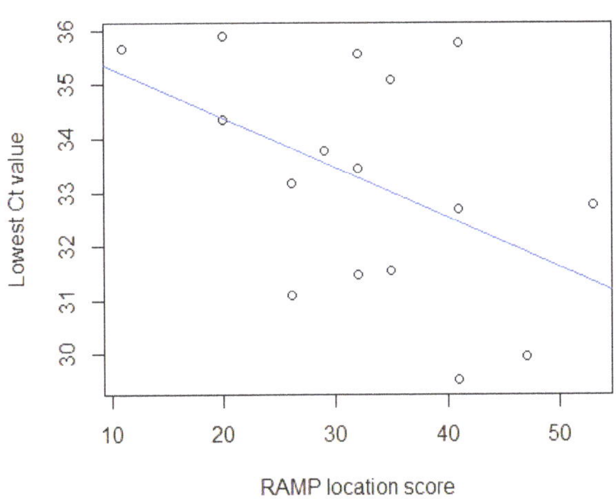

Figure 2. Scatterplot to illustrate the correlation between the lowest Ct value for a location and the RAMP score for the location.

The association between the lowest Ct value for a location and aWHP was assessed by selecting all PCR-positive sample-level results (n = 16) from all farms. There was no correlation found using Spearman's rank correlation coefficient (−0.16; p-value = 0.55)).

4. Discussion

This study presents a novel and objective strategy for monitoring the efficacy of RAMPs conducted in herds engaged in a Johne's disease control programme for reducing the risk of transmission of MAP within herds. The MAP-infected herds in this study were all registered in the IJCP, and engaged in annual RAMPs conducted by a veterinary practitioner. Therefore, it may be concerning that 46% of the herds had detectable MAP in either their calving pen, calf pen, or both locations, particularly if one assumes the bedding material usually present in these areas reduces the sensitivity of environmental sampling for MAP compared to lactating cow housing. While a degree of contamination in the calving area would be expected and accepted as inevitable in infected herds, it is concerning that 21% of herds had detectable levels of MAP in the pre-weaned calf area. The median aWHP across all herds was relatively low at 3.5%, however, the proportion of truly infected animals may be much higher than this, due to the low sensitivity of serology tests for MAP [25]. Another potential factor contributing to the high proportion of positive results could be that the biocontainment measures practiced by the farmers were inadequate, allowing contamination with MAP either through direct shedding from infectious animals (most likely in the calving area) or indirectly through fomites (most likely in pre-weaned calf area), which may include the farmers' boots. On a typical Irish spring-calving farm the compact calving season may increase the chances of contamination of calf pens with faecal material from adult cows due to the frequent movements of animals and personnel between cow housing (including the calving pen) and the calf rearing areas.

Similar rates of positivity in calving areas and calf pens have been demonstrated in other studies. Pillars et al. (2009) reported finding MAP in the calving area and pre-weaned calf area on 4/6 and 3/6 infected farms that were sampled longitudinally over four years [26]. The herds had annual RAMPs conducted for the duration of the study. The study also found significant associations between increasing within-herd prevalence and the number of positive samples obtained in the calving area and the pre-weaned calf area. Another study of the environmental distribution of MAP in herds enrolled in a Johne's disease control programme in the US found that 24% of calving pens were positive among 68 farms that had positive environmental samples, however none of the farms had detectable MAP in the pre-weaned calf area [27].

In the present study the relationship between PCR test result and aWHP or RAMP scores at farm-level, location-level or sample-level was not significant. This may be due to insufficient statistical power, and/or the limitations of using RAMP scores or aWHP to assess transmission risk. As mentioned previously, aWHP based on serology likely underestimates the true proportion of infected animals in the herd that may be contributing to contamination, while RAMP scores are limited by the subjective, point-in-time nature of the assessment [28]. In seasonal calving systems, as the majority of Irish dairy herds are, RAMP assessments are often conducted outside of the busy calving season when the calving pen and calf pens are more likely to be empty. This makes it more difficult for the veterinarian to conduct an accurate assessment of what the transmission risk is, with consequent subjectivity and high inter-observer variability affecting RA scores [29].

The moderate negative correlation (p = 0.08) found between the lowest Ct value obtained at a location and the RAMP score for the location suggests that higher RAMP scores may be associated with higher concentrations of MAP in a sample (a lower Ct value). The required significance level (p = 0.05) was not achieved, most likely due to low statistical power due to the small sample size and possibly the limitations of the RAMP as discussed above. Factors that may contribute to higher MAP concentrations in an environmental sample from a calving pen, for example, include the number of shedding animals present and the amount of MAP each animal is shedding, as well as the level of hygiene maintained in the pen. The RAMP directly assesses the latter factor, while indirect assumptions about the level of shedding in the pen can be made using aWHP and stocking density. Clumping of MAP bacteria in a sample can also affect the quantification of MAP bacteria during analysis [30]. Bedding material usually present in calving pens and calf pens also likely affects the sensitivity of environmental sampling, however most literature evaluating the sensitivity of environmental sampling for MAP has been focused on main cow housing and manure storage. Pillars et al. (2009) reported that 7% and 14% of calf pen floor and calving pen floor samples, respectively, were culture-positive for MAP, compared to 44% of lactating cow floor samples on seven infected dairy farms.

A limitation of the present study is the small number of farms that were sampled, which likely impacted the significance of the modelling results. A larger study, with longitudinal sampling, would provide further insight into the factors that affect contamination of the pre-weaned calf area and calving area on infected farms.

5. Conclusions

This research has demonstrated a potential novel application for environmental sampling as an objective test to complement the annual RAMP, to help monitor the progress of infected farms in reducing the risk of transmission of MAP to calves. However further research conducted on a larger scale is needed to support the results reported in the present study.

Author Contributions: Conceptualization, N.L.F., C.G.M. and J.F.M.; Methodology, N.L.F., C.G.M. and J.F.M.; Formal Analysis, N.L.F. and C.G.M.; Investigation, N.L.F.; Resources, N.L.F.; Data Curation, N.L.F.; Writing—Original Draft Preparation, N.L.F.; Writing—Review and Editing, N.L.F., C.G.M. and J.F.M.; Supervision, C.G.M. and J.F.M.; Project Administration, N.L.F. All authors have read and agreed to the published version of the manuscript.

Funding: This study was funded by the Irish Dairy Levy Trust through funding provided for Teagasc project MKAB1170.

Institutional Review Board Statement: This study did not involve any interventions on animals or humans so did not require ethical approval.

Informed Consent Statement: Informed consent was obtained from all subjects involved in the study.

Data Availability Statement: The data presented in this study are available in article.

Acknowledgments: We wish to acknowledge the collaboration and technical support of the microbiology PCR laboratory of the Irish Equine Centre during the planning and completion of this study.

Conflicts of Interest: The authors declare no conflict of interest.

References

1. Whittington, R.J.; Marshall, D.J.; Nicholls, P.J.; Marsh, I.B.; Reddacliff, L.A. Survival and dormancy of Mycobacterium avium subsp. paratuberculosis in the environment. *Appl. Environ. Microbiol.* **2004**, *70*, 2989–3004. [CrossRef] [PubMed]
2. Sweeney, R.W. Pathogenesis of Paratuberculosis. *Vet. Clin. N. Am. Food Anim. Pract.* **2011**, *27*, 537–546. [CrossRef] [PubMed]
3. Garcia, A.B.; Shalloo, L. Invited review: The economic impact and control of paratuberculosis in cattle. *J. Dairy Sci.* **2015**, *98*, 5019–5039. [CrossRef] [PubMed]
4. McAloon, C.G.; Whyte, P.; More, S.J.; Green, M.J.; O'Grady, L.; Garcia, A.; Doherty, M.L. The effect of paratuberculosis on milk yield—A systematic review and meta-analysis. *J. Dairy Sci.* **2016**, *99*, 1449–1460. [CrossRef]
5. Ott, S.L.; Wells, S.J.; Wagner, B.A. Herd-level economic losses associated with Johne's disease on US dairy operations. *Prev. Vet. Med.* **1999**, *40*, 179–192. [CrossRef]
6. Waddell, L.A.; RajiĆ, A.; StÄRk, K.D.C.; McEwen, S.A. The zoonotic potential of *Mycobacterium avium* ssp. paratuberculosis: A systematic review and meta-analyses of the evidence. *Epidemiol. Infect.* **2015**, *143*, 3135–3157. [CrossRef]
7. Field, N.L.; McAloon, C.G.; Gavey, L.; Mee, J.F. Mycobacterium avium subspecies paratuberculosis infection in cattle—A review in the context of seasonal pasture-based dairy herds. *Ir. Vet. J.* **2022**, *75*, 12. [CrossRef]
8. Geraghty, T.; Graham, D.A.; Mullowney, P.; More, S.J. A review of bovine Johne's disease control activities in 6 endemically infected countries. *Prev. Vet. Med.* **2014**, *116*, 1–11. [CrossRef]
9. Groenendaal, H.; Nielen, M.; Jalvingh, A.W.; Horst, S.H.; Galligan, D.T.; Hesselink, J.W. A simulation of Johne's disease control. *Prev. Vet. Med.* **2002**, *54*, 225–245. [CrossRef]

10. Kudahl, A.B.; Østergaard, S.; Sørensen, J.T.; Nielsen, S.S. A stochastic model simulating paratuberculosis in a dairy herd. *Prev. Vet. Med.* **2007**, *78*, 97–117. [CrossRef]
11. Doré, E.; Paré, J.; Côté, G.; Buczinski, S.; Labrecque, O.; Roy, J.P.; Fecteau, G. Risk factors associated with transmission of Mycobacterium avium subsp. paratuberculosis to calves within dairy herd: A systematic review. *J. Vet. Intern. Med.* **2012**, *26*, 32–45. [CrossRef] [PubMed]
12. McAloon, C.G.; Roche, S.; Ritter, C.; Barkema, H.W.; Whyte, P.; More, S.J.; O'Grady, L.; Green, M.J.; Doherty, M.L. A review of paratuberculosis in dairy herds—Part 1: Epidemiology. *Vet. J.* **2019**, *246*, 59–65. [CrossRef]
13. Khol, J.L.; Wassertheurer, M.; Sodoma, E.; Revilla-Fernandez, S.; Damoser, J.; Oserreicher, E.; Dunser, M.; Kleb, U.; Baumgartner, W. Long-term detection of Mycobacterium avium subspecies paratuberculosis in individual and bulk tank milk from a dairy herd with a low prevalence of Johne's disease. *J. Dairy Sci.* **2013**, *96*, 3517–3524. [CrossRef] [PubMed]
14. Stabel, J.R.; Bradner, L.; Robbe-Austerman, S.; Beitz, D.C. Clinical disease and stage of lactation influence shedding of Mycobacterium avium subspecies paratuberculosis into milk and colostrum of naturally infected dairy cows. *J. Dairy Sci.* **2014**, *97*, 6296–6304. [CrossRef] [PubMed]
15. Whittington, R.; Donat, K.; Weber, M.F.; Kelton, D.; Nielsen, S.S.; Eisenberg, S.; Arrigoni, N.; Juste, R.; Saez, J.L.; Dhand, N.; et al. Control of paratuberculosis: Who, why and how. A review of 48 countries. *BMC Vet. Res.* **2019**, *15*, 1–29. [CrossRef]
16. Ridge, S.; Baker, I.; Hannah, M. Effect of compliance with recommended calf-rearing practices on control of bovine Johne's disease. *Aust. Vet. J.* **2005**, *83*, 85–90. [CrossRef] [PubMed]
17. Sorge, U.; Kelton, D.; Lissemore, K.; Godkin, A.; Hendrick, S.; Wells, S. Attitudes of Canadian dairy farmers toward a voluntary Johne's disease control program. *J. Dairy Sci.* **2010**, *93*, 1491–1499. [CrossRef]
18. Gavey, L.; Citer, L.; More, S.J.; Graham, D. The Irish Johne's Control Programme. *Front. Vet. Sci.* **2021**, *8*, 703843. [CrossRef]
19. Raizman, E.A.; Wells, S.J.; Godden, S.M.; Fetrow, J.; Friendshuh, K.; Michael Oakes, J. Characterization of Minnesota dairy herds participating in a Johne's disease control program and evaluation of the program risk assessment tool. *Prev. Vet. Med.* **2006**, *75*, 22–33. [CrossRef]
20. Donat, K.; Hahn, N.; Eisenberg, T.; Schlez, K.; Köhler, H.; Wolter, W.; Rohde, M.; Pützschel, R.; Rösler, U.; Failing, K.; et al. Within-herd prevalence thresholds for the detection of Mycobacterium avium subspecies paratuberculosis-positive dairy herds using boot swabs and liquid manure samples. *Epidemiol. Infect.* **2016**, *144*, 413–424. [CrossRef]
21. Eisenberg, T.; Wolter, W.; Lenz, M.; Schlez, K.; Zschock, M. Boot swabs to collect environmental samples from common locations in dairy herds for Mycobacterium avium ssp paratuberculosis (MAP) detection. *J. Dairy Res.* **2013**, *80*, 485–489. [CrossRef]
22. Caldwell, D.J.; Hargis, B.M.; Corrier, D.E.; DeLoach, J.R. Frequency of Isolation of Salmonella from Protective Foot Covers Worn in Broiler Houses as Compared to Drag-Swab Sampling. *Avian Dis.* **1998**, *42*, 381–384. [CrossRef]
23. Talorico, A.A.; Bailey, M.A.; Munoz, L.R.; Chasteen, K.S.; Pal, A.; Krehling, J.T.; Bourassa, D.V.; Buhr, R.J.; Macklin, K.S. The use of roller swabs for Salmonella detection in poultry litter. *J. Appl. Poult. Res.* **2021**, *30*, 100163. [CrossRef]
24. Wolf, R.; Barkema, H.W.; De Buck, J.; Slomp, M.; Flaig, J.; Haupstein, D.; Pickel, C.; Orsel, K. High herd-level prevalence of Mycobacterium avium subspecies paratuberculosis in Western Canadian dairy farms, based on environmental sampling. *J. Dairy Sci.* **2014**, *97*, 6250–6259. [CrossRef]
25. Nielsen, S.S.; Toft, N. Ante mortem diagnosis of paratuberculosis: A review of accuracies of ELISA, interferon-γ assay and faecal culture techniques. *Vet. Microbiol.* **2008**, *129*, 217–235. [CrossRef] [PubMed]
26. Pillars, R.B.; Grooms, D.L.; Kaneene, J.B. Longitudinal study of the distribution of Mycobacterium avium subsp paratuberculosis in the environment of dairy herds in the Michigan Johne's disease control demonstration herd project. *Can. Vet. J. Rev. Vet. Can.* **2009**, *50*, 1039–1046.
27. Raizman, E.A.; Wells, S.J.; Godden, S.M.; Bey, R.F.; Oakes, M.J.; Bentley, D.C.; Olsen, K.E. The distribution of Mycobacterium avium ssp paratuberculosis in the environment surrounding Minnesota dairy farms. *J. Dairy Sci.* **2004**, *87*, 2959–2966. [CrossRef] [PubMed]
28. McAloon, C.G.; Whyte, P.; More, S.J.; O'Grady, L.; Doherty, M.L. Development of a HACCP-based approach to control paratuberculosis in infected Irish dairy herds. *Prev. Vet. Med.* **2015**, *120*, 152–161. [CrossRef] [PubMed]
29. Pieper, L.; DeVries, T.J.; Sorge, U.S.; Godkin, A.; Hand, K.J.; Perkins, N.R.; Imada, J.; Kelton, D.F. Variability in Risk Assessment and Management Plan (RAMP) scores completed as part of the Ontario Johne's Education and Management Assistance Program (2010–2013). *J. Dairy Sci.* **2015**, *98*, 2419–2426. [CrossRef] [PubMed]
30. Elguezabal, N.; Bastida, F.; Sevilla, I.A.; González, N.; Molina, E.; Garrido, J.M.; Juste, R.A. Estimation of Mycobacterium avium subsp. paratuberculosis growth parameters: Strain characterization and comparison of methods. *Appl. Env. Microbiol.* **2011**, *77*, 8615–8624. [CrossRef] [PubMed]

animals

MDPI

Article

Presence of Non-Tuberculous Mycobacteria Including *Mycobacterium avium* subsp. *paratuberculosis* Associated with Environmental Amoebae

Vincent Rochard [1,†], Thierry Cochard [2,†], Stéphanie Crapart [1], Vincent Delafont [1], Jean-Louis Moyen [3], Yann Héchard [1] and Franck Biet [2,*]

1 Laboratoire Ecologie et Biologie des Interactions, UMR Centre National de la Recherche Scientifique 7267, Université de Poitiers, Equipe Microbiologie de l'Eau, F-86073 Poitiers, France; vincent.delafont@univ-poitiers.fr (V.D.); yann.hechard@univ-poitiers.fr (Y.H.)

2 Institut National de Recherche pour l'Agriculture—INRAE, Université de Tours, ISP, F-37390 Nouzilly, France; thierry.cochard@inrae.fr

3 Laboratoire Départemental d'Analyse et de Recherche de la Dordogne, F-24660 Coulounieix-Chamiers, France

* Correspondence: franck.biet@inrae.fr; Tel.: +33-247-427-869

† These authors contributed equally to this work.

Simple Summary: The prevalence of Johne's disease (JD) due to *Mycobacterium avium* ssp. *paratuberculosis* (*Map*), responsible for considerable economic losses in the dairy industry worldwide, remains very high. In addition, recent studies have shown that *Map* may hamper the success of bovine tuberculosis (bTB) eradication programs due to frequent co-infection of hosts. Therefore, the risk of *Map* contamination by environmental sources must be studied given achieving improved control of this enzooty. Our previous work showed that environmental amoebae could function as reservoirs and vectors of these pathogenic mycobacteria. This study aims to describe amoeba-mycobacteria co-occurrence by sampling the water points of cattle herds monitored for *Map* and *Mycobacterium bovis* (*M. bovis*) infection. Our study showed that a wide diversity of amoebae and non-tuberculous mycobacteria (NTM) species, including *Map*, live closely together in water troughs of herds monitored for JD or bTB. The exact association between amoebae and mycobacteria calls for further investigation.

Abstract: One of the obstacles to eradicating *paratuberculosis* or Johne's Disease (JD) seems to be the persistence of *Mycobacterium avium* subsp. *paratuberculosis* (*Map*) in the environment due to its ability to survive alone or vectorized. It has been shown that *Map* is widely distributed in soils and water. Previously, we isolated amoebae associated with *Map* strains in the environment of bovines from an infected herd. This work aims to verify our working hypothesis, which suggests that amoebae may play a role in the transmission of JD. In this study, we sampled water in the vicinity of herds infected with *Map* or *Mycobacterium bovis* (*M. bovis*) and searched for amoebae and mycobacteria. Live amoebae were recovered from all samples. Among these amoebae, four isolates associated with the presence of mycobacteria were identified and characterized. *Map* and other mycobacterial species were detected by qPCR and, in some cases, by culture. This study suggests that amoebae and *Map* may be found in the same environment and might represent a risk of exposure of animals to pathogenic mycobacteria. These data open up new perspectives on the control measures to be put in place to prevent contamination by *Map*.

Keywords: amoebae; environment; Johne's disease; *Mycobacterium avium* subsp. *paratuberculosis*; transmission

1. Introduction

Johne's disease (JD), caused by *Mycobacterium avium* subsp. *paratuberculosis* (*Map*) is a chronic ruminant disease responsible for considerable economic losses in the dairy industry worldwide [1]. Despite implementing control programs in most developed countries and notwithstanding substantial financial efforts, the prevalence of Johne's disease remains very high (around 50% in European herds and 80% in the United States) [2,3].

The reasons for this worrisome situation and its negative impact on animal health and the livestock economy are multiple, both in terms of the biology of the disease (especially due to the long-term sub-clinical nature of the infection) and of the specificities of this very slowly growing mycobacteria, of which the isolation always remains a challenge, as do control policies [4].

Another concern is that *Map* may hamper bovine TB (bTB) eradication programs due to frequent co-infection of *Map* and *M. bovis*. This may lead to misdiagnosis (preventing accurate identification of infected animals), and it has been suggested that herds with "chronic" *M. bovis* infection may be more likely to present with co-infection [5]. A key factor in the difficulty of controlling these infections may be the environmental survival of these mycobacteria.

Cattle infected with *Map* shed the bacteria in their faeces. For this reason, fecal shedding by cows is the main route of transmission to calves [6]. Simulation models have confirmed the hypothesis that intermittent, low, and transient shedding animals play a major role in maintaining a low prevalent infection rate in dairy herds. As a result, *Map* is widely distributed within and beyond the boundaries of the agricultural environment, in both soil and water [7]. However, these analyses are often based on the detection of DNA and do not provide information on the general potential for survival and multiplication of ex-host *maps*, which is a key element in the dynamics of disease transmission.

Microbial hosts in the shared agricultural environment may likely serve as vectors to *Map* and contribute to the spread and persistence of JD. Environmental protozoa such as free-living amoebae (FLA) have long been considered "incubators" for intracellular bacteria in the environment [8,9]. They are a training ground for pathogenic bacteria, including several *Mycobacterium* species [10–12]. The interactions between *Map* and amoebae remain little studied to date. A previous study involving amoebae has shown that *Acanthamoeba* spp. It enables *Map* replication and enhances virulence [13]. Another study showed that *Map* ingested by *Acanthamoeba* resisted digestion for at least 24 days [14]. White et al. described the co-occurrence of *Map* and FLA in soil (2010) [15]. An in vitro study demonstrated that *Map* persisted for up to four years in the presence of *Acanthamoeba* [16].

Interestingly, a report showed that *Map* was found in *amoebae* isolated from the soil after the application of cattle manure spiked with *Map* [17]. In our previous study, we reported that various lineages of *Map* strains were able to grow within *Acanthamoeba castellanii* and that they can survive for several days within their host [18]. In addition, we showed that *Map* might be found in amoebae isolated from a cattle environment, and we also showed that the strain genotype was identical to a previous infectious strain isolated from a bovine [18]. These results strongly suggest a risk of transmission mediated by infected amoebae.

In the present study, we sought to extend these previous investigations on amoeba-mycobacteria association by sampling through water points from six herds monitored for *Map* and *M. bovis* infection status.

2. Materials and Methods

2.1. Sampling

Sampling campaigns occurred from April 2018 to April 2019 in six farms in southwestern France (Figure 1 and Table 1). For this study, we selected farms from this area for two reasons. The first is that JD is endemic in this region, and the second is that although France is one of the officially tuberculosis-free (OTF) member states of the EU, it nevertheless experiences multi-host tuberculosis enzootic situations, mainly in the southwestern region [19]. The bTB and JD status of the herds enrolled in this study is detailed in Table 1. On average, three samples were taken per herd, including troughs and outdoor water points. Figure 2 summarizes the process of sample treatment.

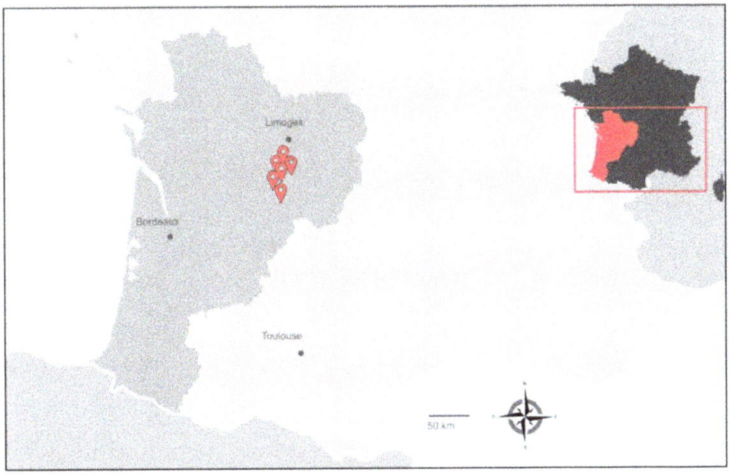

Figure 1. Location of the sampled bovine herds from southwest France Nouvelle Aquitaine region.

Table 1. Sampling details.

Herd N°	Number of Cattle	bTB Status	JD Status *	Sample Number	Nature of Water Sampling
1	68	positive 6 confirmed cows	ND	1 2 3 4	Pond Pond Pond Pond
2	82	positive Two confirmed cows	ND	5 6 7 8	Pond River Pond Pond
3	87	positive 1 confirmed cow	negative	9 10	Outdoor drinking trough Pond
4	19	positive Five confirmed cows	ND	11	Outdoor drinking trough
5	40	positive 12 confirmed cows	positive 3.5%	12 13 14 15 16 17	Outdoor drinking trough Outdoor drinking trough Outdoor drinking trough Outdoor drinking trough Outdoor drinking trough Outdoor drinking trough
6	235	positive 6 confirmed cows	positive 0.85%	18 19 20	Outdoor drinking trough Outdoor drinking trough Outdoor drinking trough

* ND; Not Determined, %; % of seroprevalence.

2.2. *Cultivation of Microorganisms*

Mycobacterial strains were grown at 37 °C in Sauton medium or Middlebrook 7H9 broth as described previously [20].

Escherichia coli strain K12 was used to isolate amoeba as described previously [18].

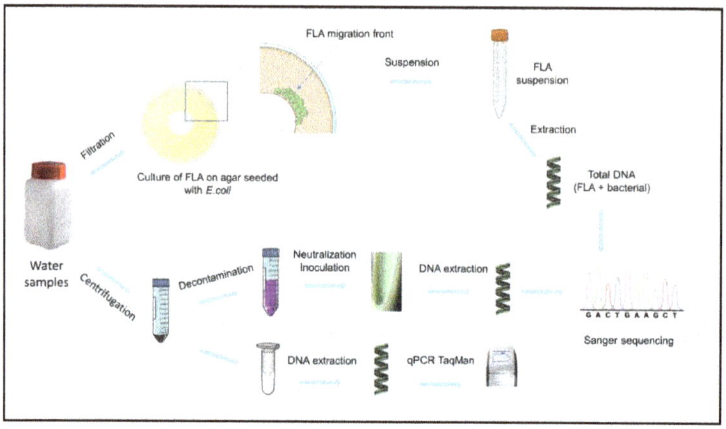

Figure 2. Experimental protocol for the analysis of water samples from cattle herds.

2.3. Isolation of Environmental Amoebae

The samples, comprising two flacons of one liter, were taken from the drinking troughs for each herd. They were treated according to the method described by Samba-Louaka et al. [18]. Briefly, water samples were filtered, and the filters were placed on agar plates seeded with *E. coli* as a food source. The plates were incubated at 20 or 37 °C, and the presence of amoeba development was followed by microscopy. To obtain a clonal culture of amoebae, they were subcultured in the same conditions or liquid culture: PAS buffer [18] spiked with *E. coli* in a 25 or 75 cm^2 flask.

2.4. DNA Extraction and Purification

The environmental samples were centrifuged at 13,000× *g* for 20 min at +4 °C. DNAs were extracted from 500 mg of pellet using the FastDNA Spin Kit for Soil, MP Biomedicals (67400 Illkirch-Graffenstaden, France) according to the manufacturer's instructions. The obtained DNAs were stored until use at −20 °C.

2.5. Detection and Quantification of Mycobacteria by Quantitative PCR (qPCR)

Bacterial genomic DNA was detected by quantitative PCR (qPCR) in a final volume of 25 µL including 5 µL of extracted DNA, 12.5 µL of TaqManTM Fast Advanced Master Mix with UNG (Applied Biosystems, Thermo Fischer Scientific, Vilnius, Lithuania), 2.5 µL of primer mix (Table 2) at 3 µM each and probed at 2.5 µM (table below) and 5 µL of ultrapure water. Experiments were carried out using a BIO-RAD CFX96 thermocycler (BIO-RAD), consisting of polymerase activation at 50 °C for 2 min, followed by an initial denaturation step at 95 °C for 10 min, followed by 40 cycles of denaturation at 95 °C for 15 s, elongation at 60 °C for 1 min. The Ct is calculated with the CFX Manager software (BIO-RAD CFX Manager 3.1). The primers used are listed in Table 2.

2.6. Identification of Mycobacteria Isolated from Samples

To isolate mycobacteria, the samples were decontaminated according to two methods in parallel. The first decontamination method used 0.9% hexadecylpyridinium chloride (HPC) (weight/volume). The environmental samples were centrifuged at 13,000× *g* for 20 min at +4 °C, and approximately 2 g of the pellet was resuspended in 25 mL of 0.9% HPC and incubated for 18 h at room temperature. The second decontamination method used 3% sodium dodecyl sulfate (SDS). Approximately 2 g of the pellet was resuspended with 30 mL of a 1% sodium hydroxide solution, 3% SDS and 25 mg/L phenol red and incubated at 37 °C for 30 min with regular gentle agitation. The suspension was then neutralized with H3PO4 until the color indicator turned.

After centrifugation for 15 min at 5000× *g* of suspensions obtained from the two decontamination methods, the pellets were resuspended with 2 mL of Middlebrook 7H9 broth (Difco Laboratories, Detroit, MI, USA) with 0.2% glycerol and supplemented with 10% Albumin Dextrose Catalase (ADC, Becton Dikinson, Le Pont de Claix, France) to inoculate two Herrold's tubes with mycobactin J and ANV (Becton Dickinson, Le Pont de Claix, France). The tubes were incubated horizontally for one week and then vertically until cultures appeared (in a maximum of 16 weeks).

Mycobacterial species were identified as previously described [21].

2.7. Amoebal Identification

The amoeba isolated from the water troughs was identified as described previously [22]. Briefly, DNA from the amoebae was isolated and amplified by PCR with universal primers targeting the 18S rRNA gene (Table 2). Then, the amplified fragments were sequenced and compared to the nucleotide database using the BLAST program.

Table 2. Primers used for mycobacteria and amoeba species identification.

Name	Primers	Sequences	Target
qPCR atpE	Forward Reverse Probe	CGGYGCCGGTATCGGYGA CGAAGACGAACARSGCCAT FAM-ACSGTGATGAAGAACGGBGTRAA-BHQ1	*Mycobacterium* ssp.
qPCR IS900	Forward Reverse Probe	CCGCTAATTGAGAGATGCGATTGG AATCAACTCCAGCAGCGCGGCCTCG FAM-TCCACGCCCGCCCAGACAGG-BHQ1	*Map*
qPCR IS6110	Forward Reverse Probe	GGTAGCAGACCTCACCTATGTGT AGGCGTCGGTGACAAAGG FAM-CACGTAGGCGAACCC-MGB	MTB complex *
qPCR IS1081	Forward Reverse Probe	CCGCCACCGTGATTTCGA GCCAGTCCGGGAAATAGCT FAM-CCGCAACCATCGACGTC-MGB	MTB complex *
qPCR IS1561	Forward Reverse Probe	GATCCAGGCCGAGAGAATCTG GGACAAAAGCTTCGCCAAAA FAM-ACGGCGTTGATCCGATTCCGC-TAMRA	MTB complex *
hsp65	Forward Reverse	CTTGTCGAACCGCATACCCT ACCAACGATGGTGTGTCCAT	*Mycobacterium* ssp.
18S rRNA	Forward 566 Reverse 1200	CAGCAGCCGCGGTAATTCC CCCGTGTTGAGTCAAATTAAGC	Amoeba

* Detection of the MTBC complex targeted two specific insertion sequences: IS6110 and IS1081. This reflects the extra sensitivity that these multicopy sequences allow. In addition, the PCR targeting IS1561 was used to discriminate members of MTB to *Mycobacterium microti* [23].

3. Results

3.1. High Diversity of Non-Tuberculous Mycobacteria from Herd Samples

We attempted to detect mycobacteria's indirect and direct presence in each sample according to the method illustrated in Figure 2. Table 3 summarizes the results of detecting DNA mycobacteria by analyses of five different qPCRs, identifying either the species or the subspecies of mycobacteria, and the results of the isolation and genetic identification of mycobacteria by culture. Overall, we observed that positive qPCR results were not confirmed in positive cultures, and conversely, samples leading to positive cultures were not always positive by qPCR analysis. Regarding qPCR analysis, the samples from herds 1, 3 and 4 did not give convincing results with signals above the positivity thresholds of Ct 40. For herd 2, only one sample was positive for *Mycobacterium* spp. For herds 5 and 6, many samples were positive for mycobacteria, including the detection of *Map* and species of the MTBC.

Table 3. Mycobacterial species were identified by qPCR on water samples and by sequencing the *hsp* 65 gene after culture.

Herd N°	Sample N°	qPCR on Water Samples *	Culture ID
1 *	1	Ct > 40	*Mycobacterium hiberniae*
	2	Ct > 40	*Mycobacterium nonchromogenicum*
	3	Ct > 40	*Mycobacterium icosiumassiliensis*
	4	Ct > 40	*Mycobacterium nonchromogenicum*
2	5	Ct > 40	*Mycobacterium hiberniae*
	6	*Myco* sp.	*Mycobacterium avium* ssp. *paratuberculosis*
	7	Ct > 40	*Mycobacterium arupensis*
	8	Ct > 40	*Mycobacterium parascrofulaceum or genavense*
3	9	Ct > 40	Contaminated
	10	Ct > 40	Contaminated
4	11	Ct > 40	Contaminated
5	12	*Myco* sp.; *Map*; MTBC	*Mycobacterium hiberniae*
	13	*Myco* sp.; *Map*; MTBC	Contaminated
	14	*Myco* sp.; *Map*; MTBC	Contaminated
	15	Ct > 40	Contaminated
	16	*Myco* sp.; *Map*; MTBC	Contaminated
	17	*Myco* sp.; *Map*; MTBC	Contaminated
6	18	Ct > 40	*Mycobacterium arupensis*
	19	*Myco* sp.; *Map*; MTBC	Contaminated
	20	*Myco* sp.; *Map*; MTBC	Contaminated

* used primers described in Table 2 and targeted specific insertion sequences IS900, IS6110, IS1081 and IS1561 to identify Map or the MTBC species.

Bacteriological analyses were not consistent with qPCR data. For the samples from herds 1 and 2, which were almost negative in qPCR, widely diverse NTM, including *Map*, was isolated. Conversely, despite positive qPCR signals for mycobacteria in herds 5 and 6 samples, few mycobacteria could ultimately be isolated in culture, mainly due to fungal contaminants that are difficult to eliminate in environmental samples.

In these samples taken, it is interesting that *Map* was frequently found by qPCR and even in culture. Other interesting results of this study highlight the diversity of NTM species presents in these samples from troughs used by cattle.

3.2. Herd Water Samples Rich in Environmental Amoebae

In a previous study, we succeeded in isolating and culturing amoebae from water samples from cattle drinkers on a farm infected with *Map* [18]. In addition, the total DNA extracted from the culture of the amoebae demonstrated the presence of *Map* DNA. These encouraging results [18], prompted us to extend more widely these investigations aimed at isolating amoebae capable of harboring pathogenic mycobacteria.

By sampling six cattle herds, we were able to collect and analyze about twenty collection points. In most samples, the amoeba culture was positive at 20 or 37 °C (Table 4).

To identify the amoeba isolates, strains were sub-cultured, and PCR targeting a portion of the 18S rRNA coding gene was performed. Sequencing of purified amplicon made it possible to identify the amoeba species. A total of 22 amoebal cultures allowed the identification of five different genera of amoeba, including, in order of frequency in the samples, *Naegleria* (59%), *Acanthamoeba* (18%), *Vermamoeba* (14%), *Filamoeba* (4.5%) and Tubulinea group (4.5%) (Figure 3). Figure 4 shows the morphological aspects of the main amoebic species isolated during this study. Positive *atpE* qPCR analyses performed on these water samples suggest an association of these amoebae isolated from trough water points with mycobacteria.

Table 4. Characterization of environmental amoebae isolated from herd water samples.

Herd N°	Sample N°	Myco qPCR Signal *	T°	FLA Identification	Best Score ID	E Value
1	1	+	20	*Naegleria*	EU377592.1	0
			37	*Filamoeba*	GU320603.1	4×10^{-154}
			37	*Naegleria*	KC164228.1	0
	3	+	20	*Acanthamoeba*	GU001160.1	0
			37	*Naegleria*	KC164228.1	0
2	5	+	20	*Naegleria*	KC164228.1	8×10^{-156}
			37	*Naegleria*	KC164228.1	0
	6	+	20	*Naegleria*	KC164228.1	0
			37	*Naegleria*	KC164228.1	0
	7	+	37	*Vermamoeba*	KX856374.1	0
	8	+	20	Tubulinea group	FN562424.1	0
			37	*Naegleria*	KC164228.1	6×10^{-132}
3	9	+	37	*Acanthamoeba*	AM408796.1	0
	10	+	20	*Naegleria*	OQ034614.1	0
			37	*Naegleria*	AY266314.1	0
4	11	+	20	*Acanthamoeba*	AY351644.1	0
			37	*Naegleria*	AY576367.1	0
5	12	+	37	*Vermamoeba*	KX856373.1	0
	14	+	37	*Naegleria*	MG699123.1	0
6	18	+	37	*Naegleria*	KC164247.1	0
	19	+	37	*Vermamoeba*	LC431240.1	0
	20	+	20	*Acanthamoeba*	HF930501.1	0

* qPCR atpE (*Mycobacterium* spp.) performed on water samples.

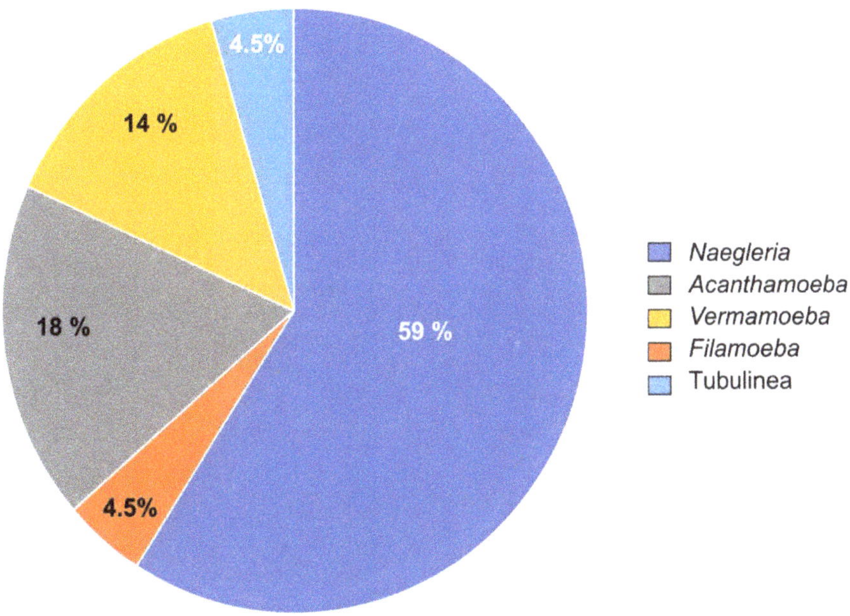

Figure 3. Pie chart of the frequency of isolated amoebic species.

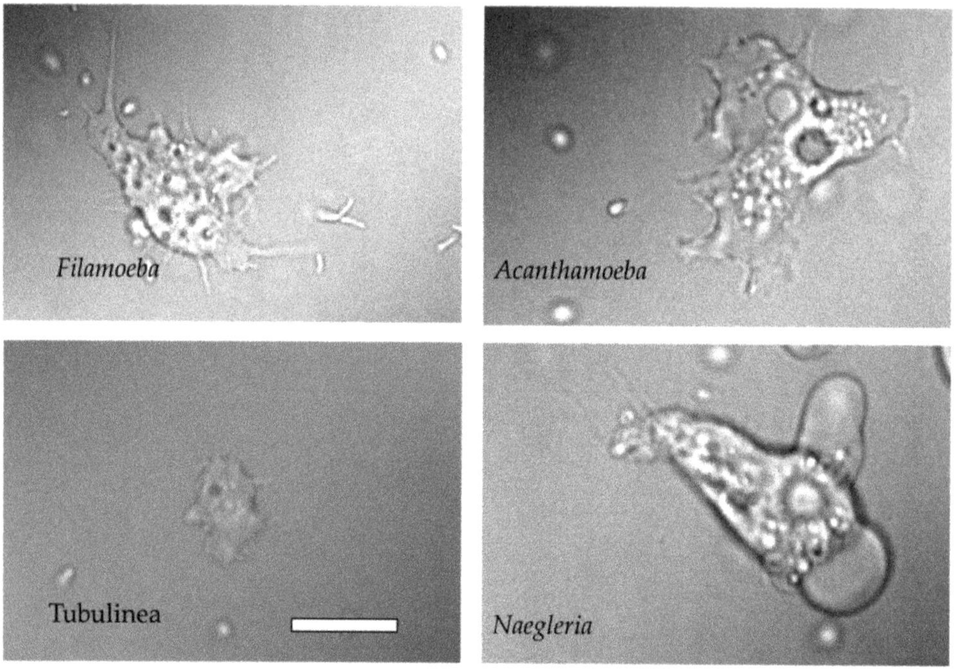

Figure 4. Morphological features highlighting the diversity of size and shape found in amoebae from the herd water sampling observed by differential interferential contrast microscopy. Bar length represents 10 μm for all panels.

4. Discussion

Livestock and wildlife spread pathogenic mycobacteria such as *Map* and *M. bovis* to pasture, soil, feed, water, and microfauna [24,25]. Field and experimental reports have indicated the persistence of these pathogenic mycobacteria under different environmental conditions representing a potential risk of infection [26].

The present study addressed the question of the presence of viable mycobacteria in the cattle environment, including environmental actors represented by free-living amoebae, protists frequently found in water and soils, to understand better the transmission routes of the disease among livestock and wildlife. For this purpose, we collected water samples from the environment of animals from six herds in southwest France. We looked for the presence of mycobacteria associated with the isolation of amoebae. The results of our study describe the presence, in all samples, of amoeba of the different genera related to positive qPCR signals of mycobacteria. Furthermore, the same samples revealed the presence of mycobacteria by qPCR and cultures, leading to the identification of different NTM species, including *Map*.

Detecting mycobacteria in environmental samples of waters containing highly diverse microbial communities, including several described and potentially undescribed NTMs [27], is always a challenge. That is why we chose two approaches, one based on detection by qPCR and the other based on the culture of mycobacteria. qPCR detection is a method of choice for mycobacteria, as it is a faster alternative to culture-based methods, hampered by the slow growth rate of mycobacteria and the high number and diversity of environmental species [28]. Additionally, PCR can detect mycobacterial DNA in environmental samples after initial contamination [29]. However, our results show that some samples, mainly from herds 1 and 2, did not show mycobacteria detection PCR signals, while at other sampling sites, we could detect mycobacteria signatures. These results could reflect different dilution rates of target mycobacteria in the samples or PCR inhibition by co-extracted contaminants. According to some studies, these effects could be minimized using an amplification facilitator (product of gene 32 of bacteriophage T4) [30]. However, there is still considerable work to be done to improve detection by PCR from complex environmental matrices.

The results of the isolation of mycobacteria by culture provided new and exciting data. Seven mycobacterial species were identified from the positive cultures. By contrast, for two sites, herds 3 and 4, we obtained neither a positive culture nor a PCR signal. This reflects the difficulty of working with field samples for which there may be hazards that we do not control. Moreover, bacteriological cultures of these samples were impossible due to fairly frequent fungal contamination despite using an antifungal cocktail. More intensive sampling or better selection could improve the isolation of mycobacteria from environmental matrices and amoebae. The seven mycobacterial species identified (*Mycobacterium hiberniae; Mycobacterium nonchromogenicum; Mycobacterium icosiumassiliensis; Map; Mycobacterium arupensis; Mycobacterium parascrofulaceum; Mycobacterium genavense*) are already well-documented NTMs that affect farm animals and wildlife. NTMs, of which 61 species have been described in publications, are also opportunistic pathogens capable of causing lymphadenitis and infections of the lungs, skin, soft tissues, bursa, joints, tendon sheath and bones [31]. NTM poses two main problems: interference with the detection of bTB and other major mycobacteriosis such as JD; and the potential to cause significant or opportunistic infections leading to considerable economic losses [32]. The present field study clearly illustrates the impact of NTM in the environment (water point) of cattle could have on herd management in JD and bTB endemic zones.

The study also showed that *Map* is frequently detected in PCR and could be isolated in cultures, in accordance with previous work that has demonstrated that *Map* is widely distributed in the environment [7]. These results may also reflect the worrisome epidemiological situation of JD in France, with very high herd prevalence (around 50% for cattle farms) [1,2]. The isolation of *Map* by culture represents a significant challenge, especially with environmental samples containing many other microorganisms (up to 10^{10} genomes per gram of soil) [28] and many culture inhibitors. To support a hypothesis of disease transmission via environmental matrices, it is essential to demonstrate the presence of living bacteria. The detection of bacteria by PCR can be significantly overestimated (studies advance a factor of 100 as compared to culture) [33] since the detection of DNA can result from the presence of DNA traces or dead bacteria or even from dormant bacteria, which have been specifically observed for *Map* [34]. Our results of environmental *Map* detection in which bTB is also present seem to confirm the recent report of Byrne et al. [5], showing robust associations between bTB herd breakdown episode risk and concurrent *Map* infections at the herd level.

Although we could detect MTBC signature (IS6110, IS1081 and IS1561 positive qPCR) in samples, no strain of the MTBC could be isolated in culture despite animals infected with *M. bovis*. This result could suggest DNA persistence is superior to cultivable bacteria, which is in agreement with the observations of Adams et al. [29]. Even though France became free of bTB in 2001, zoonosis at the interface between animals and humans remains endemic in certain areas where wildlife is also affected. One of the reasons for this would be direct transmission by inhalation or ingestion of infected environmental substances [25,35,36]. In a previous study, *M. bovis* could be detected by PCR, but not isolated in culture, in environmental samples from tuberculosis outbreaks [28]. It seems that, similarly to our study, viable isolation of *M. bovis* from the environment is difficult to obtain. Therefore, sampling should target very recent, heavily affected outbreaks.

It was in a study by Delafont et al. [37] that, for the first time, an association between amoebae and mycobacteria, more precisely NTM, was observed in drinking water network, thereby highlighting the importance of FLA in the ecology of NTMs. More recently, screening of environments surrounding *Map*-infected herds led to the recovery of a free-living amoeba harboring a *Map* strain from a drinking trough [18]. In the present study, following the same approach, we were able to isolate and identify different amoebal isolates from all sampled herds. The discovery of this species diversity enlarges our knowledge of the environmental amoebae in interactions with livestock. All in all, 22 amoebal strains were isolated from the environment. They are distributed in four genera (*Naegleria*, *Acanthamoeba*, *Vermamoeba* and *Filamoeba*) and one group (Tubulinea). The majority of isolated amoebae belong to the *Naegleria* genus.

Interestingly, in herds 1 and 2, seven isolates of *Naegleria* were similar, and the best BLAST score was the same (KC164228.1), suggesting that this isolate is highly represented in the area, which is not the case for the other herds. Furthermore, *Acanthamoeba* and *Vermamoeba* are also highly represented. This was unsurprising as these are the major amoeba genera described in environmental water [9,38]. Finally, isolates from these genera have been characterised in interaction with *Mycobacterium*, particularly NTM, mainly identified in ponds and river water [7], suggesting that it could also occur in our environment [12].

5. Conclusions

Our study showed that a wide diversity of amoebae and NTM species, including *Map*, live closely in water troughs in herds monitored for bovine tuberculosis or JD. The exact association between amoebae and mycobacteria cannot be readily determined with these data. However, we hypothesize that given the results obtained elsewhere, *Map*, other NTM and *M. bovis* could survive in the environmental matrices of cattle thanks to the amoebae capable of hosting them. Therefore, the control of bTB and JD and sanitation of herds must consider environmental sources where *Map* and *M. bovis* could survive with amoebae and infect cattle and wildlife. In addition, numerous NTMs exposed to cattle may explain the cross-reactions observed during the diagnosis of these diseases. Therefore, the frequent presence of *Map* potentially associated with amoebae appears as an underestimated obstacle for controlling JD.

Further research is required to understand the association more clearly between mycobacteria and hosting amoeba from field samples, which could help to explain their survival in a multiplicative form and to more closely study the pathobionts consisting of a large variety of amoebae and mycobacteria.

Author Contributions: Conceptualization, F.B., J.-L.M. and Y.H.; methodology, V.R., T.C., S.C., V.D. and J.-L.M.; formal analysis, resources, J.-L.M.; writing—original draft preparation, F.B. and Y.H.; writing—review and editing, all authors.; funding acquisition, F.B. and Y.H.; All authors have read and agreed to the published version of the manuscript.

Funding: This research was funded by the Ministry of Agriculture and Food Sovereignty, DGAL RSFA AMYCO, grant number 000001897.

Institutional Review Board Statement: Not applicable.

Informed Consent Statement: Not applicable.

Data Availability Statement: Not applicable.

Acknowledgments: In this section, you can acknowledge any support not covered by the author's contribution or funding sections. This may include administrative and technical support or donations in kind (e.g., materials used for experiments). We also thank Jeffrey Arsham for the English-language revision.

Conflicts of Interest: The authors declare no conflict of interest.

References

1. Whittington, R.; Donat, K.; Weber, M.F.; Kelton, D.; Nielsen, S.S.; Eisenberg, S.; Arrigoni, N.; Juste, R.; Saez, J.L.; Dhand, N.; et al. Control of paratuberculosis: Who, why and how. A review of 48 countries. *BMC Vet. Res.* **2019**, *15*, 198. [CrossRef]
2. Nielsen, S.S.; Toft, N. A review of prevalences of paratuberculosis in farmed animals in Europe. *Prev. Vet. Med.* **2009**, *88*, 1–14. [CrossRef] [PubMed]
3. Ott, S.L.; Wells, S.J.; Wagner, B.A. Herd-level economic losses associated with Johne's disease on US dairy operations. *Prev. Vet. Med.* **1999**, *40*, 179–192. [CrossRef] [PubMed]

4. Barkema, H.W.; Orsel, K.; Nielsen, S.S.; Koets, A.P.; Rutten, V.; Bannantine, J.P.; Keefe, G.P.; Kelton, D.F.; Wells, S.J.; Whittington, R.J.; et al. Knowledge gaps that hamper prevention and control of Mycobacterium avium subspecies paratuberculosis infection. *Transbound. Emerg. Dis.* **2018**, *65* (Suppl. 1), 125–148. [CrossRef]
5. Byrne, A.W.; Graham, J.; Milne, G.; Guelbenzu-Gonzalo, M.; Strain, S. Is There a Relationship Between Bovine Tuberculosis (bTB) Herd Breakdown Risk and Mycobacterium avium subsp. paratuberculosis Status? An Investigation in bTB Chronically and Non-chronically Infected Herds. *Front. Vet. Sci.* **2019**, *6*, 30. [CrossRef] [PubMed]
6. Field, N.L.; McAloon, C.G.; Gavey, L.; Mee, J.F. Mycobacterium avium subspecies paratuberculosis infection in cattle—A review in the context of seasonal pasture-based dairy herds. *Ir. Vet. J.* **2022**, *75*, 12. [CrossRef]
7. Rhodes, G.; Henrys, P.; Thomson, B.C.; Pickup, R.W. Mycobacterium avium subspecies paratuberculosis is widely distributed in British soils and waters: Implications for animal and human health. *Environ. Microbiol.* **2013**, *15*, 2761–2774. [CrossRef]
8. Allen, A.R.; Skuce, R.A.; Byrne, A.W. Bovine Tuberculosis in Britain and Ireland—A Perfect Storm? the Confluence of Potential Ecological and Epidemiological Impediments to Controlling a Chronic Infectious Disease. *Front. Vet. Sci.* **2018**, *5*, 109. [CrossRef] [PubMed]
9. Samba-Louaka, A.; Delafont, V.; Rodier, M.H.; Cateau, E.; Hechard, Y. Free-living amoebae and squatters in the wild: Ecological and molecular features. *FEMS Microbiol. Rev.* **2019**, *43*, 415–434. [CrossRef] [PubMed]
10. Molmeret, M.; Horn, M.; Wagner, M.; Santic, M.; Abu Kwaik, Y. Amoebae as training grounds for intracellular bacterial pathogens. *Appl. Environ. Microbiol.* **2005**, *71*, 20–28. [CrossRef]
11. Salah, I.B.; Ghigo, E.; Drancourt, M. Free-living amoebae, a training field for macrophage resistance of mycobacteria. *Clin. Microbiol. Infect.* **2009**, *15*, 894–905. [CrossRef] [PubMed]
12. Drancourt, M. Looking in amoebae as a source of mycobacteria. *Microb. Pathog.* **2014**, *77*, 119–124. [CrossRef] [PubMed]
13. Cirillo, J.D.; Falkow, S.; Tompkins, L.S.; Bermudez, L.E. Interaction of Mycobacterium avium with environmental amoebae enhances virulence. *Infect. Immun.* **1997**, *65*, 3759–3767. [CrossRef] [PubMed]
14. Whan, L.; Grant, I.R.; Rowe, M.T. Interaction between Mycobacterium avium subsp. paratuberculosis and environmental protozoa. *BMC Microbiol.* **2006**, *6*, 63. [CrossRef]
15. White, C.I.; Birtles, R.J.; Wigley, P.; Jones, P.H. Mycobacterium avium subspecies paratuberculosis in free-living amoebae isolated from fields not used for grazing. *Vet. Rec.* **2010**, *166*, 401–402. [CrossRef]
16. Mura, M.; Bull, T.J.; Evans, H.; Sidi-Boumedine, K.; McMinn, L.; Rhodes, G.; Pickup, R.; Hermon-Taylor, J. Replication and long-term persistence of bovine and human strains of Mycobacterium avium subsp. paratuberculosis within Acanthamoeba polyphaga. *Appl. Environ. Microbiol.* **2006**, *72*, 854–859. [CrossRef]
17. Salgado, M.; Alfaro, M.; Salazar, F.; Badilla, X.; Troncoso, E.; Zambrano, A.; Gonzalez, M.; Mitchell, R.M.; Collins, M.T. Application of cattle slurry containing Mycobacterium avium subsp. paratuberculosis (MAP) to grassland soil and its effect on the relationship between MAP and free-living amoeba. *Vet. Microbiol.* **2015**, *175*, 26–34. [CrossRef]
18. Samba-Louaka, A.; Robino, E.; Cochard, T.; Branger, M.; Delafont, V.; Aucher, W.; Wambeke, W.; Bannantine, J.P.; Biet, F.; Hechard, Y. Environmental Mycobacterium avium subsp. paratuberculosis Hosted by Free-Living Amoebae. *Front. Cell Infect. Microbiol.* **2018**, *8*, 28. [CrossRef]
19. Reveillaud, E.; Desvaux, S.; Boschiroli, M.L.; Hars, J.; Faure, E.; Fediaevsky, A.; Cavalerie, L.; Chevalier, F.; Jabert, P.; Poliak, S.; et al. Infection of Wildlife by Mycobacterium bovis in France Assessment Through a National Surveillance System, Sylvatub. *Front. Vet. Sci.* **2018**, *5*, 262. [CrossRef]
20. Lefrancois, L.H.; Bodier, C.C.; Cochard, T.; Canepa, S.; Raze, D.; Lanotte, P.; Sevilla, I.A.; Stevenson, K.; Behr, M.A.; Locht, C.; et al. Novel feature of Mycobacterium avium subsp. paratuberculosis, highlighted by characterization of the heparin-binding hemagglutinin adhesin. *J. Bacteriol.* **2013**, *195*, 4844–4853. [CrossRef]
21. Lefrancois, L.H.; Cochard, T.; Branger, M.; Peuchant, O.; Conde, C.; Pastuszka, A.; Locht, C.; Lanotte, P.; Biet, F. Feature of Adhesins Produced by Human Clinical Isolates of Mycobacterium intracellulare, Mycobacterium intracellulare subsp. chimaera and Closely Related Species. *Microorganisms* **2020**, *8*, 1154. [CrossRef] [PubMed]
22. Hadziavdic, K.; Lekang, K.; Lanzen, A.; Jonassen, I.; Thompson, E.M.; Troedsson, C. Characterization of the 18S rRNA gene for designing universal eukaryote specific primers. *PLoS ONE* **2014**, *9*, e87624. [CrossRef] [PubMed]
23. Huard, R.C.; Lazzarini, L.C.; Butler, W.R.; van Soolingen, D.; Ho, J.L. PCR-based method to differentiate the subspecies of the Mycobacterium tuberculosis complex on the basis of genomic deletions. *J. Clin. Microbiol.* **2003**, *41*, 1637–1650. [CrossRef] [PubMed]
24. Barasona, J.A.; Torres, M.J.; Aznar, J.; Gortazar, C.; Vicente, J. DNA Detection Reveals Mycobacterium tuberculosis Complex Shedding Routes in Its Wildlife Reservoir the Eurasian Wild Boar. *Transbound. Emerg. Dis.* **2017**, *64*, 906–915. [CrossRef]
25. Phillips, C.J.; Foster, C.R.; Morris, P.A.; Teverson, R. The transmission of Mycobacterium bovis infection to cattle. *Res. Vet. Sci.* **2003**, *74*, 1–15. [CrossRef] [PubMed]
26. Allen, A.R.; Ford, T.; Skuce, R.A. Does Mycobacterium tuberculosis var. bovis Survival in the Environment Confound Bovine Tuberculosis Control and Eradication? A Literature Review. *Vet. Med. Int.* **2021**, *2021*, 8812898. [CrossRef]
27. Falkinham, J.O., 3rd. Surrounded by mycobacteria: Nontuberculous mycobacteria in the human environment. *J. Appl. Microbiol.* **2009**, *107*, 356–367. [CrossRef]

28. Barbier, E.; Boschiroli, M.L.; Gueneau, E.; Rochelet, M.; Payne, A.; de Cruz, K.; Blieux, A.L.; Fossot, C.; Hartmann, A. First molecular detection of Mycobacterium bovis in environmental samples from a French region with endemic bovine tuberculosis. *J. Appl. Microbiol.* **2016**, *120*, 1193–1207. [CrossRef]
29. Adams, A.P.; Bolin, S.R.; Fine, A.E.; Bolin, C.A.; Kaneene, J.B. Comparison of PCR versus culture for detection of Mycobacterium bovis after experimental inoculation of various matrices held under environmental conditions for extended periods. *Appl. Environ. Microbiol.* **2013**, *79*, 6501–6506. [CrossRef]
30. Opel, K.L.; Chung, D.; McCord, B.R. A study of PCR inhibition mechanisms using real time PCR. *J. Forensic Sci.* **2010**, *55*, 25–33. [CrossRef]
31. Falkinham, J.O., 3rd. Epidemiology of infection by nontuberculous mycobacteria. *Clin. Microbiol. Rev.* **1996**, *9*, 177–215. [CrossRef] [PubMed]
32. Biet, F.; Boschiroli, M.L. Non-tuberculous mycobacterial infections of veterinary relevance. *Res. Vet. Sci.* **2014**, *97*, S69–S77. [CrossRef] [PubMed]
33. Betelli, L.; Duquenne, P.; Grenouillet, F.; Simon, X.; Scherer, E.; Gehin, E.; Hartmann, A. Development and evaluation of a method for the quantification of airborne Thermoactinomyces vulgaris by real-time PCR. *J. Microbiol. Methods.* **2013**, *92*, 25–32. [CrossRef] [PubMed]
34. Whittington, R.J.; Marsh, I.B.; Reddacliff, L.A. Survival of Mycobacterium avium subsp. paratuberculosis in dam water and sediment. *Appl. Environ. Microbiol.* **2005**, *71*, 5304–5308. [CrossRef] [PubMed]
35. Palmer, M.V.; Waters, W.R.; Whipple, D.L. Investigation of the transmission of Mycobacterium bovis from deer to cattle through indirect contact. *Am. J. Vet. Res.* **2004**, *65*, 1483–1489. [CrossRef]
36. Vicente, J.; Hofle, U.; Garrido, J.M.; Fernandez-de-Mera, I.G.; Acevedo, P.; Juste, R.; Barral, M.; Gortazar, C. Risk factors associated with the prevalence of tuberculosis-like lesions in fenced wild boar and red deer in south central Spain. *Vet. Res.* **2007**, *38*, 451–464. [CrossRef]
37. Delafont, V.; Mougari, F.; Cambau, E.; Joyeux, M.; Bouchon, D.; Hechard, Y.; Moulin, L. First evidence of amoebae-mycobacteria association in drinking water network. *Environ. Sci. Technol.* **2014**, *48*, 11872–11882. [CrossRef]
38. Rodriguez-Zaragoza, S. Ecology of free-living amoebae. *Crit. Rev. Microbiol.* **1994**, *20*, 225–241. [CrossRef]

animals MDPI

Article

An ELISA Using Synthetic Mycolic Acid-Based Antigens with DIVA Potential for Diagnosing Johne's Disease in Cattle

Paul S. Mason [1], Thomas Holder [2], Natasha Robinson [3], Brendan Smith [4], Rwoa'a T. Hameed [5,6], Juma'a R. Al Dulayymi [5], Valerie Hughes [7], Karen Stevenson [7], Gareth J. Jones [2], H. Martin Vordermeier [2], Shawn Mc Kenna [3] and Mark S. Baird [1,5,*]

1 Diagnostig Ltd., M-SParc, Gaerwen, Anglesey LL60 6AG, Wales, UK; mason@diagnostig.com
2 Animal and Plant Health Agency, Addlestone KT15 3NB, Surrey, UK; thomas.holder@apha.gov.uk (T.H.); gareth.jones@apha.gov.uk (G.J.J.); martin.vordermeier@apha.gov.uk (H.M.V.)
3 Department of Health Management, Atlantic Veterinary College, Charlottetown, PE C1A 4P3, Canada; ngrobinson@upei.ca (N.R.); slmckenna@upei.ca (S.M.K.)
4 Bimeda Animal Health Ltd., Airton Close, Airton Road, Tallaght, D24 FH9V Dublin, Ireland; bsmith@bimeda.com
5 School of Natural and Environmental Sciences, Bangor University, Bangor LL57 2UW, Wales, UK; rwoaatareq@uomosul.edu.iq (R.T.H.); jumaaldulayymi@gmail.com (J.R.A.D.)
6 College of Science, University of Mosul, Mosul 41002, Iraq
7 Moredun Research Institute, Penicuik EH26 0PZ, Midlothian, UK; val.hughes@moredun.ac.uk (V.H.)
* Correspondence: baird@diagnostig.com; Tel.: +44-7984203180

Simple Summary: Johne's disease, caused by *Mycobacterium avium* subsp. *paratuberculosis* (MAP), is endemic in cattle herds in the UK and throughout the world. It is a wasting disease with a protracted incubation period, during which clinical signs are not apparent, and is responsible for considerable economic losses. Infection can be transmitted silently from animal to animal, and MAP can find its way into the food chain through the products of infected animals. Ante-mortem diagnosis of the disease is normally achieved through detection of the organism (by faecal culture or PCR) or serological tests, but agreement as to a positive diagnosis is often poor and the certainty of identifying all positive animals is low, particularly in early-stage infections. Mycobacterial cells contain high-molecular-weight lipids, mycolic acids (MAs), which are recognised by antibodies in active infections. Here, we show that a simple assay using different classes of single synthetic MAs can detect antibodies in serum from cattle with MAP infections and distinguish uninfected animals from animals positive both to faecal culture and a commercial serodiagnostic assay. It does so without interference from vaccination either for MAP (a Gudair vaccination) or for bovine TB (BCG vaccination). When the assay is optimised to distinguish negative samples from those positive only by faecal PCR, it identifies additional positives that are not seen using a commercial serodiagnostic assay; this suggests that, with further development, it may provide a method for detecting infections before the commercial serodiagnostic assay can.

Abstract: The problem: Ante-mortem diagnosis of Johne's disease, caused by *Mycobacterium avium* subsp. *paratuberculosis* (MAP), is normally achieved through faecal culture, PCR, or serological tests, but agreement as to which samples are positive for Johne's disease is often poor and sensitivities are low, particularly in early-stage infections. The potential solution: Mycobacterial cells contain very complex characteristic mixtures of mycolic acid derivatives that elicit antibodies during infection; this has been used to detect infections in humans. Here, we explore its application in providing an assay differentiating infected from vaccinated animals (DIVA assay) for Johne's disease in cattle. Method: Antibody responses to different classes of mycolic acid derivatives were measured using ELISA for serum from cattle positive for MAP by both faecal PCR and commercial serum ELISA, or just by PCR, and from animals from herds with no history of Johne's disease, bovine tuberculosis reactors, BCG-vaccinated, BCG-vaccinated and *M. bovis*-infected, and Gudair-vaccinated animals. Results: The best-performing antigens, ZAM295 and ST123—the latter a molecule present in the cells of MAP but not of *Mycobacterium bovis*—achieved a sensitivity of 75% and 62.5%, respectively, for serum from animals positive by both faecal PCR and a commercial MAP serum ELISA, at a specificity of 94% compared to 80 no-history negatives. Combining the results of separate assays with two antigens (ST123 and JRRR121) increased the sensitivity/specificity to 75/97.5%. At the same cut-offs, animals vaccinated with Gudair or BCG vaccines and bTB reactors showed a similar specificity. The specificity in BCG-vaccinated but *M. bovis*-infected animals dropped to 85%. Combining the results of two antigens gave a sensitivity/specificity of 37.5/97.5% for the full set of 80 PCR-positive samples, detecting 30 positives compared 16 for IDEXX. Conclusion: Serum ELISA using synthetic lipids distinguishes effectively between MAP-negative cattle samples and those positive by both PCR and a commercial MAP serodiagnostic, without interference by Gudair or BCG vaccination. It identified almost twice as many PCR positives as the commercial serodiagnostic, offering the possibility of earlier detection of infection.

Keywords: MAP; mycolic acids; serodiagnosis; TDM; TMM; GMM; DIVA; lipid ELISA; *Mycobacterium avium* subspecies *paratuberculosis*; Johne's disease; vaccine

1. Introduction

Mycobacterium avium subsp. *paratuberculosis* (MAP) causes disease in ruminants, camelids, and leporids and can infect a wide range of animals [1–6]. MAP causes Johne's disease in cattle, characterised by weight loss and diarrhoea and is associated with lower economic yields in infected animals. The organism is present in the milk of infected animals and may be incompletely inactivated by pasteurisation. There appear to be increased instances of the presence of MAP in patients with Crohn's disease, although causality has not been proven. Although Johne's disease is not reportable in Great Britain and many other countries, it is notifiable in some countries and in certain species. There is growing pressure to ensure food products are from 'Johne's-free' animals [1].

The organism is primarily transferred via ingestion of infected milk or food and water contaminated with faeces from infected animals. It can survive for long periods in the environment, 6–18 months in water and 5–8 months on pasture [7]. Not all animals exposed to MAP become infected or go on to develop disease. The disease develops slowly in cattle and there are a number of models of disease development. In one, three stages are described: (i) subclinical infection, defined as MAP infection without demonstrable pathology in tissues; (ii) subclinical disease, defined as the presence of pathology without weight loss or diarrhoea; (iii) clinical disease, defined as the presence of pathology with weight loss and/or diarrhoea [7]. Gross pathology, and more specifically histopathology, provides the most definitive diagnosis of the disease stages defined above, and since it can only be determined post-mortem, differentiating subclinical infection and subclinical disease using ante-mortem tests is not possible. Thus, there are no routine, reliable methods for the ante-mortem diagnosis of early-stage infection [8], and later stages of disease are normally diagnosed by either faecal culture or PCR, or a serodiagnostic assay detecting disease antibodies. Elevated MAP antigen induced interferon (IFN)-γ production in subclinically infected animals has been used in diagnosis [9], but this cytokine declines in the clinical stage [10].

Detection of the organism in faeces by culture or PCR is often used as evidence of infection. However, while it is certainly evidence of exposure, it may not always be indicative of infection but rather the result of the passage of contaminated material through the gut (passive shedding) [11]. Furthermore, infected animals may exhibit intermittent shedding and may return to a more active phase of disease with frequent shedding, possibly triggered by stress during lactation or parturition. Whatever the trigger, it has been suggested that the progression from subclinical to clinical disease coincides with a shift in immune function from a Th1 to a Th2 response. Because there are no other widely used, reliable methods for the detection of early infection, the balance of these possibilities remains unclear. There is therefore often a poor correlation between faecal culture or

PCR and a serodiagnostic assay, particularly in early-stage disease [7,8,12,13]. The current widely used commercial serodiagnostic assays generally provide excellent specificity, but recommended cut-offs often lead to very low sensitivities, certainly compared to culture or PCR. Moreover, specificity may be low in geographical regions where animals are exposed to environmental mycobacteria [14]. Although predictive models have been developed [15,16], the use of faecal PCR on its own must also be treated with caution, and the recommended maximum number of quantification cycles that can safely be interpreted as a positive diagnosis was reduced in one recent study from <37 to <33, with any value from 33 to 40 potentially being a false positive and requiring supporting evidence [17]. It is worth noting that false results could be caused by inhomogeneity of sampling, by passive shedding, by intermittent shedding, or by changes in PCR quantification cycles when appropriate controls are not applied.

A major challenge in controlling Johne's disease is therefore the ability to detect infected cattle prior to appearance of disease signs, such as diarrhoea and heavy faecal shedding of MAP. Cattle that appear healthy may not be routinely evaluated using serial test bleeds and analysis unless they are part of a cattle health scheme. An assay that detects responses in blood or serum at an earlier point in disease development, and one that can help to determine whether a greater number of positive PCR do indeed correspond to disease and not exposure and direct pass-through, is therefore highly desirable [18].

A further complication is the need for diagnostic methods the results of which are not affected by vaccination for MAP [19,20], or by developing vaccines for bovine TB [21], and indeed by the statutory skin test for bovine tuberculosis (bTB).

Like other mycobacteria, the cells of MAP contain complex mixtures of very long chain hydroxy-fatty acids, mycolic acids (MA) (Figure 1), and of their sugar esters, in particular trehalose mono- and di-mycolates (TMM and TDM) [22]. The natural mixtures comprise molecules with a variety of groups X and Y and chain lengths a–d. The detailed pattern of these molecules forms a fingerprint which can be used directly to identify a particular mycobacterium, for example by mass spectrometry. In *Mycobacterium tuberculosis* and *Mycobacterium bovis* (*M. bovis* throughout this manuscript), the MAs are largely from three classes, alpha (X includes a cyclopropane), keto (X includes a ketone with an adjacent methyl branch), and methoxy (X includes a methoxy group with an adjacent methyl branch); commonly, group Y includes a cyclopropane fragment. In *Mycobacterium avium* complex (MAC), the keto class is replaced by 'wax esters' containing an additional oxygen in group X [23]. Other non-tuberculous mycobacteria (NTM) often contain much simpler groups X and Y containing either *cis*- or *trans*-alkenes. Within an infection such as tuberculosis (TB), the balance of specific MA classes (and the stereochemistry within the classes) is known to change with disease progression [22].

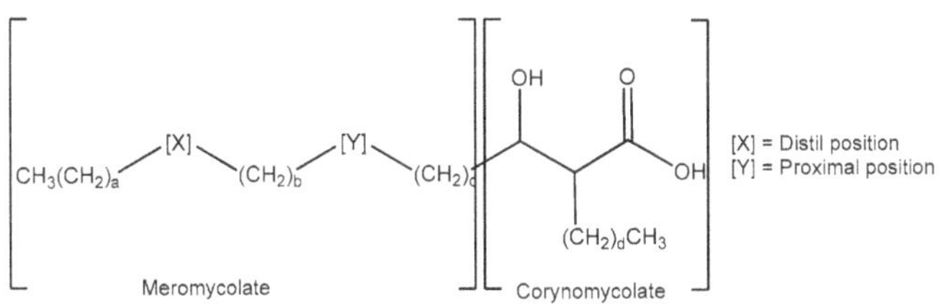

Figure 1. The generic structure of mycolic acids (MAs).

Mycolic acid and mycolate ester extracts of cells of *Mycobacterium tuberculosis* are antigenic to antibodies in the serum of patients with active tuberculosis. They have therefore been evaluated in the diagnosis of the disease, using ELISA [24–26]. In the same way, complex mixtures of trehalose mycolates extracted from the cells of *Mycobacterium avium* can be used to distinguish MAC infection from MTb infection [27]. Single synthetic mycolic acids and their sugar esters have also been used in TB diagnosis; in these studies, a multiplex assay was used in which the results of individual assays with several antigens were combined statistically to optimise sensitivity and specificity [28]. A similar method has been used in the diagnosis of infections by *Mycobacterium abscessus* in immune-compromised patients [29].

We now report the evaluation of unique synthetic mycolic acids and their sugar esters as antigens in ELISA for the detection of antibodies generated to these lipids in MAP infection in cattle, including control animals, and those that had been vaccinated for bTB or for MAP. A key aim was to determine whether specific antigens or antigen combinations would be recognised by antibodies to mycolic acid-based lipids, and if so, whether the responses observed could be correlated with some PCR data better than those to standard antigens in currently used serodiagnostics. This might then allow for the confirmation of developing disease at an earlier time and represent the initial stage of the rigorous process of evaluation of an improved diagnostic assay [30]. Early diagnosis has also been the target for serodiagnostics using arrays of protein antigens [31].

2. Materials and Methods

2.1. Serum Samples

Cattle serum samples were provided either from Atlantic Veterinary College, Prince Edward Island (PEI) in Canada or from Moredun Research Institute (MRI) or Animal and Plant Health Agency (APHA) in the UK.

2.1.1. Positive Serum Samples

Faecal PCR-positive serum samples (n = 80) were provided by PEI. Serum samples were from animals from MAP-infected herds. Animals were considered infected with MAP if they tested positive by faecal PCR and/or a commercial paraTB ELISA manufactured by IDEXX. Of 80 animals, 16 were positive by both faecal PCR and IDEXX paraTB ELISA (PEI++ cohort). The remaining 64 samples were faecal PCR-positive but IDEXX ELISA-negative (PEI+− cohort). In this study, the whole herd was first screened by faecal PCR and positive samples were then tested by IDEXX ELISA so no samples in this set were IDEXX ELISA-positive and PCR negative. The PCR quantification cycles for all these samples were within the limits set by the commercial assay used (see below) and these and the IDEXX ELISA results are presented in Supplementary Table S1.

2.1.2. Negative Serum Samples

(i) From Prince Edward Island (PEI) (n = 80)

Two sets of 40 sera from negative animals from two farms (F5 and F6) were provided by PEI. Both farms had participated in annual herd testing for at least 10 years with no history of infection. These animals were faecal PCR-negative and IDEXX ELISA-negative (PEI−−). Animals from herd F5 were kept outdoors on pasture during suitable times of the year; those from herd F6 were housed exclusively indoors year-round. The PCR and IDEXX ELISA results for all the negative samples are presented in Supplementary Table S1.

(ii) From Moredun Research Institute (MRI) (n = 9)

A further set of negative sera were provided by MRI. The animals providing the serum samples were from herds participating in the national paratuberculosis control programme (CHECS-Technical-document-Johnes-disease-rule-change-released-220621-date-01-11-2021.pdf, accessed on 28 December 2023) [32] and the herd had consistently tested negative for MAP by serological testing using ID Screen Paratuberculosis Indirect ELISA (Innovative Diagnostics, Grabels France) for at least the previous 3 years. The farm had no history of bTB and is located in Scotland, which is designated an 'Officially bovine Tuberculosis-Free' (OTF) region.

(iii) From Animal and Public Health Authority (APHA) (n = 20)

Five sets of samples were provided by APHA from field animals from a number of farms:

(a) Serum Samples from Skin-Test (SICCT)-Negative Animals (n = 20)

Serum samples from 20 SICCT-negative animals were provided by APHA (Control 1–20). Serum samples were taken 10 days post skin test.

(b) Serum samples from Skin-Test (SICCT)-Positive Animals (bTB reactors) (n = 20)

Serum samples from Skin-Test (SICCT)-Positive Animals were provided by APHA (Reactor 1–20). These animals had a natural *M. bovis* infection, which was subsequently confirmed by both the commercial Interferon-gamma Release Assay (IGRA) and post-mortem examination. Serum samples were taken 8 weeks after SICCT.

(c) Serum Samples from BCG-Vaccinated Animals (n = 20)

Serum samples from BCG-vaccinated (but not SICCT-tested) animals were provided by APHA (BCG Vac 1–20). Animals were inoculated with 1–4 $\times$ 10^6 CFU of BCG SSI via the subcutaneous route. Serum samples used in this study were taken 6 weeks post vaccination.

(d) Serum Samples from Gudair-Vaccinated Animals (n = 20)

Serum samples from Gudair-vaccinated animals were provided by APHA (Gudair 1–20). Animals received 2 ml Gudair vaccine (CZ Veterinaria, Spain) via the subcutaneous route. Serum samples were taken 8 weeks post vaccination.

(e) Serum Samples from BCG-Vaccinated, *M. bovis*-infected Animals (n = 20)

Serum samples from BCG-vaccinated, *M. bovis*-infected animals were provided by APHA (BCG Vac-Inf 1–20). These animals were BCG-vaccinated as above, and then experimentally infected 10 weeks post vaccination with *M. bovis* strain AF2122/97. Serum samples used in this study were taken 22 weeks post vaccination.

(iv) Longitudinal negatives from APHA: 90 serum samples from 10 IDEXX Para TBserum ELISA negative and *M. bovis*-free animals over 9 time points.

These animals had been imported from a bTB-free country at the age of 35–50 days. Serial serum samples were taken at nine timepoints over a period of 47 weeks. Serum samples tested consistently negative by IDEXX ParaTB serum ELISA (Supplementary Table S2) and within assay limits by Bovigam IGRA for *M. bovis* infection, though the PPDA values for this assay did show some increased values (Supplementary Table S2).

2.2. Antigens

Six antigens were evaluated in this work. Their structures are presented in Supplementary Figure S1. Antigen MOD171 is based on a cell-wall-bound mycolic acid fragment, a triarabinose dimycolate, incorporating a major α-mycolic acid present in many mycobacteria including MAC [33]. ZAM295 is a trehalose dimycolate of a diene-mycolic acid, a subgroup commonly present in NTM [34]. ST123 is a trehalose dimycolate of a wax ester; this class of molecule is present in MAC, replacing the keto-mycolic acid class present in many other mycobacteria [22,35]. SMP70 is a glucose monomycolate of a common methoxymycolic acid [36]. RT237F2 is an ester of a major α-mycolic acid with an amino-sugar; it is not a known component of mycobacteria, but a simpler analogue of a mycolyl muramyl dipeptide [37]. JRRR121 is a free methoxy-mycolic acid, incorporating a *trans*-cyclopropane fragment; such *trans*-cyclopropanes are major components of the methoxy-mycolic acid class present in MAC [38].

2.3. Faecal PCR

DNA was extracted from 2 g of frozen faecal sample using the MAP Extraction System TC-9014-100 (Tetracore, Rockville, MD, USA) according to manufacturer's instructions. Real-time PCR was carried out using the VetAlert Johne's Real-Time PCR kit TC-9828-100 (Tetracore, Rockville, MD, USA) in accordance with manufacturer instructions. The qPCR was performed in a Cepheid SmartCycler II Thermocycler (Cepheid, Sunnyvale,

CA, USA). After an enzyme activation step (95 °C), a 2-step cycling reaction (95 °C and 62 °C) was used. The cut-off value for the positive control was set to between 20 and 26 cycle threshold values. Samples that were positive within 42 cycles were considered MAP positive as recommended in this assay. Of the 16 PEI++ samples (positive to both PCR and IDEXX), 12 gave quantification cycles below 31, and the highest count was 36.2; none of the 64 PEI+− samples (positive to PCR, negative in IDEXX) gave quantification cycles below 33.

2.4. IDEXX ELISA

The IDEXX MAP Ab ELISA test was performed according to the manufacturer's instructions using a recommended cut-off of 0.55 for a positive.

2.5. Lipid ELISA

Responses in ELISA were measured using six individual synthetic lipid antigens, comprising different classes of mycolic acid and sugar ester, as described in Section 2.2.

The ELISA was carried out on 96-well microplates and the purified antigens were dissolved in THF/n-hexane (1:50, 5 mL) at a concentration of 5.6 μM; 50 μL of this antigen solution was placed in each well. The plates were left to dry at room temperature overnight. In order to reduce non-specific hybridisations, the plates were blocked with casein/PBS containing 0.5% *w*/*v* casein (300 μL/well) and left to incubate at 25 °C for 30 min. The solution was aspirated using the LT-3500 plate washer and the plates flicked dry. The serum (1:40 dilution in casein/PBS) was added to the plate (50 μL/well) and left to incubate at 25 °C for 1 h. The serum was aspirated and the plates washed three times with casein/PBS (400 μL/well) using the same plate washer and the plates were again flicked dry. The secondary antibody (Peroxidase-AffiniPure Goat Anti-Bovine IgG, Stratech Fc; 1:1000 dilution in casein/PBS) was added to the plate (50 μL/well) and left to incubate at 25 °C for 30 min. The plates were washed three times again with casein/PBS (400 μL/well) and dried. OPD substrate (0.2 mg/mL) was added and the plates (50 μL/well) incubated at 25 °C for 30 min; the reaction was then stopped by the addition of 2.5 M H_2SO_4 (50 μL/well) and the absorbances of each well were read straight away at 492, 450, and 620 nm using an LT-4000 plate reader; the values at 492 nm were used for the current analysis.

Typically, two replicate samples were analysed. The average optical densities (492 nm) of the replicates and their standard deviations were recorded. On each plate, a common pooled positive control (using samples from animals positive to both faecal PCR and IDEXX ELISA) and a PBS/casein negative control were run alongside test serum samples. Whenever a pooled positive control had to be updated, it was standardised against the existing control. It was also compared to a commercial positive standard obtained from GD Animal Health. The positive control was used to normalise optical density reading on different plates by converting OD readings for individual wells into a percentage of the pooled positive response. Average optical densities were converted into percentage of the positive control for each plate and are recorded in Supplementary Tables S1–S3.

2.6. Statistical Analyses

The current data were analysed in two ways. In the first, the PEI++ samples were used as the positive set as those best corroborated as positive using current assays. In the second, the data were analysed to provide the best fit to the complete set of 80 PCR positive samples.

The significance of differences between the data for each cohort of samples compared either to the PEI++ or PEI+ samples was determined using the Tukey test. In each case, optimal cut-offs for each antigen were initially determined by ROC analysis using the 80 F5 and F6 as negatives. These cut-offs were then applied to analyse the other sets of samples. With some antigens, this gave a low specificity unsuitable for Johne's diagnosis; in these cases, cut-offs were determined to provide the maximum accuracy (average of sensitivity and specificity) for positive/negative distinction, except that a minimum specificity of 93% was applied, as described in Table 1.

Table 1. Median responses for each set of samples with each of six antigens as % positive control for that antigen.

	Antigen or Antigen Combination Using R	MOD 171	JRRR 121	RT237 F2	SMP 70	ZAM 295	ST 123	ST123 + ZAM295	ST123 + JRRR121	ST123 + JRRR121 + ZAM295	JRRR121 + SMP70 + ZAM295	All Six Antigens
		% Pooled Pos						Probability of positive (0 to 1) with antigen combinations				
1	PEI++ n = 16 median	27.3	38.8	56.5	88.4	96.3	79.0	0.9	0.9	0.9	0.9	0.8
2	F5 plus F6 negative, n = 80 median	24.7	17.4	48.8	54.2	50.7	23.7	0.1	0.1	0.1	0.1	0.1
3	p * (F5 plus F6 compared to PEI++)	ns *	<0.0001	ns	<0.0001	<0.0001	<0.0001	<0.0001	<0.0001	<0.0001	<0.0001	<0.0001
		Sensitivity analysis for single antigens						**Sensitivity/specificity analysis for antigens combined in R**				
4	**Optimal cut-off for >93% specificity**	57	40	76	93	91	68	0.8	0.63	0.73	0.63	0.78
5	Optimal sensitivity %	25.0	50.0	6.3	37.5	75.0	62.5	56.3	75.0	68.8	68.8	68.8
6	Optimal specificity %	95.0	96.3	93.8	95.0	93.8	93.8	93.8	97.5	97.5	97.5	100.0
7	Accuracy % (sens + spec)/2	60.0	73.1	50.0	66.3	84.4	78.1	75.0	86.3	83.1	83.1	84.4
	Data for PEI++ relative to just F5 or F6 at cut-offs for above sensitivity											
8	F5 Specificity %	97.5	92.5	92.5	95	95	97.5	97.5	100	97.5	100	100
9	p (F5 compared to PEI++)	<0.05	<0.0001	ns	<0.0001	<0.0001	<0.0001	<0.0001	<0.0001	<0.0001	<0.0001	<0.0001
10	F6 Specificity %	92.5	100	95	95	92.5	90	90	95	97.5	95	100
11	p (F6 compared to PEI++)	ns	<0.0001	ns	<0.01	<0.0001	<0.0001	<0.0001	<0.0001	<0.0001	<0.0001	<0.0001

Median responses for each set of samples with each of six antigens as % positive control for that antigen. Optimal cut-offs are based on 16 PEI PCR- and IDEXX-positive samples (PEI++) compared to all 80 PEI F5 and F6 negatives (again as % positive control), and sensitivity and specificity of each group are based on those cut-offs. In some cases, such as for MOD171, the optimal cut-off by ROC analysis led to a very low specificity; cut-offs were then identified to give maximum accuracy (average of sensitivity and specificity) at >93% specificity. The data for F5 and F6 samples are also presented separately, using the same cut-offs. The final five columns show the median calculated negative or positive status (on 0 to 1 scale) of PEI++ compared to all 80 F5 and F6 negative samples, combining results with several specified antigens using R statistics; the calculated values for each serum sample have then been analysed by ROC and optimal cut-offs determined to provide the highest accuracy (average of sensitivity and specificity). * Significance values were calculated using the Tukey test; ns = not significant.

In each case, the results for various combinations of individual antigens were combined using the R statistics package, identifying the serum samples as known negative or positive sets and generating the best statistical +/− separation. This separated the results along two axes, resulting in a two-dimensional scatter plot. It also provided a prediction of negative or positive status of each sample on a scale from 0 (negative) to 1 (positive). The optimal cut-off for this predicted status was again determined as that giving the maximum accuracy, but with a minimum 93% specificity (Table 1). The resulting combination of responses to various antigens, and the cut-offs established were then applied to predict the status of any other sample or sample set. Box-and-whisker plots were produced in the R statistics package.

3. Results

The ELISA results for all the samples with each of the six antigens are presented as % positive control in Supplementary Tables S1–S3.

3.1. *Distinguishing PEI PCR- and IDEXX-Positive Samples and MAP Negatives*

Initial analysis was carried out comparing the responses for the 16 PEI++ serum samples with those for the 80 PEI 'no-history' negative serum samples (F5 plus F6). Table 1 shows the medians for the PEI++ sample set, together with those for the combined and separate F5 and F6 PEI−− sets with each antigen and the significance values for the differences between the sets, together with optimal combinations of specificity and sensitivity. The final five columns show the corresponding results when the data for individual serum samples and defined antigens were combined using the R statistics package, optimising the positive/negative separation for the 16 PEI++ sera compared to all 80 negatives. Further information, including ROC analysis for each antigen or antigen combination, is provided in Supplementary Table S4 and Figure S2. The distributions of responses for each cohort of cattle are shown as box-and-whisker plots in Figure S3.

Antigens MOD171, a model mycobacterial cell-wall fragment, and RT237F2, a model amino-sugar mycolate, gave only a very low sensitivity at >90% specificity. Compared with the complete set of 80 PEI−−-negative sera, the free mycolic acid JRRR121, and glucose mycolate SMP70 gave an optimal specificity and sensitivity of 96/50 and 95/38, respectively, but trehalose dimycolates ZAM295 and ST123 both gave higher sensitivity/specificity combinations (94/75 and 94/63), respectively. The results for each serum sample with sets of antigens were combined using the R statistics package; this gave a specificity of 97.5% and sensitivity of 75% with two antigens (ST123 and JRRR121) and 100/69 if all six antigens were combined.

3.2. *Optimising the Distinction of All PCR-Positive Samples from PEI MAP Negatives*

Given that the commercial serodiagnostic only gave a positive response with 16 of 80 PCR-positive samples, it was important to determine whether antibodies to mycolic acid-based lipids could be detected in more of the PCR-positive sera. The responses to the lipid antigens (Supplementary Table S1) were therefore re-analysed compared to the complete set of PCR positives. The optimal cut-offs for each antigen determined in this way and the optimal antigen combinations derived in R to provide the best distinction between the complete set of 80 PEI PCR-positive sera (PEI++ and PEI+− samples) and the 80 F5 and F6 'no-history' negatives are given in Table 2. This shows the sensitivity for the 80 PCR-positive animals for each antigen, the specificity in each case being set to optimise the accuracy of match to PCR. The final two rows show the number of serum samples with each antigen or antigen combination at a fixed specificity of 97.5%. In this case, sensitivity is the percentage of all 80 PCR positive samples predicted to be positive in the lipid assay, and specificity is the percentage of no-history negatives identified as negative in the lipid assay (thus, for reference, the commercial IDEXX assay has a value of 20/100). The results with individual antigens were then combined in R so as to optimise the distinction between 80 PCR positives and 80 PCR-negative F5 and F6 samples.

Table 2. Optimisation of results from lipid assay to predict status all 80 PEI PCR-positive samples compared to 80 F5 and F6 negatives, with single antigens and combinations of antigens in R.

		Antigen						Antigen Combination					
		MOD 171	**JRRR 121**	**RT 237 F2**	**SMP 70**	**ZAM 295**	**ST 123**	**ST123 + ZAM295**	**ST123 + JRRR121**	**SMP70 + JRRR121 + ZAM295**	**ST123 + JRRR121 + ZAM295**	**All Six Antigens**	**MAP IDEXX**
		% Pooled Pos						**Samples Meeting Cut-Off for Positive Diagnosis**					
PCR+ n = 80	Sens %	10.0	20.0	8.8	12.5	22.5	26.3	22.5	37.5	38.8	38.8	36.3	20
Number matching PCR		8	16	7	10	18	21	18	30	31	31	29	16
PEI 'no-history' negatives													
F5 + F6 n = 80	Spec %	92.5	92.5	96.3	91.3	95.0	91.3	95.0	97.5	95.0	95.0	95.0	100
Compared to PCR+		ns	ns	ns	ns	ns	ns	ns	ns	ns	ns	ns	ns
F5 n = 40	Spec %	95	97.5	95	95	92.5	87.5	92.5	97.5	92.5	92.5	95	100
p (compared to PEI+)		ns	<0.05	ns	<0.0001	<0.001	<0.05	<0.001	<0.001	<0.0001	0.0001	<0.0001	0.05
F6 n = 40	Spec %	90	87.5	97.5	87.5	97.5	95	97.5	97.5	95	97.5	95	100
p (compared to PEI+)		ns	<0.05	<0.05	ns	<0.001	ns	<0.05	<0.05	<0.001	<0.05	<0.05	<0.05

Table 2. *Cont.*

		Antigen						Antigen Combination					
		MOD 171	JRRR 121	RT 237 F2	SMP 70	ZAM 295	ST 123	ST123 + ZAM295	ST123 + JRRR121	SMP70 + JRRR121 + ZAM295	ST123 + JRRR121 + ZAM295	All Six Antigens	MAP IDEXX
		% Pooled Pos						Samples Meeting Cut-Off for Positive Diagnosis					
		Using a fixed specificity of 97.5%											
PCR+ sens.	Sens %	5	17.5	7.5	7.5	11.3	17.5	17.5	37.5	33.8	18.9	30	21.3
Number matching PCR PEI+		4	14	6	6	9	14	14	30	27	15	24	17

The optimal sensitivity and specificity, assuming the 80 PCR positive samples are true positives, for the two sets of 'no-history' negatives from farms F5 and F6, and the 80 combined F5 plus F6 PEI samples. Data for sets of antigens are derived by optimising the combinations of antigen responses to provide the best distinction between 80 PCR positive and PCR negative samples using the R statistics package. Significance values were calculated using the Tukey test; ns = not significant. The final column is the result with a commercial IDEXX assay; the number of predicted positives in IDEXX, in comparison to the lipid assay, at 97.5% specificity was calculated by reducing the cut-off to reach that specificity with the 80 F5 and F6 negatives using the data in Table S1. An analysis of the data for other sample cohorts using these cut-offs and antigen combinations optimised for all 80 PCR positives is shown in Table S5.

Thus, only two of the individual antigens identified more PCR-positive samples than IDEXX MAP ELISA: ZAM295, picking up two more but with a specificity reduced to 95%, and ST123, picking up five more but with a specificity reduced to 91%. However, the combination of the results with ST123 and JRRR121 identified 30 of the 80 PCR positive samples as positive in the assay, compared to just 16 using the IDEXX ELISA (the 16 PEI++ samples); in this case, the specificity was only 97.5%, compared to 100% for IDEXX ELISA. Of the 30 samples identified as positive in this way, 11 had PCR counts of 25.5–33 (n = 13), 7 from 33 to 37 (n = 23), 5 from 37 to 38.5 (n = 24), and 7 from 38.5 to 40 (n = 18).

An analysis of the fit of the above prediction of the lipid ELISA optimised using all 80 PCR positives to each sub-group of serum samples is presented in Table S6.

3.3. Determining Possible Effect of SICCT and M. bovis Infection and Vaccination on Lipid ELISA Performance

The optimal cut-offs for each antigen, as determined in Table 1, and the optimal antigen combinations derived in R (giving the results shown in Table 1) were applied to predict the positive or negative status of the additional sets of samples in Section 2.1 (ii, iii a–e, and iv); the results are presented in Table 3.

The distributions of the responses for each cohort of samples, including the F5 and F6 negatives and the PEI++ samples using single antigens ZAM295 and ST123 and two combinations of antigens, obtained using R statistics are shown in Figure 2. The corresponding distributions for other antigens and antigen combinations are shown in Supplementary Figure S4.

Compared to the PEI++ MAP-positive samples, the MRI-negative sera gave a very low median, and all were predicted to be negative with all antigens except RT237F2 and all selected antigen combinations.

For the SICCT-negative animals (APHA control 1–20), the lipid ELISA showed 100% specificity for all of the antigens except for RT237F2, and for all selected combinations of the antigens in R statistics. Similar results were obtained for the bTB reactors, except that one animal in twenty was false-positive with antigens ST123 and MOD171, and two with the ST123/JRRR121 combination.

Of the twenty Gudair-vaccinated cattle, one was detected as a false positive with antigen ST123, six with MOD171, and four with RT237F2. In this case, R values for specificity with the selected antigen combinations were 90–100%.

Table 3. Median responses (as % positive control) and sensitivity/specificity for Moredun 'no-history' samples and for each cohort of APHA serum samples compared to 16 PEI PCR- and IDEXX-positive samples using the same cut-offs as determined in Table 1.

			Antigen						Antigen Combination Using R				
			MOD 171	**JRRR 121**	**RT237 F2**	**SMP 70**	**ZAM 295**	**ST 123**	**ST123 + ZAM295**	**ST123 + JRRR121**	**ST123 + JRRR121 + ZAM295**	**JRRR121 + SMP70 + ZAM295**	**All Six Antigens**
		Cut-off	57	40	76	93	91	68	0.8	0.63	0.73	0.63	0.78
PEI++ (as per Table 1)		Sens %	25.0	50.0	6.3	37.5	75.0	62.5	56.3	75.0	68.8	68.8	68.8
	n	Group in Section 2.1	Specificity based on single antigens						Specificity based on R combinations				
Moredun serum	9	(ii)	100.0	100.0	88.9	100.0	100.0	100.0	100.0	100.0	100.0	100.0	100.0
p (compared to PEI++)			ns	<0.05	ns	<0.001	<0.001	<0.001	0.0001	<0.001	0.0001	0.0001	<0.001
APHA controls	20	(iii) (a)	100.0	100.0	75.0	100.0	100.0	100.0	100.0	100.0	100.0	95.0	100.0
p (compared to PEI++)			ns	<0.01	ns	<0.001	<0.0001	<0.0001	<0.0001	<0.0001	<0.0001	<0.0001	<0.0001
bTB reactors	20	(iii) (b)	95.0	100.0	95.0	100.0	100.0	95.0	100.0	90.0	100.0	100.0	100.0
p (compared to PEI++)			ns	<0.001	<0.01	<0.001	<0.0001	<0.0001	<0.0001	<0.0001	<0.0001	<0.0001	<0.0001
Gudair-vaccinated	20	(iii) (c)	70.0	100.0	80.0	100.0	100.0	95.0	100.0	90.0	95.0	90.0	100.0
p (compared to PEI++)			ns	<0.01	ns	<0.001	<0.0001	<0.0001	<0.0001	<0.0001	<0.0001	<0.0001	<0.0001
BCG-vaccinated	20	(iii) (d)	95.0	100.0	85.0	95.0	100.0	100.0	100.0	100.0	100.0	100.0	100.0
p (compared to PEI++)			ns	<0.001	ns	<0.001	<0.0001	<0.0001	<0.0001	<0.0001	<0.0001	<0.0001	<0.0001
BCG vac-inf	20	(iii) (e)	90.0	100.0	85.0	75.0	90.0	85.0	90.0	90.0	85.0	85.0	90.0
p (compared to PEI++)			ns	<0.01	ns	<0.05	<0.0001	<0.001	<0.0001	<0.0001	<0.0001	<0.0001	<0.0001
Longitudinal samples: 10 ParaTB IDEXX-negative animals, each over 9 times (total 47 weeks)	90	(iv)	97.8	98.9	93.3	96.7	91.1	96.7	94.4	90.0	93.3	85.6	97.8
p (compared to PEI++)			<0.05	<0.0001	ns	<0.0001	<0.0001	<0.0001	<0.0001	<0.0001	<0.0001	<0.0001	<0.0001

Median responses (as % positive control) and sensitivity/specificity for Moredun 'no-history' samples and for each cohort of APHA serum samples (Section 2.1), using the same cut-offs as determined in Table 1. PEI++ data show the corresponding sensitivity with the 16 PEI samples both positive to PCR and giving IDEXX responses above the recommended cut-off for MAP infection. Specificity values for single antigens are determined from Tables S2 and S3 using the cut-offs from Table 1. Specificity values for combinations of antigens using R for each cohort are calculated using the antigen combinations determined in Table 1; the cut-offs for R calculations are as in Table 1 and are on a scale 0 (negative) to 1 (positive). Significance values were calculated using the Tukey test; ns = not significant. Individual cohorts are Moredun 'no-history' MAP negatives, APHA controls, BTB reactor cattle, Gudair-vaccinated cattle, BCG-vaccinated cattle, and BCG-vaccinated but M. bovis-infected cattle. The final set of 90 'longitudinal' serum samples are from 10 control animals, each at 9 times over 47 weeks, and negative by MAP IDEXX at each time.

For the BCG-vaccinated cattle, the lipid ELISA showed 100% specificity for three antigens (JRRR121, ZAM295, and ST123) and all selected antigen combinations in R. However, the BCG-vaccinated animals experimentally infected with *M. bovis* showed a somewhat reduced specificity (75–90%), although with antigen JRRR121, the specificity for MAP was 100%. Three antigen combinations using R statistics also showed 2–3 false positives. It is worth noting that ZAM295 on its own gave 100% specificity with all these sample sets except that from the vaccinated infected animals.

With the serial serum samples from 10 MAP- and *M. bovis*-free animals taken over nine time-points ('longitudinal' samples), the lipid ELISA overall gave a specificity of 97% at the cut-off for PEI++ samples giving 62.5% sensitivity using antigen ST123, 91% specificity and 75% sensitivity with ZAM295, and 98% with the six-antigen combination in R using the cut-off to give 69% sensitivity. The individual responses are shown for each animal in Figure 3 for antigen ST123. Although very few of the responses are actually above the cut-off for a positive prediction, three of the animals do show persistent high responses. In two cases, AP2169 and AP2176, these animals also show somewhat elevated Bovigam PPDA responses (around 4 and 6 times, respectively, the averages for the remaining eight animals) (Supplementary Table S2); whether there is any link is not clear.

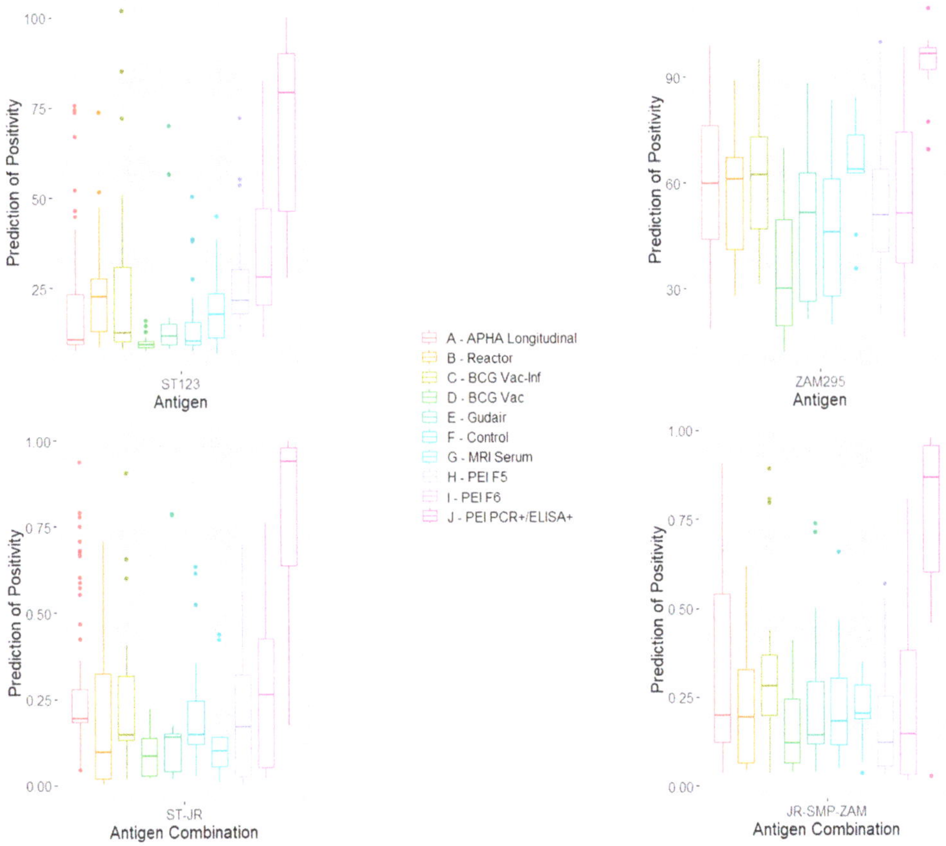

Figure 2. Box-and-whisker plots of the distribution of antigen responses with each cohort of MAP-negative samples and PEI++ samples (positive to PCR and IDEXX) to individual antigens ST123 and ZAM295, together with those to combinations of antigens ST123/JRRR121 and JRRR121/SMP70/ZAM295 using the R statistics package. A–J refer to columns from left to right.

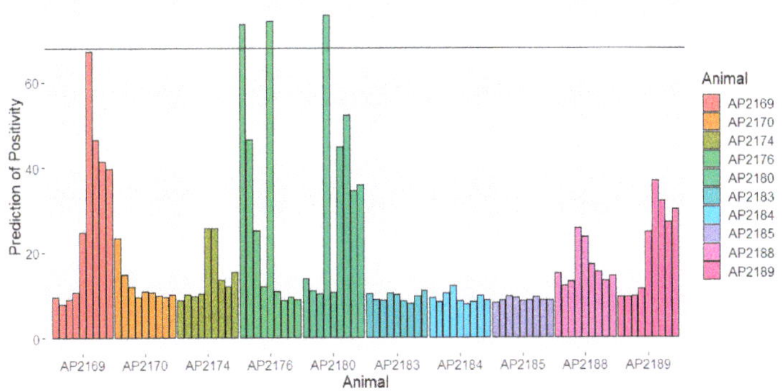

Figure 3. Responses for 10 Bovigam- and MAP IDEXX-negative animals in ELISA with antigen ST123, over 47 weeks (9 samples each) showing optimal cut-off for distinguishing PEI++ samples from 80 F5 and F6 negatives. The horizontal line corresponds to the cut-off from Table 3 for a positive diagnosis with this antigen.

4. Discussion

The diagnosis of Johne's disease in cattle, particularly at an early stage, has proven challenging, not least because of the very slow development of the symptoms of clinical infection. Thus, the use of faecal PCR, or indeed faecal culture, is complicated by intermittent shedding, by sampling issues, by direct pass-through, and, arguably, by animals exposed but apparently able to resolve the initial infection. Indeed, a cut-off <33 PCR cycles of quantification for a confirmed infection has recently been suggested [17]. Serology has conventionally used quite high cut-offs in ELISA, leading to high specificity, but often to a rather low sensitivity. As seen from the 80 faecal PCR MAP-positive animals in this study, only 16 of which were IDEXX MAP-positive, there can be very poor agreement between these methods as to which animals are positive, though that agreement is better when PCR quantification cycles are low. In this case, the average cycle count for PEI++ samples (PCR-positive, MAP IDEXX-positive) was 27.6 (13 of 16 below 33), whereas for PEI+− samples (PCR-positive, MAP IDEXX-negative), it was 37.6 (none below 33). The use of histopathology at slaughter in following disease development and herd management is clearly limited. Analysis of results is further complicated by the different approaches to using these methods from country to country and between research groups, by herd infection and disease progression levels, and by inter-animal and husbandry variations.

In this study, we set up an ELISA utilising six individual synthetic lipid antigens for the serodiagnosis of MAP infection. In the initial analysis comparing the serological responses of 16 MAP-infected animals positive both by PCR and IDEXX ELISA (PEI++) with 80 test-negative animals from two farms (F5 and F6), individual antigens ZAM295 and ST123 gave the best combination of specificity and sensitivity, with specificities and sensitivities of 93.8/75 and 93.8/62.5, respectively. Interestingly, combining the individual assay results using R statistics for these two antigens resulted in a reduced sensitivity of 56.3% at the same specificity of 93.8%; this may reflect the particular combination of false-positive responses, as the optimal cut-off was higher than with other combinations. However, the combination of the results for ST123 and JRRR121 gave the best specificity at 97.5% and sensitivity of 75%. Combining the results from all six individual antigens gave a higher specificity of 100% but lower sensitivity of 68.8%. It is possible that similar improvements might be achieved by mixing two or more of the antigens in a single ELISA; however, keeping the assays separate and then combining the results in a multiplex approach allows selective changes, such as different dominant antibody responses in different stages of infection, and indeed different infections to be recognised. Antigens MOD171 and RT237F2 generally gave low sensitivity and were less discriminative between MAP-infected animals, bTB reactors, Gudair- and BCG-vaccinated animals, and BCG-vaccinated and *M. bovis*-infected animals.

The results show that there are some differences in response between the test-negative animals on the two PEI farms with no history of Johne's disease. The animals from farm 5 were primarily housed outside, whereas those from farm 6 were housed inside exclusively, and therefore in closer proximity to each other and with different environmental exposures. Although farm 6 was a total confinement herd with no outdoor exposure, this farm did have a history of a large number of resident birds in the rafters of the barn and had a lower level of overall hygiene with increased stocking density. This particular barn had not been deep-cleaned for several years and there was significant detritus in the rafters that would occasionally drop down into the cattle living area, including the feed bunks and water troughs. It is therefore proposed that there may have been some cross-reactivity of the ELISAs with other environmental mycobacteria species that affected the observed specificity of the lipid ELISA in this herd.

Trehalose dimycolate ST123 gave the highest combination of sensitivity and specificity with the samples from farm 5; this antigen is a wax ester mycolic acid characteristic of MAC and not present in organisms such as *M. bovis* [35]. This antigen performed well with the samples from farm 6, though in this case, the best antigen was trehalose dimycolate ZAM295, more typical of faster-growing environmental mycobacteria. Antigen JRRR121 gave high specificity with both sets of MAP negative samples, albeit with only modest sensitivity for the PEI++ samples; this antigen is a *trans*-cyclopropane containing mycolic acid, again a sub-structure common in MAC [38]. The other three antigens showed rather lower sensitivities for the PEI++ samples at cut-offs giving as low as 91% specificity for F5 and F6 negatives. Antigen MOD171 is a model fragment from a mycobacterial cell wall, unlike the other non-wall-bound antigens used in this study, and may be less antigenic for that reason. The glucose ester SMP70 incorporates a methoxymycolic acid that is common in many mycobacterial cells and modest performance in the assay may reflect cross-reactivity with antibodies raised to other mycobacterial exposure/infection; it is worth noting that this class of mycolic acid is generally reduced in BCG vaccines [22]. Although antigen RT237F2 does contain a widely distributed mycolic acid, it also includes an amino-sugar and is not a known natural product; nonetheless, it was included in this study because it did produce very strong responses with some PCR-positive samples. Antigens SMP70, ST123, and ZAM295 are all examples of different classes of non-wall-bound trehalose mycolates, which are known to be strongly antigenic [22,23,25–28]. JRRR121 is a common class of free mycolic acid; these molecules are known to be antigenic, though responses are generally weaker than with trehalose or glucose esters [22,23,28].

It has been reported previously that the correlation between faecal PCR and serological assays may be poor at individual animal [39] or herd level [40], particularly in disease that has not reached a late stage. There have been similar studies on the agreement between faecal culture and serology [12,13]. With the current 80 PEI PCR-positive serum samples, there is a clear correlation between high IDEXX responses and low PCR cycle numbers—twelve of the sixteen samples that had positive IDEXX responses also had quantification cycles below 31, and none of those with IDEXX responses below the cut-off had quantification cycles below 33. There could be a number of reasons for the PCR-positive and IDEXX-negative samples, including direct pass-through of the organism. It was of interest, therefore, to see whether the lipid assay would provide a better, or different, fit than IDEXX to the complete set of PCR-positive data; in other words, whether the lipid assay gives more responses above a particular cut-off for the complete PCR-positive set than the 16 from IDEXX ELISA. In this study, two of the individual lipid antigens did identify more PCR-positive samples than the commercial ELISA: ZAM295 detected two more but with a specificity reduced to 95%, and ST123 detected five more but with a specificity reduced to 91%. However, the combination of the results with ST123 and JRRR121 identified 30 of the PCR positive samples as positive in the lipid ELISA assay, compared to just 16 using the IDEXX ELISA (the 16 PEI++ samples used in the initial analysis). In this case, the specificity was only 97.5% compared to 100% for the IDEXX ELISA; however, if the cut-off in the IDEXX ELISA was reduced to give 97.5% specificity, it only picked up 17 positives. The lipid ELISA could therefore be a good alternative to the commercial IDEXX ELISA for routine screening of cattle, particularly in the early stages of disease progression [41], and this would be worth investigating further. In the commercial PCR used in this study, samples were considered MAP-positive within 42 cycles. Of the 80 PCR-positive samples utilised in the analyses, 13 gave PCR quantification cycles <33, 38 between 33 and 38, 27 between 38 and 40, and 2 gave >40. A recent publication recommends that the maximum number of quantification cycles that can safely be interpreted as 'infection' is <33 and that a PCR cycle count of >33 and <40 should be regarded as inconclusive and requires a second confirmatory test [17]. Of the 30 samples identified as positive using the combination of the results for ST123 and JRRR121 in the R statistics package, only 11 had PCR counts of 25.5–33, and the others were spread across all quantification cycles from 33 to 40. Thus, the lipid ELISA identifies antibody responses in serum that appear to support infection or developing disease in a number of the high-cycle-count PCR positives.

All antigens except for RT237F2 correctly identified the APHA control animals as negative for MAP infection. Since these animals had been tested ten days previously with the statutory bTB skin test, this suggests that responses to these lipid antigens are unaffected by the skin test. The number of animals tested in this study is small, so further investigations are necessary to confirm this. Currently, cattle are not tested using a commercial serum ELISA within three months of a bTB skin test to avoid any cross-reactivity.

Responses to individual antigens JRRR121, SMP70, and ZAM295 and either a combination of all the antigens or ST123/ZAM295 were unaffected by Gudair vaccination eight weeks post vaccination. This suggests that the lipid ELISA has potential as a DIVA diagnostic. However, individual antigens MOD171, RT237, and ST123 and combinations ST123/JRRR121, ST123/JRRR121/ZAM295, and JRRR121/SMP70/ZAM295 were less specific, detecting a few false-positive animals. This may reflect changes in immune responses caused by vaccination selectively activating antibodies to other non-tuberculous mycobacteria that are recognised by these antigens, or could represent antibodies against MAP directly generated by vaccination with Gudair vaccine (inactivated strain of *M. paratuberculosis*) [22].

Responses to individual antigens ST123, ZAM295, and JRRR121 and to all selected antigen combinations were unaffected by vaccination with BCG. BCG vaccination is being investigated as a potential vaccine for bTB [41], and if found to be effective, it would likely be used extensively in countries with endemic bTB. It is therefore important to develop new MAP diagnostics with this in mind.

JRRR121 was found to have 100% specificity for animals infected with *M. bovis* (bTB reactors and BCG-vaccinated/*M. bovis*-infected animals). Other lipid antigens and selected antigen combinations gave reduced specificity: 90–95% in bTB reactors and 75–90% in BCG-vaccinated/*M. bovis*-infected animals. There were differences in the responses between these two groups, for which there could be several reasons. The bTB reactors represent natural *M. bovis* infections contracted in the field, whereas the animals in the BCG-vaccinated/*M. bovis*-infected group were infected experimentally with high doses of *M. bovis*. The animals in the bTB reactor group will be at different stages of infection to the experimentally infected animals. Additionally, in the BCG-vaccinated/*M. bovis*-infected group, the vaccination may have changed the animals' immune response to infection, which may have resulted in a change in the antibody responses to the lipid antigens.

The lipid ELISA was used to evaluate the responses to the antigens over the course of a year in 10 animals, all of which remained negative for MAP infection, as determined by the IDEXX ELISA, and negative for *M. bovis*, as indicated by low IGRA responses. The specificity of the lipid ELISA was somewhat reduced. Using the combination of ST123/JRRR121, 93% specificity corresponded to six false positives; however, four of these were from a single animal at several time-points, possibly reflecting an early infection (by MAP or other mycobacteria) not detected by the IDEXX ELISA. This requires further investigation.

These results suggest that assays detecting the binding of antibodies generated in MAP disease can be detected in ELISA with specific lipid antigens, or combinations of those assays, in more PCR-positive samples than when using a current commercial serodiagnostic device. Moreover, the assay is not interfered with by vaccination for MAP or bovine TB. On the basis that not all PCR positives are caused by direct pass-through, and some represent real disease, an assay that can detect which ones correspond to developing disease would be ideal for a diagnostic test. These initial results appear to support that possibility, but clearly require validation with much larger sample sets and with longitudinal sets of serum from animals known to be exposed to and infected with MAP.

5. Conclusions

This study provides evidence that ELISA using synthetic lipid antigens identical to individual components of mycobacterial cells can identify cattle that are positive for MAP infection by a combination of PCR and commercial MAP ELISA. The best antigens from the six evaluated here were a trehalose ester of a wax ester mycolic acid (ST123) characteristic of *Mycobacterium avium* complex, a free mycolic acid containing a *trans*-cyclopropane group common in *Mycobacteium avium* complex (JRRR121), and a trehalose ester of a diene containing mycolic acid class common in non-tuberculous mycobacteria (ZAM295). The assay is not interfered with by BCG vaccination for bTB, Gudair vaccination for MAP, or by a bTB skin test, and is not compromised by active bTB. Thus, it may find applications as a DIVA assay. It also provides a better correlation between PCR positivity and serodiagnosis than the commercial MAP IDEXX ELISA and may provide a useful alternative for routine screening of cattle for Johne's disease, particularly in the early stages of disease.

Supplementary Materials: The following supporting information can be downloaded at: https://www.mdpi.com/article/10.3390/ani14060848/s1. Table S1: ELISA responses to six synthetic lipid antigens and sample information for PEI samples, together with faecal PCR cycle counts and responses in a commercial IDEXX MAP assay (0.55 recommended cut-off for positive). Table S2: ELISA responses (% positive control) to six synthetic lipid antigens and sample information for APHA timeline samples (10 animals at 9 times). Columns 2 and 3 are results for commercial Bovigam IGRA and ParaTB IDEXX assays. Table S3: ELISA responses (% positive control) to six synthetic lipid antigens for APHA and Moredun samples. Table S3: ELISA responses (% positive control) to six synthetic lipid antigens for APHA and Moredun samples. Table S4: Median ELISA responses for each set of samples with each of six antigens as % positive control for that antigen. Figure S1: Structures of synthetic antigens used in this work. Figure S2: ROC analysis of ELISA responses for PEI++ samples compared to 80 F5 and F6 negatives. Figure S3: Box plots showing distribution of ELISA responses for each group of samples PEI++, F5 and F6 when combined in R as in Tables 1 and S1. Table S5: Median ELISA responses (as % positive control) and sensitivity/specificity for Moredun 'no history' samples and for each cohort of APHA serum samples, using the same cut-offs as determined in Table 1. Figure S4: Box plots showing distribution of ELISA responses for each group of samples PEI++, F5 and F6 compared to Moredun and APHA cohorts. Table S6: Analysis of all 80 PEI PCR samples giving non-zero cycle counts (Table S1) with cut-offs optimised for maximum accuracy compared to all 80 F5 and F6 negatives.

Author Contributions: Conceptualisation, M.S.B., B.S., K.S., V.H., S.M.K. and H.M.V.; methodology, P.S.M., V.H., N.R., R.T.H. and J.R.A.D.; software, P.S.M.; validation, P.S.M., N.R. and T.H.; formal analysis, M.S.B., P.S.M., T.H. and G.J.J.; investigation, R.T.H., J.R.A.D. and P.S.M.; resources, M.S.B. and B.S.; data curation, M.S.B. and P.S.M.; writing—original draft preparation, M.S.B.; writing—review and editing, M.S.B., V.H., K.S., G.J.J., T.H. and S.M.K.; visualisation, M.S.B.; supervision, M.S.B., H.M.V. and S.M.K.; project administration, M.S.B., S.M.K. and G.J.J.; funding acquisition, M.S.B. and B.S. All authors have read and agreed to the published version of the manuscript.

Funding: This research received partial support from the Welsh Government Smart Programme (Grant: 2018/ED/007).

Institutional Review Board Statement: Not applicable as there were no human or animal interventions.

Informed Consent Statement: Not applicable.

Data Availability Statement: All data are provided as part of the Supplementary Material.

Acknowledgments: We thank D.A. Thomas for the statistical analysis of the data.

Conflicts of Interest: Prof. Mark Stephen Baird is a director of Diagnostig Ltd., who hold background IP on the applications of synthetic mycolic acid based liqids. That IP was generated before the start of the work that is currently published. Paul S. Mason worked for Diagnostig Ltd. during this project but is no longer employed by the company and has no conflict interest. Brendan Smith is a consultant for Bimeda Animal Health Ltd.; Bimeda have an interest in the development of improved assays for infections in farm animals.

References

1. Mallikarjunapa, S.; Brito, L.F.; Pant, S.D.; Shenkel, F.S.; Meade, K.G.; Karrow, N.A. Johne's Disease in Dairy Cattle: An Immunogenetic Perspective. *Front. Vet. Sci.* **2021**, *8*, 718987. [CrossRef]
2. Whittington, R.; Donat, K.; Weber, M.F.; Kelton, D.; Nielsen, S.S.; Eisenberg, S.; Arrigoni, N.; Juste, R.; Sáez, J.L.; Dhand, N.; et al. Control of paratuberculosis: Who, why and how. A review of 48 countries. *BMC Vet. Res.* **2019**, *15*, 198. [CrossRef] [PubMed]
3. Coussens, P.; Lamont, E.A.; Kabara, E.; Sreevatsan, S. Host-pathogen interactions and intracellular survival of *Mycobacterium avium* subsp. paratuberculosis. In *Paratuberculosis: Organism, Disease, Control*; CAB International: Wallingford, UK, 2020; pp. 109–125. [CrossRef]

4. Wu, C.W.; Livesey, M.; Schmoller, S.K.; Manning, E.J.B.; Steinberg, H.; Davis, W.C.; Hamilton, M.J.; Talaat, A.M. Invasion and persistence of *Mycobacterium avium* subsp. paratuberculosis during early stages of Johne's disease in calves. *Infect. Immun.* **2007**, *75*, 2110–2119. [CrossRef] [PubMed]
5. O'Brien, R.; Mackintosh, C.; Griffin, F. Paratuberculosis in deer, camelids and other ruminants. In *Paratuberculosis: Organism, Disease, Control*, 2nd ed.; Behr, M., Stevenson, K., Kapur, V., Eds.; CAB International: Wallingford, UK, 2020; pp. 188–199. [CrossRef]
6. Fox, N.J.; Smith, L.A.; Stevenson, K.; Davidson, R.; Marion, G.; Hutchings, M. Infection of non-ruminant wildlife by *Mycobacterium avium* subsp. paratuberculosis. In *Paratuberculosis: Organism, Disease, Control*, 2nd ed.; Behr, M., Stevenson, K., Kapur, V., Eds.; CAB International: Wallingford, UK, 2020; pp. 200–212.
7. Whittington, R.J.; Begg, D.J.; de Silva, K.; Purdie, A.C.; Dhand, N.K.; Plain, K.M. Case definition terminology for paratuberculosis (Johne's disease). *BMC Vet. Res.* **2017**, *13*, 328. [CrossRef] [PubMed]
8. National Researcg Council (US) Committee on Diagnosis and Control of Johne's Disease. *Diagnosis and Control of Johne's Disease*; National Academies Press (US): Washington, DC, USA, 2003.
9. Jungersen, G.; Mikkelsen, H.; Grell, S.N. Use of the johnin PPD interferon-gamma assay in control of bovine paratuberculosis. *Vet. Immunol. Immunopathol.* **2012**, *148*, 48–54. [CrossRef] [PubMed]
10. Hussain, T.; Shah, S.Z.A.; Zhao, D.; Sreevatsan, S.; Zhou, X. The role of IL-10 in *Mycobacterium avium* subsp. *paratuberculosis* infection. *Cell Commun. Signal.* **2016**, *14*, 29. [CrossRef] [PubMed]
11. Beaver, A.; Sweeney, R.W.; Hovingh, E.; Wolfgang, D.R.; Grohn, Y.T.; Schukken, Y.H. Longitudinal relationship between fecal culture, fecal quantitative PCR, and milk ELISA in *Mycobacterium avium* ssp. paratuberculosis-infected cows from low-prevalence dairy herds. *J. Dairy Sci.* **2017**, *100*, 7507–7521. [CrossRef] [PubMed]
12. Whitlock, R.H.; Wells, S.J.; Sweeney, R.W.; Van Tiem, J. ELISA and fecal culture for paratuberculosis (Johne's disease): Sensitivity and specificity of each method. *Vet. Microbiol.* **2000**, *77*, 387–398. [CrossRef] [PubMed]
13. Neilsen, S.S.; Toft, N. Ante mortem diagnosis of paratuberculosis: A review of accuracies of ELISA, interferon-g assay and faecal culture techniques. *Vet. Microbiol.* **2008**, *129*, 217–235. [CrossRef]
14. Pitt, D.J.; Pinch, D.S.; Janmaat, A.; Condron, R.J. An estimate of specificity for a Johne's disease absorbed ELISA in northern Australian cattle. *Aust. Vet. J.* **2002**, *80*, 57–60. [CrossRef]
15. Kralik, P.; Slana, I.; Kralova, A.; Babak, V.; Whitlock, R.H.; Pavlik, I. Development of a predictive model for detection of *Mycobacterium avium* subsp. paratuberculosis in faeces by quantitative real time PCR. *Vet. Microbiol.* **2011**, *149*, 133–138. [CrossRef] [PubMed]
16. Russo, S.; Cortimiglia, C.; Filippi, A.; Palladini, G.; Garbarino, C.; Massella, E.; Ricchi, M. Validation of digital PCR assay for the quantification of *Mycobacterium avium* subsp. paratuberculosis in bovine faeces according to the ISO 20395:2019. *J. Microbiol. Methods* **2023**, *213*, 106825. [CrossRef] [PubMed]
17. Caldow, G.L.; Douglas, L.; Thacker, J.; Mackay, E.; Berezowski, J. Estimation of the specificity of an antibody ELISA for paratuberculosis generated from a sector of the UK cattle population using results from a paratuberculosis control programme. *Prev. Vet. Med.* **2023**, *215*, 105910. [CrossRef] [PubMed]
18. Bannantine, J.P.; Bayles, D.O.; Waters, W.R.; Palmer, M.V.; Stabel, J.R.; Paustian, M.L. Early antibodyresponse against *Mycobacterium avium* subspecies *paratuberculosis* antigens in subclinical cattle. *Proteome Sci.* **2008**, *6*, 5. [CrossRef] [PubMed]
19. Muskens, J.; van Zijderveld, F.; Eger, A.; Bakker, D. Evaluation of the long term immune response in cattle after vaccination against paratuberculosis in two Dutch dairy herds. *Vet. Microbiol.* **2002**, *86*, 269–278. [CrossRef] [PubMed]
20. Middleton, S.; Steinbach, S.; Coad, M.; McGill, K.; Brady, C.; Duignan, A.; Wiseman, J.; Gormley, E.; Jones, G.J.; Vordermeier, H.M. A molecularly defined skin test reagent for the diagnosis of bovine tuberculosis compatible with vaccination against Johne's Disease. *Sci. Rep.* **2021**, *11*, 2929. [CrossRef] [PubMed]
21. Vordermeier, H.M.; Jones, G.J.; Buddle, B.M.; Hewison, R.G.; Villareal, B. Bovine tuberculosis in cattle: Vaccines, DIVA tests, and host biomarker discovery. *Annu. Rev. Anim. Biosci.* **2016**, *4*, 87–109. [CrossRef]
22. Verschoor, J.A.; Baird, M.S.; Grooten, J. Towards understanding the functional diversity of cell wall mycolic acids of *Mycobacterium tuberculosis*. *Prog. Lipid Res.* **2012**, *51*, 325–339. [CrossRef]
23. Jones, A.; Lee, O.Y.C.; Minnikin, D.E.; Baird, M.S.; Al Dulayymi, J.R. A re-investigation of the mycolic acids of Mycobacterium avium. *Chem. Phys. Lipids* **2020**, *230*, 104928. [CrossRef]
24. Schleicher, G.K.; Feldman, C.; Vermaak, Y.; Verschoor, J.A. Prevalence of anti-mycolic acid antibodies in patients with pulmonary tuberculosis co-infected with HIV. *Clin. Chem. Lab. Med.* **2002**, *40*, 882–887. [CrossRef]
25. Maekura, R.; Nakagawa, M.; Nakamura, Y.; Hiraga, T.; Yamamura, Y.; Ito, M.; Ueda, E.; Yano, S.; He, H.; Oka, S.; et al. Clinical-Evaluation of Rapid Serodiagnosis of Pulmonary Tuberculosis by Elisa with Cord Factor (Trehalose-6,6′-Dimycolate) as Antigen Purified from Mycobacterium-Tuberculosis. *Am. Rev. Respir. Dis.* **1993**, *148*, 997–1001. [CrossRef] [PubMed]
26. Fujita, Y.; Ogata, H.; Yano, I. Clinical evaluation of serodiagnosis of active tuberculosis by multiple-antigen ELISA using lipids from *Mycobacterium bovis* BCG Tokyo 172. *Clin. Chem. Lab. Med.* **2005**, *43*, 1253–1262. [CrossRef] [PubMed]
27. Fujita, Y.; Doi, T.; Maekura, R.; Ito, T.; Yano, I. Differences in serological responses to specific glycopeptidolipid-core and common lipid antigens in patients with pulmonary disease due to *Mycobacterium tuberculosis* and *Mycobacterium avium* complex. *J. Med. Microbiol.* **2006**, *55*, 189–199. [CrossRef] [PubMed]

28. Jones, A.; Pitts, M.; Al Dulayymi, J.R.; Gibbons, J.; Ramsay, A.; Goletti, D.; Gwenin, C.D.; Baird, M.S. New synthetic lipid antigens for rapid serological diagnosis of tuberculosis. *PLoS ONE* **2017**, *12*, 0181414. [CrossRef] [PubMed]
29. Bashford, J.; Flowers, W.; Haworth, C.; Ryan, J.; Cervi, A.; Al Dulayymi, J.R.; Mason, P.S.; Plank, A.; Baird, M.; Floto, A. Evaluation of a novel ELISA test using synthetic mycolic acid antigens for serodiagnosis of non-tuberculous mycobacterial (NTM) infections. *Thorax* **2023**, *78*, 309–312. [CrossRef] [PubMed]
30. Diaz, F. OIE Standard on principles and methods of validation of diagnostic assays for infectious diseases. In Proceedings of the OIE Regional Workshop for OIE National Focal Points for Veterinary Products, Ottawa, ON, Canada, 25–27 August 2014; pp. 3–5.
31. Li, L.; Bannantine, J.P.; Campo, J.J.; Randall, A.; Grohn, Y.T.; Schilling, M.A.; Katani, R.; Radzio-Basu, J.; Easterling, L.; Kapur, V. Identification of Sero-Diagnostic Antigens for the Early Diagnosis of Johne's Disease Using MAP Protein Microarrays. *Sci. Rep.* **2019**, *9*, 17573. [CrossRef]
32. Premium Cattle Health Scheme, SRUC Veterinary Services, St. Boswells, Roxburghshire. CHECS.co.uk: Setting Industry Standards in Cattle Disease Control. Available online: https://checs.co.uk/wp-content/uploads/2021/11/CHECS-Technical-document-Johnes-disease-rule-change-released-220621-date-01-11-2021.pdf (accessed on 28 December 2023).
33. Ali, O.T.; Mohammed, M.O.; Al Dulayymi, J.R.; Baird, M.S. Synthesis of a Di-Mycoloyl Tri-Arabinofuranosyl Glycerol Fragment of the Mycobacterial Cell Wall, Based on Synthetic Mycolic Acids. *Molecules* **2019**, *24*, 3596. [CrossRef]
34. Alhuwaymil, Z.S.; Al-Araj, I.Q.M.; Al Dulayymi, A.R.; Jones, A.; Gates, P.J.; Valero-Guillen, P.L.; Baird, M.S.; Al Dulayymi, J.R. Mycobacterium alvei (omega-1)-methoxy mycolic acids: Absolute stereochemistry and synthesis. *Chem. Phys. Lipids* **2020**, *233*, 104977. [CrossRef]
35. Taher, S.G.; Al Dulayymi, J.R.; Tima, H.G.; Ali, H.M.; Romano, M.; Baird, M.S. Synthesis of wax esters and related trehalose esters from *Mycobacterium avium* and other mycobacteria. *Tetrahedron* **2016**, *72*, 3863–3876. [CrossRef]
36. Sahb, M.M.; Al Dulayymi, J.R.; Baird, M.S. Glucose monomycolates based on single synthetic mycolic acids. *Chem. Phys. Lipids* **2015**, *190*, 9–14. [CrossRef] [PubMed]
37. Hameed, R.T. Synthesis of Mycobacterial Cord Factors and Related Compounds. Ph.D. Thesis, Bangor University, Bangor, UK, 2014.
38. Koza, G.; Muzael, M.; Schubert-Rowles, R.R.; Theunissen, C.; Al Dulayymi, J.R.; Baird, M.S. The synthesis of methoxy and keto mycolic acids containing methyl-trans-cyclopropanes. *Tetrahedron* **2013**, *69*, 6285–6296. [CrossRef]
39. Geisbauer, E.; Khol, J.L.; Wassertheurer, M.; Damoser, J.; Osterreicher, E.; Dunser, M.; Revilla-Fernandez, S.; Baumgartner, W. Longterm investigation in an Austrian dairy herd with low prevalence of paratuberculosis detection of antibodies in blood and milk. *Vet. Q.* **2007**, *29*, 138–148. [CrossRef]
40. Field, N.L.; Mee, J.F.; McAloon, C.G. Characteristics (sensitivity and specificity) of herd-level diagnostic tests for *Mycobacterium avium* subspecies paratuberculosis in cattle—A systematic review. *Vet. J.* **2022**, *27*, 105786. [CrossRef] [PubMed]
41. Buddle, B.M.; Denis, M.; Aldwell, F.E.; Vordermeier, M.H.; Hewinson, G.R.; Wedlock, N.D. Vaccination of cattle with *Mycobacterium bovis* BCG by a combination of systemic and oral routes. *Tuberculosis* **2008**, *88*, 595–600. [CrossRef] [PubMed]

animals

MDPI

Article

Comparison of Machine Learning Tree-Based Algorithms to Predict Future Paratuberculosis ELISA Results Using Repeat Milk Tests

Jamie Imada [1], Juan Carlos Arango-Sabogal [2], Cathy Bauman [1], Steven Roche [1,3] and David Kelton [1,*]

[1] Department of Population Medicine, University of Guelph, Guelph, ON N1G 2W1, Canada; jamiebimada@gmail.com (J.I.); cbauman@uoguelph.ca (C.B.); sroche@acerconsult.ca (S.R.)
[2] Département de Pathologie et Microbiologie, Faculté de Médecine Vétérinaire, Université de Montréal, Saint-Hyacinthe, QC J2S 2M2, Canada; juan.carlos.arango.sabogal@umontreal.ca
[3] ACER Consulting, 100 Stone Rd West #101, Guelph, ON N1G 5L3, Canada
* Correspondence: dkelton@ovc.uoguelph.ca

Simple Summary: Johne's disease is a chronic progressive gastrointestinal disease of ruminants. The diagnosis and control of this disease remains a challenge for cattle producers and herd veterinarians. Machine learning is a broad class of algorithms and statistical analysis belonging to the area of artificial intelligence. These techniques allow for the recognition of patterns in large datasets that are not found by traditional analytical and statistical methods. While artificial intelligence and machine learning have been used in human medicine and in various aspects of animal husbandry, they have only just started to be used in Johne's disease research. By using results from milk component tests and past Johne's individual test results, tree-based models, a type of machine learning, were used to predict future Johne's results. This type of predictive algorithm may help producers and veterinarians more reliably identify those animals most likely to yield a positive result. This will, in turn, help improve control efforts and reduce the testing burden for herds working towards disease control and eradication.

Abstract: Machine learning algorithms have been applied to various animal husbandry and veterinary-related problems; however, its use in Johne's disease diagnosis and control is still in its infancy. The following proof-of-concept study explores the application of tree-based (decision trees and random forest) algorithms to analyze repeat milk testing data from 1197 Canadian dairy cows and the algorithms' ability to predict future Johne's test results. The random forest models using milk component testing results alongside past Johne's results demonstrated a good predictive performance for a future Johne's ELISA result with a dichotomous outcome (positive vs. negative). The final random forest model yielded a kappa of 0.626, a roc AUC of 0.915, a sensitivity of 72%, and a specificity of 98%. The positive predictive and negative predictive values were 0.81 and 0.97, respectively. The decision tree models provided an interpretable alternative to the random forest algorithms with a slight decrease in model sensitivity. The results of this research suggest a promising avenue for future targeted Johne's testing schemes. Further research is needed to validate these techniques in real-world settings and explore their incorporation in prevention and control programs.

Keywords: paratuberculosis; Johne's disease; disease control; machine learning; random forest; decision tree; diagnostics; dairy farming; cattle

1. Introduction

Mycobacterium avium subspecies paratuberculosis (MAP) is the causative agent of Johne's disease (JD). This untreatable, chronic, and progressive disease of ruminants results in economic losses of up to USD 28 million for the Canadian dairy industry annually [1]. Global approaches to on-farm JD control are typically centered around the implementation of management practices that minimize the introduction or within herd spread of MAP and the identification and removal of MAP-positive animals from the herd [2].

The test and cull control method is hampered by the long latency/incubation of the disease before clinical signs, the large proportion of infected populations existing as asymptomatic infectious carriers, and the absence of accurate and reliable diagnostic tests [3,4]. Historically, JD tests have demonstrated a low sensitivity and high specificity. False negatives occur due to the biological characteristics of the disease wherein the shedding of the pathogen and/or the circulation of detectable antibodies occur late in the stage of infection and can be inconsistent over time [5,6]. False positives are much less likely to occur, but, when they do, they can often be attributed to the cross-reactivity of the antigens being identified with molecular tests [6] or due to the "pass-through" phenomenon in the case of organism detection tests [7]. It has also been suggested that animals exposed to environmental non-MAP mycobacterium may generate cross-reacting antibodies, which can produce false positive results for bovine paratuberculosis serologic tests [8]. When tests have an imperfect sensitivity and specificity, producers may have difficulty making culling decisions. It can be difficult to cull a valuable animal which tests positive on a JD test in the absence of clinical signs or frustrating when an animal which tested negative previously demonstrates clinical signs at a later date.

Compounding this frustration is interpreting these imperfect tests over time. Herds which participate in long-term serologic or milk testing frequently encounter conflicting results (i.e., a cow undergoing testing every time they are dried off may have two positive tests followed by a negative test) [9]. Interpreting results that are sequentially inconsistent requires a systematic decision-making process.

To date, this process of interpreting inconsistent sequential results is beyond the ability of researchers with the analytical tools currently available to them. With the advent of machine learning (ML), a subfield of artificial intelligence (AI), the classification of data based on pattern-recognizing algorithms has become possible. The goal of ML is to simplify the data analysis process, increase the reliability of the insight gained, and improve the decision-making process. While ML-guided decision support systems have been used in human medicine since the 1970s, their application in veterinary medicine has been few and far between [10]. Most applications of AI in the veterinary and animal husbandry sector have been focused on solving animal health and welfare problems through analyzing animal behavior [11,12] with very few studies addressing disease detection [11]. In a review by Basran and Appleby [13], the authors highlight the unmet potential for AI in veterinary medicine; however, the published work on AI in veterinary medicine has mainly been centered on medical image analysis (e.g., radiographs, CT, MRI, ultrasound images, and teat images on dairy cattle).

There are many different ML approaches, each with their own unique benefits, depending on the data being used [14]. The ML algorithms are broadly classified as either supervised or unsupervised, depending on whether the data are labeled or not [15]. The supervised ML algorithms, such as decision trees (DTs), random forests (RFs), and neural networks (NNs), are methodologies that can be used to analyze data to help predict or classify future results. Algorithms of the same type may also perform differently. As Matvieiev et al. [16] noted in their research involving artificial neural networks (ANNs) to predict milk yield and milk components, different ANN models were better at predicting different outcomes (i.e., one model that performed well predicting milk yield was not as good as another ANN at predicting protein content).

Researchers have attempted to address various aspects of companion animal medicine using ML and AI, from the evaluation of the criteria in structured histology reports [17] to the differentiation between inflammatory bowel disease and alimentary lymphoma in cats [18] to the early prediction of canine cancer through blood serum analysis [19] and the short- and medium-term survival of acute-on-chronic kidney disease in cats using DT models [20]. In their work using ML to predict leptospirosis, Reagan et al. [21] cautioned that models should have some form of post-implementation evaluation for each new population that the model is applied to. Furthermore, the models developed to predict diseases function as predictive algorithms rather than diagnostic tools, and are best used for identifying animals at higher risk, who should thus be prioritized for further analysis [22,23].

Much of the AI and ML research in dairy farming has focused on behavior classification, early lameness, estrus detection, fertility predictions, and disease detection (mostly subclinical mastitis and ketosis) [24]. In their review of 142 publications covering ML in dairy farming, Lokhorst et al. [25] found that, among the 134 studies that used supervised learning, 64% used it for classification. This finding is also reported by Slob et al. [26] and Wang et al. [27], who found that decision tree-based algorithms, a type of classification approach, are by far the most utilized followed by ANNs and support vector machines (SVMs).

Schmeling et al. [28] used the data from accelerometers, magnetometers, and gyroscopes to predict cow lying behavior. The researchers used RF, decision tree, SVM and naive Bayes to model the data with the RF achieving the highest accuracy. Zhou et al. [29] utilized collar data alongside milking system data with eight different ML algorithms to predict common disorders in dairy cattle with SVM, DT, and RF performing the best (accuracy >80%). The benefit that these ML methods have in dairy disease predictions (and disease detection in general) is that they can discover complex, latent patterns between predictor variables and the trait of interest (the disease status) even if this relationship is non-linear. These methods do not need to have a preconceived hypothesis compared with other more traditional statistical methods as their purpose can be solely to identify patterns [30]. However, a major hurdle in the use of AI and ML on dairy farms is that dairy-related data exists in silos that are not connected [24].

Another challenge with some machine learning algorithms is their lack of interpretability. To leverage the power of machine learning methods while avoiding this lack of interpretability, Sykes et al. [31] used explainable AI models to assess the impact of on-farm biosecurity practices on the predicted risk of a Porcine Reproductive and Respiratory Syndrome (PRRSV) outbreak. The researchers utilized SVMs, RFs, and gradient boosted machines in their study and performed a variable importance analysis to identify the most influential factors within the models. The generated models have the potential to inform individualized biosecurity assessments, informing which practices are most important at a specific farm.

While the current research with ML being applied to disease surveillance and risk assessment has been limited, ML does show promise as a useful tool. Pereira et al. [32] used 77 predictors related to the type of farm, type of lactation, number of animals on the property, and gradient boosted trees to predict which farms were at the highest risk for tuberculosis infection. Stański et al. [33] used various ML techniques such as NNs, RFs, gradient boosted trees, and support vector classifiers to improve the predictions of herd-level bovine tuberculosis beyond that of the traditional herd-level single intra-dermal comparative cervical tuberculin test. By using data on the farm location, animal movements, links to infected farms, past testing history, and land cover around the farm, the researchers were able to demonstrate an increase in herd sensitivity from 61 to 67% and herd specificity from 90 to 92%.

Various groups have attempted to use ML methods to evaluate, classify, and predict udder and teat health. Researchers have used data such as milking parameters [34], cytometric fingerprints [35], and SCC [36] to predict udder health and mastitis. Moving beyond classifying the health status of the udder, Heald et al. [37] used ANNs to classify the bacteriologic causes of mastitis. Others have attempted to utilize image data, such as Porter et al. [38] who used convolutional neural networks (CNNs) to classify images of teat ends achieving an AUC of 77%. The authors noted that subtle differences between some classes may have limited the classification performance of their model. Researchers have also utilized the variable importance of ML models to identify those factors which are the best predictors of disease [39,40] or, in the case of Johne's research, to identify the most influential control measures for disease prevention [41].

Machine learning is a family of statistical analysis techniques that are relatively new to MAP research. Volatile organic compounds (VOCs) have been explored as a possible marker for infection with MAP [42,43], demonstrating the possible potential for detecting infection prior to bacterial culture results [42]. Meanwhile, others have attempted to utilize gut microbiota composition to predict the infection status and shedding severity [44,45]. Johne's ML research continues to expand to explore new avenues in the detection of MAP. The changes to sera microRNA [46] and metabolomic changes in polyunsaturated fatty acids and eicosanoids [47] have been used to differentiate between MAP infected and non-infected animals.

Researchers have highlighted that there are still challenges with the application of ML. Obtaining quality data from multiple sources remains one of the biggest barriers [48,49]. Furthermore, the datasets need to be accurately labelled for supervised learning techniques to be effective [50]. As in most modeling, the quality of the model is dependent on the quality and representativeness of the datasets used by the ML algorithms [48,51,52]. In their review of the applications of ML in animal and veterinary public health surveillance, Guitian et al. [51] note that ML algorithms can and have been used to increase the likelihood of pathogen detection, allowing for the prioritization of samples most likely to yield a positive result. However, few studies have explored the efficient and practical deployment of ML algorithms to real-world scenarios [52,53]. Hennessey et al. [49] concluded that the establishment of large, labeled, open-source data sets and the incentivization of ML researchers to take an interest in the veterinary space will be essential for progress to continue, as it has in human medicine.

2. Objective and Hypothesis

The main objective of this study was to assess the ability of tree-based classification models to predict a Johne's test result based on concurrent milk components testing results, cow demographic data, historical Johne's testing of the cow, and the Johne's positivity rate of the herd. As a secondary objective, we compared the results of decision tree (DT) and random forest (RF) models to illustrate the trade-off between complexity and interpretability commonly faced when fitting ML models. We hypothesize that tree-based classification models can incorporate milk component test results with past Johne's test results to provide accurate predictions on future test results. Further, DTs might have a comparable performance to RFs with results that are more easily interpreted by stakeholders.

3. Material and Methods

3.1. Data Preparation

The data for the current project were obtained from a longitudinal study conducted on Ontario dairy farms from 2010 to 2014. The farms enrolled in this study were conducting frequent (2–4 times per year) whole herd individual cow JD testing and were offered financial compensation for the tests in exchange for access to the test results. These cow-level results consisted of 4407 IDEXX milk ELISA (Idexx laboratories, Westbrook, ME, USA) or Parachek® milk ELISA (Prionics USA, Inc., Omaha, NE, USA) results and their corresponding interpretation based on the manufacturer's instructions (interpretation) (i.e., positive, negative, or suspect) with an additional interpretation category (i.e., high) that represented strong positive ELISA results (>1.0 optical density) [54] for 1197 cows. This "high" classification was part of the Ontario Johne's control program "http://www.johnes.ca/" (accessed 2 November 2021), with the rationale being to target and cull "highly positive" animals to remove animals shedding high amounts of pathogen. All herds in this study were enrolled in milk recording through Lactanet, and the JD tests were conducted on samples collected as part of the herd's routine milk testing. The demographic data and milk testing results—lactation number (lact), cow breed (breed), days in milk (dim), 24 h milk yield (m24), 24 h fat yield (f24), 24 h protein yield (p24), predicted 305-day milk yield (m305), predicted 305-day protein yield (p305), predicted 305-day fat yield (f305), somatic cell count (scc), somatic cell count linear score (sccls), and milk urea nitrogen (mun)—from the corresponding test dates were retrieved from the dairy herd improvement records as collected by Lactanet (n = 4750). The ELISA results from the longitudinal study were combined with the milk testing data and matched by test day and animal identification number (n = 4008). The 4008 JD ELISA results with matched milk testing data represented the results from 1059 animals (70% Holstein, 30% Jersey, average scc = 253.72, average m305 = 8372) across eight different herds, of which there were 164 cows with only one Johne's test result. A total of 895 cows had a minimum of two JD test results. Seven of these animals had a maximum of 10 test results (Figure 1).

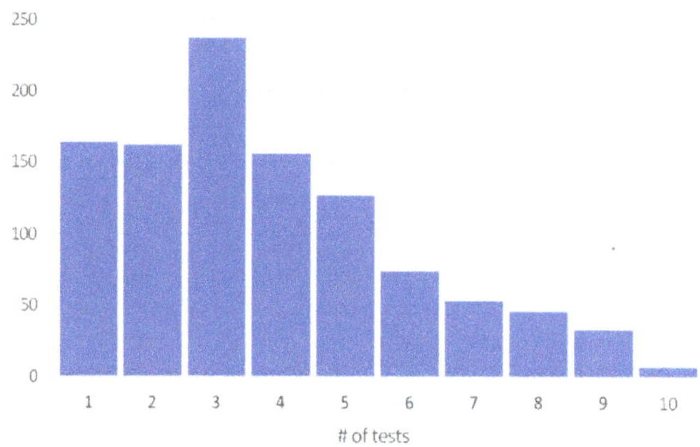

Figure 1. Distribution of the number of tests per cow for 1059 dairy cows undergoing serial Johne's disease milk ELISA testing on eight Ontario farms.

While over 84% of the cows were tested for JD multiple times, each JD test result was considered as an independent observation in the dataset (n = 4008), because each result represented an animal in different lactations, stages of lactation, and involved different protein yields, fat levels, and milk yields, so accommodation was not made for within cow correlation. For each observation, additional variables were created to capture previous Johne's test results for that particular animal (i.e., the number of previous tests (prevtests); the number of tests including the current test that were negative (neg), positive (pos), suspect (susp), or high (high); the number of previous tests that were negative (prevneg), suspect (prevsusp), or high (prevhigh); the proportion of previous tests that were positive (proppos); and the average ELISA (avELISA) score of previous tests). Both the proppos and avELISA were created from all test results except the most current result. The proppos was created by dividing the number of previous positive tests by the total number of previous tests. The avELISA was created by calculating the average ELISA OD from the previous JD tests. In addition, two variables (herd.pos.rt and herd.pos.rt.and.susp) were created that described the context of the herd Johne's positivity in which the animal originated (i.e., the herd positivity rate). This herd positivity variable was created using the whole herd results for the specific test day, with the number of positive results on test day divided by the total number of animals tested. The herd positivity rate was defined as low (L) if less than 5% of the herd tested positive, medium (M) if 5–15% of the herd tested positive, and high (H) if greater than 15% of the herd tested positive on a specific test date. This herd positivity rate was calculated using those results that were reported as positive (positive + high) (herd.pos.rt) and calculated again with the inclusion of suspects in the numerator (suspects + positives + highs) (herd.pos.rt.and.susp).

The data processing and analysis were performed using R Studio Version 1.3.1093. The variables with few or no observations were identified through subsetting and plotting using the heatmap function of the stats package. Variables missing more than 3500 or 80% of their observations were excluded from further analysis. Due to the nature of the RF models, the correlation between the variables may not significantly affect the model outcomes, due to the fact that variables are selected at random for each tree of the RF. However, the remaining variables (after removal of those variables missing 80% of their observations) were checked for collinearity using the Spearman rank correlation coefficient and the cor function in the stats package. The highly correlated variables were identified as those with values greater than 0.8. Of the highly correlated pairs of variables (>0.8), if the variables represented a similar biological measurement, the variable with the most relevance was retained in the model, otherwise the highly correlated pairs of variables both remained in the models. The categorical variables were evaluated for correlation through the predictive power score function pps from the ppsr package.

After removal of the highly correlated variables representing a similar biological measurement and those missing large amounts of data, four different data subsets were generated from the full dataset to test the RF model with different variables. For the first and third data subsets, the JD test results (e.g., propos, avELISA, prevtests, etc.) were removed as predictors to assess how milk components performed as predictors on their own. Within the JD-related predictors, the variables avELISA and proppos were only applicable to animals with repeat tests; therefore, the second and fourth subsets consisted of only observations that contained at least one previous test result. This was achieved by removing the first test for each animal (as the animal at this test had no previous results), as well as animals with only one test. The third and fourth subsets were created to address the concern that there were only 257 'positive', 49 'high', and 37 'suspect' results, as well as 3665 'negative' results, and the low number of high and suspect classifications may limit the prediction accuracy for the model. Therefore, for the third and fourth subsets, the outcome (interpretation) of the first and second subsets was changed from multiclass (negative, suspect, positive, and high) to dichotomous (negative and positive) with the 'high' and 'positive' results being classified in the positive category and the 'suspect' and 'negative' results pooled in the negative category (Table 1).

Table 1. Data subsets used for the training and testing of random forest models used to predict results of Johne's milk ELISA.

Dataset	Number of Observations	Number of Cows [1]	Outcome (Levels)	# of Predictors	Model Performance
Subset 1	4008	1059	Interpretation (negative, suspect, positive, high)	10	κ: 0.054, AUC: 0.514
Subset 2	2949	895	Interpretation (negative, suspect, positive, high)	18	κ: 0.518, AUC: 0.763
Subset 3	4008	1059	Interpretation (negative, positive)	10	κ: 0.038, AUC: 0.372
Subset 4	2949	895	Interpretation (negative, positive)	16	κ: 0.626, AUC: 0.915

[1] The number of cows represented in each data subset. The data used for each of the four subsets contained the Johne's milk ELISA results from cows with repeat Johne's tests from eight dairy farms. Subsets 1 and 3 utilized the milk component and demographic data as predictors; subsets 2 and 4 utilized the milk component, demographic, and previous Johne's test results as predictors. κ: cohen's kappa; and AUC: area under the curve for receiver operating characteristic curve.

3.2. Model Building

Training and testing datasets were created by splitting each of the four data subsets 3:1 (training: testing) using the initial_split function in the rsample package. Due to the fact that the ELISA interpretation results were unbalanced in the data subsets (with many more negative values than positive, suspect, or high results), the data split was stratified by interpretation so that both the training and testing datasets had a similar proportion of positive, suspect, high, and negative observations (Table 2). Using the tidymodels package, the baseline versions of the RF models were trained using the training dataset from each data subset, through bootstrap aggregation, with a default of 500 trees and a random selection of the predictor variables for each node.

Table 2. Comparison of the training and test dataset splits for the four data subsets of 4008 Johne's test results on 1059 Ontario dairy cows used in random forest models to predict the results of Johne's milk ELISA.

	Training Dataset				Test Dataset			
Data Subset	1	2	3	4	1	2	3	4
Number of observations	3006	2211	3006	2211	1002	738	1002	738
Proportion of negative results	0.92	0.9	0.93	0.92	0.91	0.91	0.92	0.91
Proportion of suspect results	0.01	0.01	NA [1]	NA[1]	0.01	0.01	NA [1]	NA [1]
Proportion of positive results	0.06	0.07	0.07	0.08	0.07	0.07	0.08	0.09
Proportion of high results	0.01	0.01	NA [1]	NA [1]	0.01	0.01	NA [1]	NA [1]

[1] Training and test datasets 3 and 4 had a dichotomous outcome and therefore the suspect and high categories were not applicable in the split. The splits were stratified by outcome to ensure equal proportions of outcome classes between the training and testing datasets.

The models were assessed using accuracy measures (accuracy and kappa) and the Variable Importance Plots (VIP) function (Table 3). In the case of accuracy, it is the proportion of data that is correctly predicted, whereas kappa is the proportion of data that is correctly predicted while adjusting for the agreement expected from chance alone. Once both the training and testing were completed on the basic model, three hyperparameters (the number of trees [tree], the minimum number of observations to split at any node [min_n], and the number of variables sampled at each split [mtry]) were utilized for tuning using the tuning function to achieve the highest possible accuracy. This was completed through the creation of a tuning grid and testing the various hyperparameters on five separate cross-validated folds of the training data.

Using the combination of hyperparameters with the highest accuracy (kappa) obtained in the cross-fold validation of each training dataset, the final RF models for each subset were trained using the training dataset, and predictions were generated using the test dataset. The models were again analyzed using the VIP function. The performance of the model was assessed through accuracy measures (kappa and roc area under the curve (AUC)), out-of-bag error, plotting gain curve, and a confusion matrix. Gain curves, also known as cumulative gain curves, are a method for evaluating the performance of a classification model. They demonstrate the percentage of observations in a given category when considering a certain percentage of the observations with the highest probability to be the target category (according to the model). The assessment of overfitting was performed through comparison of the kappa of the trained model and the kappa of the model applied to the test dataset.

The most accurate model generated from the four data subsets was further analyzed through a confusion matrix to describe the model's sensitivity, specificity, positive predictive value, and negative predictive value. Additionally, an exploration of the hyperparameter combinations was completed on the model to demonstrate the effects on kappa.

Decision tree (DT) models were then built using the data subset of the most accurate RF model and the rpart function of the rpart package to assess if a gain in interpretability can be achieved without compromising accuracy when compared to the best performing RF model. The DT was first built with a training subset and the performance was assessed on the testing subset of data. Similar to the approach taken with the RF models, the DT hyperparameters (cost_complexity and tree_depth) were utilized for tuning using the tuning function to achieve the highest possible accuracy. This was completed through the creation of a tuning grid and testing the various hyperparameter combinations on five separate cross-validated folds of the training data. The performance (accuracy, kappa, sensitivity, specificity, positive predictive value, negative predictive value, AUC, and gain curve) of the DT was also evaluated.

Table 3. Variable importance values for the top 10 variables for the four final random forest models used to predict the result of Johne's milk ELISA.

Model 1									
(mtry = 5, min_n = 2, and trees = 1)									
m305	dim	m24	f305	p305	f24	sccls	p24	lact	breed
72.77	66.52	62.21	58.38	52.01	48.93	43.69	41.22	10.04	5.71
Model 2									
(mtry = 18, min_n = 21, and trees = 1000)									
proppos	avELISA	f24	p24	m305	dim	m24	f305	p305	sccls
82.27	78.94	16.26	15.42	14.04	12.03	11.55	11.54	11.12	9.43
Model 3									
(mtry = 1, min_n = 21, and trees = 1)									
f305	dim	p24	m305	sccls	p305	lact	m24	f24	breed
31.36	28.37	16.92	15.82	15.72	13.88	12.71	11.64	9.69	2.92
Model 4									
(mtry = 8, min_n = 2, and trees = 2000)									
avELISA	proppos	prevneg	f24	dim	p24	m305	f305	m24	p305
97.15	50.25	13.15	8.47	7.80	7.19	6.82	5.96	5.94	4.83

The values for variable importance were calculated using the Gini index. Models 1 and 3 used the data from milk components testing only. Models 2 and 4 used the milk component testing results, as well as the Johne's test results. Models 1 and 2 made predictions on a multiclass outcome (negative, suspect, positive, or high) while models 3 and 4 predicted a binary outcome (negative or positive). Mtry = number of variables sampled at each split; trees = number of decision trees; min_n = the minimum number of data points in a node that is required for the node to be split further; m24, f24, and p24: 24 h milk, fat, and protein yields, respectively; m305, f305, and p305: the predicted 305-day milk, fat, and protein yields, respectively; dim: days in milk; lact: lactation number; breed: cow breed; sccls: somatic cell count linear score; prevneg: number of previous Johne's tests that were negative; proppos: proportion of previous Johne's tests that were positive; and avELISA: the average ELISA optical density of previous Johne's test.

4. Results

Among the 895 animals with more than one test result, 92 had a different interpretation result for their last test compared to the result of their first test. There were five animals that changed from positive on their first test to negative on their last test. In total, 87 changed from negative on their first test to positive on their last. Of the 803 animals with an unchanged test result for their last test compared to their first test result (770 negative cows, 33 positive cows), 13 had at least one intervening test that was opposite to the first and last test (e.g., a cow that tests negative on first test, positive on second test, and negative on third test). The variable scc was removed due to a high correlation with sccls and to simplify the model, while mun was dropped due to a high number of missing values (n = 3507).

4.1. Model 1

The first model was trained and tested on data subset 1 (*n* = 4008 observations).

The data contained 10 predictor variables representing the results for milk component testing and an outcome variable with four levels (negative, suspect, positive, and high). In the baseline model with 500 trees, the most important variable in the model was m305. This model yielded an accuracy of 0.913 and a kappa of 0.021. Tuning the hyperparameters (mtry, trees, and min_n) for kappa yielded a model with mtry = 5, trees = 1, and min_n = 2. Over the 5-fold cross-validation, this combination of hyperparameters averaged a kappa of 0.075 and an out-of-bag error of 0.162. The variable m305 remained the most important variable in the optimized final model for subset 1. When the model was applied to the test dataset, a kappa of 0.054 and a roc AUC of 0.514 was achieved (Table 1). There was evidence of overfitting as the model performed better on the trained data (kappa = 0.075) than on the test data (kappa = 0.054). Plotting the gain curve for the model demonstrated a poor predictive ability across all four levels of the outcome (interpretation) (Figure 2).

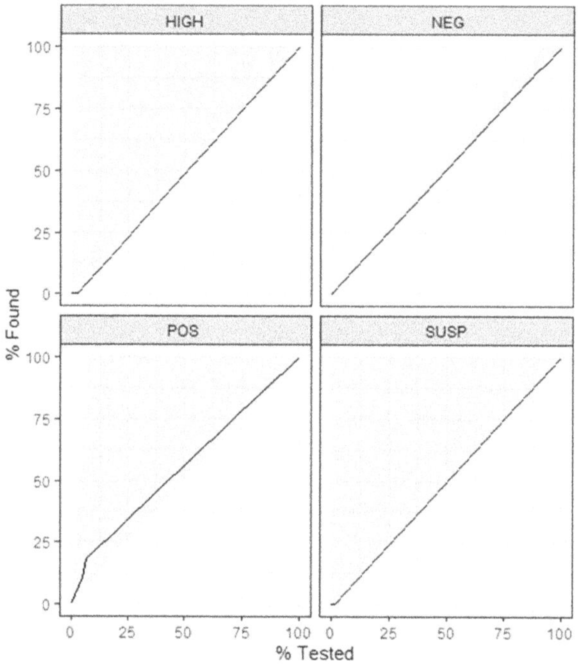

Figure 2. Gain curve for random forest model 1 using the milk component testing results and demographic data to predict the Johne's milk ELISA result across four levels. The grey shading represents the perfect gain curve. Model 1 demonstrates a poor predictive accuracy across all four levels of the outcome.

4.2. Model 2

The second model was trained and tested on data subset 2. This subset consisted of 2949 observations with 18 predictor variables and an outcome variable with four levels (negative, suspect, positive, and high). The most important variable for this baseline model (500 trees) was avELISA. The model yielded an accuracy of 0.924 and a kappa of 0.508. Optimizing the hyperparameters of the model for kappa generated a model with mtry = 18, trees = 1000, min_n = 21, and an out-of-bag error of 0.063. The average kappa across five folds of the training data was 0.543. The variable propPOS was the most important variable within the model with the optimized hyperparameters. When the model was applied to the test dataset, a kappa of 0.518 and a roc auc of 0.763 was achieved (Table 1). There was no clear evidence of overfitting as the model performed only marginally better on the trained data (kappa = 0.543) than on the test data (kappa = 0.518). Plotting the gain curve for the model demonstrated a better performance across all outcomes than with the model generated from just the milk component testing results (Figure 3).

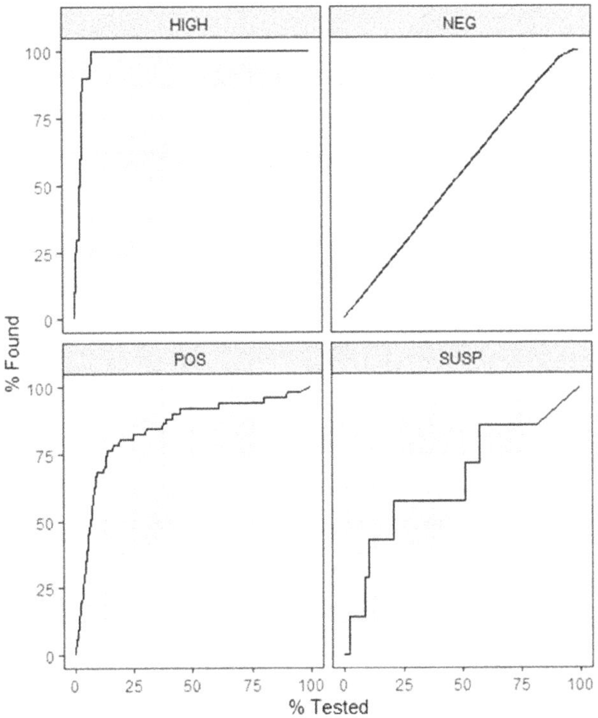

Figure 3. Gain curve for random forest model 2 using the milk component testing results, demographic data, and Johne's test results to predict the Johne's milk ELISA result across four levels. Model 2 demonstrates an improved prediction accuracy with the addition of Johne's test results; however, the model performs poorly for 'suspect' test results.

4.3. Model 3

The data subset used for model 3 contained the same number of observations (and predictors) as data subset 1; however, the outcome variable was dichotomous (positive, negative). The most important variable for the baseline model 3 was m305, similar to the baseline model 1. Applying this baseline model 3 to the test data demonstrated an accuracy of 0.915 and a kappa of 0.019. The hyperparameters optimized for kappa were mtry = 1, trees = 1, and min_n = 21 with an average kappa of 0.076 and an out-of-bag error of 0.084. The most important variable in this optimized model was f305. Applied to the test dataset, the model had a kappa of 0.038 and a roc AUC of 0.372 (Table 1). As in model 1, there was evidence of overfitting as the model performed better on the trained data (kappa = 0.076) than on the test data (kappa = 0.038). The gain curve (Figure 4) demonstrated that less than 50% of the positive tests within the testing dataset are in the top 25% of observations with the highest predicted probability of testing positive.

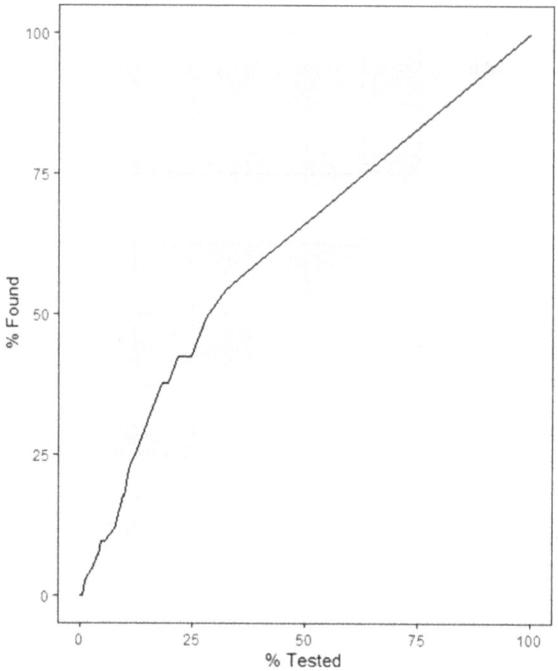

Figure 4. Gain curve for random forest model 3 using the milk component testing results and demographic data to predict the positive test results (Johne's milk ELISA). Less than 50% of the positive test results were correctly predicted within the top 25% of the observations with the highest predicted probability of testing positive.

4.4. Model 4

The final model utilized data subset 4, which contained the same 2949 observations as data subset 2, with 16 predictor variables and a dichotomous outcome variable (positive, negative). The most important variable in baseline model 4 was the avELISA, and the model achieved an accuracy of 0.943 and a kappa of 0.614 when applied to the training data. The optimal (for kappa) combination of hyperparameters for model 4 was mtry = 8, trees = 2000, and min_n = 2, with an out-of-bag error of 0.038. This combination of hyperparameters achieved a kappa of 0.616 across five folds of the training dataset. Similar to the baseline model, the avELISA was the most important predictor in the model. When applied to the test data, the optimized model 4 demonstrated a kappa of 0.626 and a roc AUC of 0.915 (Table 1). Similar to what was seen in model 2, the performance of model 4 applied to the training data was similar to that of test data, and therefore no evidence of overfitting was present. Through plotting the gain curve (Figure 5), it was seen that over 80% of the positive test results were within the top 25% of the observations with the highest predicted probability of testing positive. Finally, a confusion matrix was tabulated with the predicted classes vs. true classifications. From this, the sensitivity and specificity of the model was calculated with a sensitivity of 72% and a specificity of 98%. There was a positive predictive value of 0.81 and a negative predictive value of 0.97.

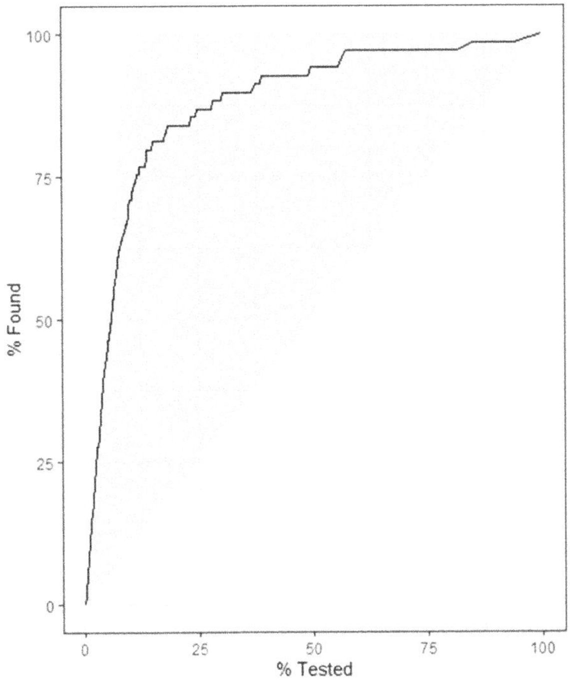

Figure 5. Gain curve for random forest model 4 using the milk component testing results, demographic data, and Johne's test results to predict the positive test results (Johne's milk ELISA). Over 80% of the positive results were correctly predicted to be positive in the model's top 25% of the observations with the highest predicted probability of testing positive.

To further explore the changes in the kappa associated with different combinations of hyperparameters, six separate RF models were trained and tested on data subset 4 without the use of a tuning grid (Table 4). Model 4-D (mtry = 8, min_n = 2, and trees = 1000) yielded a kappa of 73.7% and an AUC of 0.947 (roc curve in Figure 6). In comparison to model 4-A (mtry = 8, tree = 2000, and min_n = 2) that correctly predicted 84% of positive tests in the top 25% of test results with the highest predicted probability, model 4-D was able to correctly predict approximately 90% of the positive tests. The confusion matrix of model 4-D yielded the same sensitivity (72%), specificity (98%), positive predictive value (0.81) and negative predictive value (0.97) as model 4-A.

Table 4. Kappa measures for various combinations of hyperparameter settings applied to model 4 for predicting the results of the Johne's milk ELISA.

Model	Mtry	Trees	Min_n	Kappa	Out-of-Bag Error
4-A [1]	8	2000	2	0.727	0.042
4-B	16	2000	2	0.724	0.043
4-C	8	2000	21	0.737	0.041
4-D	8	1000	2	0.737	0.042
4-E	8	1000	40	0.717	0.041
4-F	16	1000	40	0.727	0.042

[1] The hyperparameters of model 4-A were determined to be the model parameters with the best kappa score according to the tuning grid using a 5-fold cross-validation of the training dataset. Mtry = number of variables sampled at each split, trees = number of decision trees, and min_n = the minimum number of data points in a node that is required for the node to be split further.

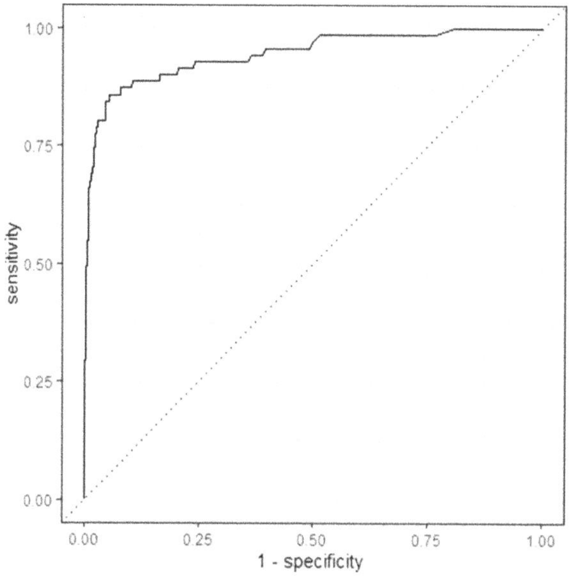

Figure 6. Receiver operating characteristic curve for model 4-D (random forest model predicting the result of the Johne's milk ELISA) with an area under the curve of 0.947. Model 4-D = random forest model with the following hyperparameters: mtry = 8, trees = 1000, and min_n = 2. Mtry = number of variables sampled at each split, trees = number of decision trees, and min_n = the minimum number of data points in a node that is required for the node to be split further.

4.5. Decision Tree

A DT was built using data from subset 4 as this was the same data used in the best-performing RF model. As seen in Figure 7, the most important variable within the tree was the avELISA. According to the DT, less than 3% of animals with an avELISA less than 0.046 were classified as positive. However, 92% of animals with an avELISA equal or greater than 0.11 and more than one third of their previous tests being positive were classified as positive (Figure 7). The DT achieved an accuracy of 0.951, a kappa of 0.629, a sensitivity of 53%, a specificity of 99%, a positive predictive value of 0.85, and a negative predictive value of 0.96. The gain curve of the DT (Figure 8) demonstrates that less than 70% of the positive test results were within the top 25% of the observations with the highest predicted probability of testing positive. Figure 9 provides the output of the receiver operator curve which resulted in an AUC of 0.803. When the DT hyperparameters were optimized for kappa, an AUC of 0.780 was achieved. This DT contained only one split, avELISA < 0.046.

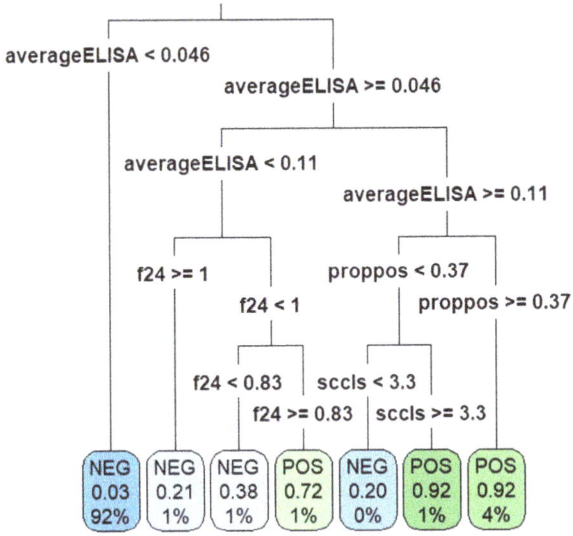

Figure 7. Decision tree to predict test results (Johne's milk ELISA). The bottom (terminal leaf) nodes of the tree are labelled with the end classification, the proportion of results within that node that are positive, and the percentage of observations from the whole dataset present within that node. For example: 92% of the observations had an avELISA less than 0.046 and were classified as negative by the DT; within these observations, 3% were positive.

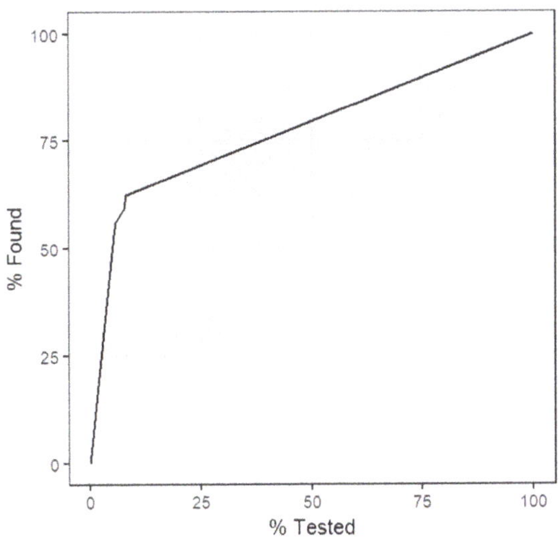

Figure 8. Gain curve for the decision tree using the milk component testing results, demographic data, and Johne's test results to predict the positive test results (Johne's milk ELISA). Less than 70% of the positive test results were within the top 25% of the observations with the highest predicted probability of testing positive.

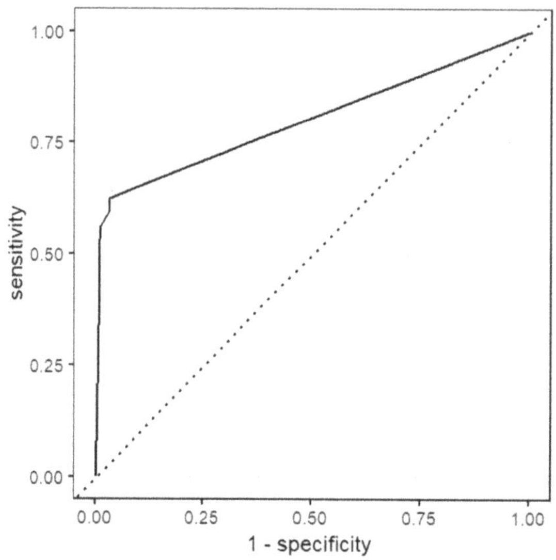

Figure 9. Receiver operating characteristic curve for the decision tree predicting the result of the Johnes milk ELISA with an area under the curve of 0.803.

5. Discussion

The tree-based models evaluated in this study (models 2 and 4) accurately predicted JD milk test results. As has been previously described [55,56], the uncertainty of JD test results, due to poor diagnostic performance, makes management decisions based on these results challenging. There were 105 animals in our dataset that had results that changed over the course of their testing history. These animals represented 12% of all animals with multiple tests; however, they illustrate an important issue in terms of confidently selecting animals to be potentially removed from the herd to limit the possible spread of JD.

Model 1 and model 3 showed evidence of overfitting; however, there was no evidence of overfitting in model 4 as the kappa on the testing dataset was higher than that achieved on the training data, and the kappa for model 2 applied to the test data was only slightly lower than the kappa on the training data [57]. This observation could be due to both the data subsets 1 and 3 containing too few observations for the number of predictor variables. Efforts were made to minimize overfitting by using cross-fold validation of the training data for the tuning of the hyperparameters and bootstrap aggregation for training the optimized models. The number of variables tried at each split was low to help promote randomness within the models.

The model including the concurrent milk component testing results, cow demographic data, historical JD testing of the cow, and JD's positivity rate of the herd (model 4 with 16 predictor variables with a dichotomous outcome) showed the highest accuracy in its predictive probability. This model may provide a tool to help producers target cows for follow-up testing. For example, the final RF model 4-D was able to capture 90% of the positive tests in the top 23% of its predicted classifications. This means that focusing on the results of 173 tests out of 738 total testing dataset results would identify 64 of the 71 positive tests. This suggests that a strategy based on the results of this model could potentially replace whole herd testing as a screening method, by only targeting those animals most likely to generate a positive result. This targeted or weighted testing strategy would, in effect, more efficiently identify candidates for removal from the herd.

The strategy would be based on the predictions generated from the 16 predictor variables included in the model (the most influential variables for model 4 can be found in Figure 10). As seen in the analysis, the variables avELISA, representing the average ELISA

OD of previous tests, and proppos, or the proportion of previous tests that were positive, were the most influential variables in the final model 4 (Figure 10). This finding is not surprising as the more consecutive positive results from previous tests from an animal, the more confident we are that the most recent positive ELISA is not a spurious result. Other researchers have demonstrated that individual animals with high ELISA S/P ratios on previous tests were more likely to test positive on their next test [9,55].

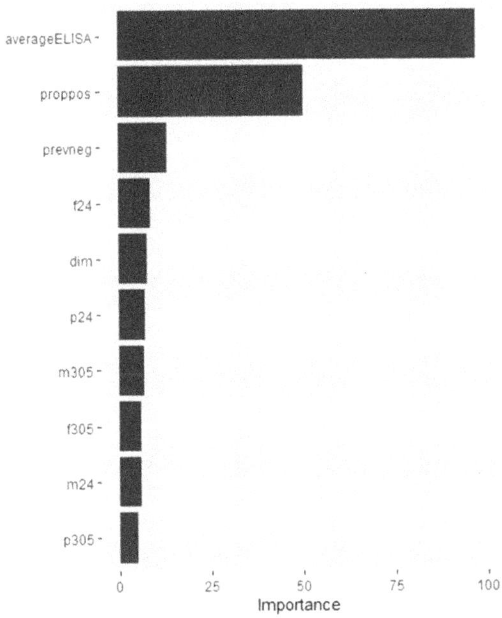

Figure 10. Variable importance plot of the top 10 variables for random forest model 4. m24, f24, and p24: 24 h milk, fat, and protein yields, respectively; m305, f305, and p305: the predicted 305-day milk, fat, and protein yields, respectively; dim: days in milk; prevneg: number of previous Johne's tests that were negative; proppos: proportion of previous Johne's tests that were positive; and averageELISA: the average ELISA optical density of previous Johne's test. The values for variable importance were calculated using the Gini index.

It should be noted that an important input of this model is previous Johne's test results, and therefore it is likely that whole herd testing results would still be required to implement this methodology into a control program. However, in the case of whole herd repeat testing, one round could be reserved for this targeted approach (i.e., a producer doing quarterly whole herd testing could elect to use a targeted approach for one or two of these quarterly tests).

As seen in Table 3, out of the 18 variables that were used to create the RF models, the milk protein and fat concentrations, milk yield, DIM, and sccls were the predictors that were within the 10 most important factors across the four final RF models. The lactation number and herd positivity rates were also important predictors in the RF models. Unfortunately, the interpretation of the variables, beyond their importance in predicting the outcome, is difficult with RF algorithms, often referred to as 'black box' algorithms [58]. However, we can see that the variables that had a large influence in our models' predictions were also significant in other research studies. The association between JD antibody milk positivity and DIM has been demonstrated by Nielsen et al. [59], with cows at the beginning of their lactation being more likely to test positive. Similarly, Nielsen and Toft [60] found that

DIM and milk yield had a significant effect on the odds an animal would test positive on milk ELISA.

Fitting a DT on the same data allows for an interpretable model (Figure 7) at the cost of prediction accuracy and model performance (Figures 8 and 9). This result is not unexpected as random forest models are generally more accurate than individual decision trees because they are an ensemble classifier utilizing multiple trees. This classification results in less overfitting, which provides a better predictive performance [14]. Pruning the DT through hyperparameter optimization resulted in a DT with one node, which results in a much simpler approach to classification with the drawback of having fewer "pure" leaves (e.g., in the group of animals classified as positive, there will be a higher proportion of animals misclassified). Using the DT approach allows a more "hands-on" option for the producer or veterinarian to make decisions surrounding culling or further testing for Johne's management. In comparison, the use of the RF allows for better sensitivity; however, it requires more technical expertise and does not allow for a directly interpretable approach to classification.

It is important to consider that the models generated here are predicting a positive test result of an imperfect test; therefore, it may be possible that some cows predicted as positive are not. However, when considering the most important variables used in predicting the positive result, it is most likely that they are MAP-infected. The ML models used in this study were mainly focused on kappa as the main measure of model performance and did not rely heavily on accuracy. Previous work has shown that accuracy was often an overestimate of model performance when there was a disproportionate number of negative results [57]. Kappa also appears to be a better performance measure for multi-class classification since it allows for the appropriate assessment of how many overall results were classified correctly above chance alone [57].

The classification task may also be more challenging with an increase in classes. As Viazzi et al. [61] demonstrated, a multi-class classification model did not perform as well as their dichotomous classification model. One of the challenges with RF models is that the training and testing of the models requires an adequately large amount of data. As seen in the comparison of the performance of the model using dichotomous data vs. the model trained and tested on data with four outcomes, the sparsity of results with suspect and high cases results in a poor overall performance of the model (AUC of 0.763 for model 2 with four levels of outcome vs. AUC of 0.915 for model 4 with two levels of outcome). There were over 139,000 cows tested over a 4-year period as part of the Ontario Johne's Education and Management Assistance Plan "http://johnes.ca" (accessed on 2 November 2021). The Ontario Animal Health Network reported that 3509 ELISA tests were performed per year in Ontario alone (2021 Q3 OAHN Bovine Veterinary Report, "https://www.oahn.ca" (accessed on 23 February 2022)). This illustrates the potential opportunity to explore machine learning if more detailed data could be collected with these tests.

A possible limitation of the above study was the assumption that the observations used for the RF models were considered independent. While the assumption of independence is not required to build RF models, the lack of independence in our dataset may result in a skewed prediction accuracy. Dependence at the animal level as well as at the herd level could potentially create a model that is overfitted to the data. Another limitation in the current dataset was that while the ELISA results were available in both an interpreted result (i.e., positive and negative) and as an optical density (OD), two different commercial tests had been used. The two tests did not have the same classifications, and thus OD considered on one scale may have led to misclassifications on certain observations, particularly those with a low positive OD. The presented modelling approach should be reevaluated when we have data that are based on the results of one test.

While the dataset initially contained many variables to use as potential predictors, a number of variables were not viable in some of the models and adjustments were required to the datasets and models. For example, the values for the variables proppos and avELISA could not be calculated for animals with no previous test. The variables neg, pos, susp, and high captured the total number of negative, positive, suspect, and high tests, respectively; however, this variable also counted the outcome of the current result. This resulted in a falsely elevated prediction accuracy, and these variables were subsequently excluded. Wagner et al. [62] found that the day-to-day and animal-to-animal variability of activity limited the predictive performance of their models (predicting subacute ruminal acidosis), emphasizing that the nature of the predictors being used in the models must be considered. This importance of predictor selection is echoed by Naqvi et al. [63], where simulated models demonstrated that data quality and within data variability impacts the model performance. This suggests that model performance may be limited if predictors are only chosen from existing or routinely collected data rather than features/predictors selected a priori based on prior knowledge (based on the associations between outcome and predictors). Researchers designing studies to develop deep learning models should therefore ensure that predictors are selected in such a way as to reduce as much as possible the unexplained farm- and cow-level variability to maximize model performance, whereas the between-farm differences in disease recording and sensor availability seem to have a smaller impact on the performance [63].

6. Conclusions

Machine learning is a tool that shows great promise in helping utilize patterns in data to predict outcomes. Its application in dairy science, dairy cow management, and veterinary medicine is still in its infancy. In our study, RF models provided a higher accuracy compared to DT models to predict JD milk ELISA positive results. However, the latter had the benefit of being easily interpreted by end users having to make culling decisions. Very few paratuberculosis studies have utilized these methods. A challenge with these approaches is understanding exactly how best to apply them to certain datasets. Random forest models allow us to incorporate measurements that may not be clearly associated with our outcome of interest but lend important information to predictive modelling. The RF models generated in our study show the potential to reduce the burden of testing for paratuberculosis. The use of milk testing data was of some importance to the models but was only of minor importance when in the presence of prior JD test data. Ideally, researchers should plan their future studies so that the data collected can accommodate ML analyses. The collection of large amounts of data, both as independent results as well as multiple variables per result, will assist us in generating better predictive models. Future opportunities exist with the collection and analysis of more detailed data, specifically, a standardized ELISA OD, the trend data for milk components, metabolites such as BHB and MUN, and the use of average ELISA OD as predictors.

Author Contributions: Conceptualization, J.I., D.K. and J.C.A.-S.; methodology, J.I.; software, J.I.; validation, J.I., C.B. and J.C.A.-S.; formal analysis, J.I.; investigation, J.I. and S.R.; resources, D.K. and S.R.; data curation, J.I. and S.R.; writing—original draft preparation, J.I.; writing—review and editing, J.I., C.B. and J.C.A.-S.; visualization, J.I.; supervision, D.K., S.R. and C.B.; project administration, D.K.; and funding acquisition, D.K. All authors have read and agreed to the published version of the manuscript.

Funding: This work was supported by the Ontario Agri-Food Alliance (Guelph, ON, Canada). The primary author received a graduate student stipend through the Ontario Veterinary College (Guelph, ON, Canada) Fellowship program. Arango-Sabogal's position is funded by the Dairy Farmers of Canada, Novalait, and the Agri-Food Innovation Partnership Program under the Canadian Agricultural Partnership, an agreement between the governments of Canada and Québec.

Institutional Review Board Statement: Not applicable.

Informed Consent Statement: Not applicable.

Data Availability Statement: The data are contained within the article.

Acknowledgments: The authors extend a sincere thank you to Zvonimir Poljak who provided guidance and a starting point for this exploration into machine learning algorithms.

Conflicts of Interest: The authors declare no conflicts of interest.

References

1. Rasmussen, P.; Barkema, H.W.; Mason, S.; Beaulieu, E.; Hall, D.C. Economic losses due to Johne's disease (paratuberculosis) in dairy cattle. *J. Dairy Sci.* **2021**, *104*, 3123–3143. [CrossRef] [PubMed]
2. Whittington, R.; Donat, K.; Weber, M.F.; Kelton, D.F.; Nielsen, S.S.; Eisenberg, S.; Arrigoni, N.; Juste, R.; Sáez, J.L.; Dhand, N.K.; et al. Control of paratuberculosis: Who, why and how. A review of 48 countries. *BMC Vet. Res.* **2019**, *15*, 1–29. [CrossRef] [PubMed]
3. McAloon, C.G.; Roche, S.; Ritter, C.; Barkema, H.W.; Whyte, P.; More, S.J.; O'Grady, L.; Green, M.J.; Doherty, M.L. A review of paratuberculosis in dairy herds–Part 1: Epidemiology. *Vet. J.* **2019**, *246*, 59–65. [CrossRef] [PubMed]
4. McAloon, C.G.; Roche, S.; Ritter, C.; Barkema, H.W.; Whyte, P.; More, S.J.; O'Grady, L.; Green, M.J.; Doherty, M.L. A review of paratuberculosis in dairy herds–Part 2: On-farm control. *Vet. J.* **2019**, *246*, 54–58. [CrossRef] [PubMed]
5. Eisenberg SW, F.; Veldman, E.; Rutten, V.P.; Koets, A.P. A longitudinal study of factors influencing the result of a *Mycobacterium avium* ssp. paratuberculosis antibody ELISA in milk of dairy cows. *J. Dairy Sci.* **2015**, *98*, 2345–2355. [CrossRef] [PubMed]
6. McAloon, C.G.; O'Grady, L.; Botaro, B.; More, S.J.; Doherty, M.; Whyte, P.; Saxmose Nielsen, S.; Citer, L.; Kenny, K.; Graham, D.; et al. Individual and herd-level milk ELISA test status for Johne's disease in Ireland after correcting for non-disease-associated variables. *J. Dairy Sci.* **2020**, *103*, 9345–9354. [CrossRef] [PubMed]
7. Fecteau, M.E.; Whitlock, R.H.; Buergelt, C.D.; Sweeney, R.W. Exposure of young dairy cattle to *Mycobacterium avium* subsp. paratuberculosis (MAP) through intensive grazing of contaminated pastures in a herd positive for Johne's disease. *Can. Vet. J.* **2010**, *51*, 198. [PubMed]
8. Norby, B.; Fosgate, G.T.; Manning, E.J.; Collins, M.T.; Roussel, A.J. Environmental mycobacteria in soil and water on beef ranches: Association between presence of cultivable mycobacteria and soil and water physicochemical characteristics. *Vet. Microbiol.* **2007**, *124*, 153–159. [CrossRef] [PubMed]
9. Hirst, H.L.; Garry, F.B.; Salman, M.D. Assessment of test results when using a commercial enzyme-linked immunosorbent assay for diagnosis of paratuberculosis in repeated samples collected from adult dairy cattle. *J. Am. Vet. Med. Assoc.* **2002**, *220*, 1685–1689. [CrossRef] [PubMed]
10. Awaysheh, A.; Wilcke, J.; Elvinger, F.; Rees, L.; Fan, W.; Zimmerman, K.L. Review of medical decision support and machine-learning methods. *Vet. Pathol.* **2019**, *56*, 512–525. [CrossRef] [PubMed]
11. Aguilar-Lazcano, C.A.; Espinosa-Curiel, I.E.; Ríos-Martínez, J.A.; Madera-Ramírez, F.A.; Pérez-Espinosa, H. Machine Learning-Based Sensor Data Fusion for Animal Monitoring: Scoping Review. *Sensors* **2023**, *23*, 5732. [CrossRef] [PubMed]
12. Bao, J.; Xie, Q. Artificial intelligence in animal farming: A systematic literature review. *J. Clean. Prod.* **2022**, *331*, 129956. [CrossRef]
13. Basran, P.S.; Appleby, R.B. The unmet potential of artificial intelligence in veterinary medicine. *Am. J. Vet. Res.* **2022**, *83*, 385–392. [CrossRef] [PubMed]
14. Gunakala, A.; Shahid, A.H. A comparative study on performance of basic and ensemble classifiers with various datasets. *Appl. Comput. Sci.* **2023**, *19*, 107–132. [CrossRef]
15. Osisanwo, F.Y.; Akinsola, J.E.; Awodele, O.; Hinmikaiye, J.O.; Olakanmi, O.; Akinjobi, J. Supervised Machine Learning Algorithms: Classification and Comparison. *Int. J. Comput. Trends Technol.* **2017**, *48*, 128–138. [CrossRef]
16. Matvieiev, M.; Romasevych, Y.; Getya, A. The Use of Artificial Neural Networks for Prediction of Milk Productivity of Cows in Ukraine. *Kafkas Üniversitesi Veteriner Fakültesi Dergisi* **2023**, *29*, 289.
17. Awaysheh, A.; Wilcke, J.; Elvinger, F.; Rees, L.; Fan, W.; Zimmerman, K. Identifying free-text features to improve automated classification of structured histopathology reports for feline small intestinal disease. *J. Vet. Diagn. Investig.* **2018**, *30*, 211–217. [CrossRef] [PubMed]
18. Awaysheh, A.; Wilcke, J.; Elvinger, F.; Rees, L.; Fan, W.; Zimmerman, K.L. Evaluation of supervised machine-learning algorithms to distinguish between inflammatory bowel disease and alimentary lymphoma in cats. *J. Vet. Diagn. Investig.* **2016**, *28*, 679–687. [CrossRef] [PubMed]
19. Flory, A.; Kruglyak, K.M.; Tynan, J.A.; McLennan, L.M.; Rafalko, J.M.; Fiaux, P.C.; Hernandez, G.E.; Marass, F.; Nakashe, P.; Ruiz-Perez, C.A.; et al. Clinical validation of a next-generation sequencing-based multi-cancer early detection "liquid biopsy" blood test in over 1,000 dogs using an independent testing set: The CANcer Detection in Dogs (CANDiD) study. *PLoS ONE* **2022**, *17*, e0266623. [CrossRef] [PubMed]
20. Renard, J.; Faucher, M.R.; Combes, A.; Concordet, D.; Reynolds, B.S. Machine-learning algorithm as a prognostic tool in non-obstructive acute-on-chronic kidney disease in the cat. *J. Feline Med. Surg.* **2021**, *23*, 1140–1148. [CrossRef] [PubMed]
21. Reagan, K.L.; Deng, S.; Sheng, J.; Sebastian, J.; Wang, Z.; Huebner, S.N.; Wenke, L.A.; Michalak, S.R.; Strohmer, T.; Sykes, J.E. Use of machine-learning algorithms to aid in the early detection of leptospirosis in dogs. *J. Vet. Diagn. Investig.* **2022**, *34*, 612–621. [CrossRef] [PubMed]
22. Biourge, V.; Delmotte, S.; Feugier, A.; Bradley, R.; McAllister, M.; Elliott, J. An artificial neural network-based model to predict chronic kidney disease in aged cats. *J. Vet. Intern. Med.* **2020**, *34*, 1920–1931. [CrossRef] [PubMed]
23. Ferreira, T.S.; Santana, E.E.C.; Junior, A.F.L.J.; Junior, P.F.S.; Bastos, L.S.; Silva, A.L.A.; Melo, S.A.; Cruz, C.A.M.; Aquino, V.S.; Castro, L.S.O.; et al. Diagnostic Classification of Cases of Canine Leishmaniasis Using Machine Learning. *Sensors* **2022**, *22*, 3128. [CrossRef] [PubMed]

24. De Vries, A.; Bliznyuk, N.; Pinedo, P. Invited Review: Examples and opportunities for artificial intelligence (AI) in dairy farms. *Appl. Anim. Sci.* **2023**, *39*, 14–22. [CrossRef]
25. Lokhorst, C.; De Mol, R.M.; Kamphuis, C. Invited review: Big Data in precision dairy farming. *Animal* **2019**, *13*, 1519–1528. [CrossRef] [PubMed]
26. Slob, N.; Catal, C.; Kassahun, A. Application of machine learning to improve dairy farm management: A systematic literature review. *Prev. Vet. Med.* **2021**, *187*, 105237. [CrossRef] [PubMed]
27. Wang, Y.; Li, Q.; Chu, M.; Kang, X.; Liu, G. Application of infrared thermography and machine learning techniques in cattle health assessments: A review. *Biosyst. Eng.* **2023**, *230*, 361–387. [CrossRef]
28. Schmeling, L.; Elmamooz, G.; Hoang, P.T.; Kozar, A.; Nicklas, D.; Sünkel, M.; Thurner, S.; Rauch, E. Training and validating a machine learning model for the sensor-based monitoring of lying behavior in dairy cows on pasture and in the barn. *Animals* **2021**, *11*, 2660. [CrossRef] [PubMed]
29. Zhou, X.; Xu, C.; Wang, H.; Xu, W.; Zhao, Z.; Chen, M.; Jia, B.; Huang, B. The Early Prediction of Common Disorders in Dairy Cows Monitored by Automatic Systems with Machine Learning Algorithms. *Animals* **2022**, *12*, 1251. [CrossRef]
30. Hernandez Brenda Contla Lopez-Villalobos, N.; Vignes, M. Identifying Health Status in Grazing Dairy Cows from Milk Mid-Infrared Spectroscopy by Using Machine Learning Methods. *Animals* **2021**, *11*, 2154. [CrossRef] [PubMed]
31. Sykes, A.L.; Silva, G.S.; Holtkamp, D.J.; Mauch, B.W.; Osemeke, O.; Linhares, D.C.; Machado, G. Interpretable machine learning applied to on-farm biosecurity and porcine reproductive and respiratory syndrome virus. *Transbound. Emerg. Dis.* **2022**, *69*, e916–e930. [CrossRef] [PubMed]
32. Pereira, L.E.C.; Ferraudo, A.S.; Panosso, A.R.; Carvalho AA, B.; Mathias, L.A.; Saches, A.C.; Hellwig, K.S.; Ancêncio, R.A. Machine Learning to predict tuberculosis in cattle from the state of Sao Paulo, Brazil. *Eur. J. Public Health* **2020**, *30* (Suppl. 5), ckaa166.849. [CrossRef]
33. Stański, K.; Lycett, S.; Porphyre, T.; Bronsvoort, B.D.C. Using machine learning improves predictions of herd-level bovine tuberculosis breakdowns in Great Britain. *Sci. Rep.* **2021**, *11*, 2208. [CrossRef] [PubMed]
34. Ebrahimie, E.; Ebrahimi, F.; Ebrahimi, M.; Tomlinson, S.; Petrovski, K.R. Hierarchical pattern recognition in milking parameters predicts mastitis prevalence. *Comput. Electron. Agric.* **2018**, *147*, 6–11. [CrossRef]
35. Dhoble, A.S.; Ryan, K.T.; Lahiri, P.; Chen, M.; Pang, X.; Cardoso, F.C.; Bhalerao, K.D. Cytometric fingerprinting and machine learning (CFML): A novel label-free, objective method for routine mastitis screening. *Comput. Electron. Agric.* **2019**, *162*, 505–513. [CrossRef]
36. Bobbo, T.; Biffani, S.; Taccioli, C.; Penasa, M.; Cassandro, M. Comparison of machine learning methods to predict udder health status based on somatic cell counts in dairy cows. *Sci. Rep.* **2021**, *11*, 13642. [CrossRef] [PubMed]
37. Heald, C.W.; Kim, T.; Sischo, W.M.; Cooper, J.B.; Wolfgang, D.R. A computerized mastitis decision aid using farm-based records: An artificial neural network approach. *J. Dairy Sci.* **2000**, *83*, 711–720. [CrossRef] [PubMed]
38. Porter, I.R.; Wieland, M.; Basran, P.S. Feasibility of the use of deep learning classification of teat-end condition in Holstein cattle. *J. Dairy Sci.* **2021**, *104*, 4529–4536. [CrossRef] [PubMed]
39. Machado, G.; Mendoza, M.R.; Corbellini, L.G. What variables are important in predicting bovine viral diarrhea virus? A random forest approach. *Vet. Res.* **2015**, *46*, 1–15. [CrossRef] [PubMed]
40. Punyapornwithaya, V.; Klaharn, K.; Arjkumpa, O.; Sansamur, C. Exploring the predictive capability of machine learning models in identifying foot and mouth disease outbreak occurrences in cattle farms in an endemic setting of Thailand. *Prev. Vet. Med.* **2022**, *207*, 105706. [CrossRef] [PubMed]
41. Camanes, G.; Joly, A.; Fourichon, C.; Ben Romdhane, R.; Ezanno, P. Control measures to prevent the increase of paratuberculosis prevalence in dairy cattle herds: An individual-based modelling approach. *Vet. Res.* **2018**, *49*, 1–13. [CrossRef] [PubMed]
42. Vitense, P.; Kasbohm, E.; Klassen, A.; Gierschner, P.; Trefz, P.; Weber, M.; Miekisch, W.; Schubert, J.K.; Möbius, P.; Reinhold, P.; et al. Detection of *Mycobacterium avium* ssp. paratuberculosis in cultures from fecal and tissue samples using VOC analysis and machine learning tools. *Front. Vet. Sci.* **2021**, *8*, 620327. [CrossRef] [PubMed]
43. Weber, M.; Gierschner, P.; Klassen, A.; Kasbohm, E.; Schubert, J.K.; Miekisch, W.; Reinhold, P.; Köhler, H. Detection of paratuberculosis in dairy herds by analyzing the scent of feces, alveolar gas, and stable air. *Molecules* **2021**, *26*, 2854. [CrossRef] [PubMed]
44. Lee, S.M.; Park, H.T.; Park, S.; Lee, J.H.; Kim, D.; Yoo, H.S.; Kim, D. A Machine Learning Approach Reveals a Microbiota Signature for Infection with *Mycobacterium avium* subsp. paratuberculosis in Cattle. *Microbiol. Spectr.* **2023**, *11*, e03134-22. [CrossRef] [PubMed]
45. Umanets, A.; Dinkla, A.; Vastenhouw, S.; Ravesloot, L.; Koets, A.P. Classification and prediction of *Mycobacterium avium* subsp. Paratuberculosis (MAP) shedding severity in cattle based on young stock heifer faecal microbiota composition using random forest algorithms. *Anim. Microbiome* **2021**, *3*, 1–13. [CrossRef] [PubMed]
46. Capewell, P.; Lowe, A.; Athanasiadou, S.; Wilson, D.; Hanks, E.; Coultous, R.; Hutchings, M.R.; Palarea-Albaladejo, J. A microRNA-based Johne's disease diagnostic predictive system: Preliminary results. *bioRxiv* **2023**. [CrossRef]
47. Taylor, E.N.; Beckmann, M.; Hewinson, G.; Rooke, D.; Mur, L.A.; Koets, A.P. Metabolomic changes in polyunsaturated fatty acids and eicosanoids as diagnostic biomarkers in *Mycobacterium avium* ssp. paratuberculosis (MAP)-inoculated Holstein–Friesian heifers. *Vet. Res.* **2022**, *53*, 1–13. [CrossRef] [PubMed]

48. Ezanno, P.; Picault, S.; Beaunée, G.; Bailly, X.; Muñoz, F.; Duboz, R.; Monod, H.; Guégan, J.F. Research perspectives on animal health in the era of artificial intelligence. *Vet. Res.* **2021**, *52*, 1–15. [CrossRef] [PubMed]
49. Hennessey, E.; DiFazio, M.; Hennessey, R.; Cassel, N. Artificial intelligence in veterinary diagnostic imaging: A literature review. *Vet. Radiol. Ultrasound* **2022**, *63*, 851–870. [CrossRef] [PubMed]
50. Kour, S.; Agrawal, R.; Sharma, N.; Tikoo, A.; Pande, N.; Sawhney, A. Artificial Intelligence and its Application in Animal Disease Diagnosis. *J. Anim. Res.* **2022**, *12*, 1–10.
51. Guitian, J.; Arnold, M.; Chang, Y.; Snary, E.L. Applications of machine learning in animal and veterinary public health surveillance. *Rev. Sci. Tech. (Int. Off. Epizoot.)* **2023**, *42*, 230–241. [CrossRef]
52. Hassan, F.A.; Moawed, S.A.; El-Araby, I.E.; Gouda, H.F. Machine Learning Based Prediction for Solving Veterinary Data Problems: A Review. *J. Adv. Vet. Res.* **2022**, *12*, 798–802.
53. Fuentes, S.; Viejo, C.G.; Tongson, E.; Dunshea, F.R. The livestock farming digital transformation: Implementation of new and emerging technologies using artificial intelligence. *Anim. Health Res. Rev.* **2022**, *23*, 59–71. [CrossRef]
54. Kelton, D.F.; Von Konigslow, T.E.; Perkins, N.; Godkin, A.; MacNaughton, G.; Cantin, R. Quantifying the cost of removing fecals shedders in a voluntary Johne's disease control program. In Proceedings of the 12th International Colloquium on Paratuberculosis, Parma, Italy, 22–26 June 2014; p. 120.
55. Sweeney, R.W.; Whitlock, R.H.; McAdams, S.; Fyock, T. Longitudinal study of ELISA seroreactivity to *Mycobacterium avium* subsp. paratuberculosis in infected cattle and culture-negative herd mates. *J. Vet. Diagn. Investig.* **2006**, *18*, 2–6. [CrossRef] [PubMed]
56. Roche, S.M.; Jones-Bitton, A.; Meehan, M.; Von Massow, M.; Kelton, D.F. Evaluating the effect of Focus Farms on Ontario dairy producers' knowledge, attitudes, and behavior toward control of Johne's disease. *J. Dairy Sci.* **2015**, *98*, 5222–5240. [CrossRef] [PubMed]
57. Montesinos López, O.A.; Montesinos López, A.; Crossa, J. Multivariate Statistical Machine Learning Methods for Genomic Prediction. In *Multivariate Statistical Machine Learning Methods for Genomic Prediction*; Springer Nature: Cham, Switzerland, 2022. [CrossRef]
58. Breiman, L. Random forests. *Mach. Learn.* **2001**, *45*, 5–32. [CrossRef]
59. Nielsen, S.S.; Enevoldsen, C.; Gröhn, Y.T. The *Mycobacterium avium* subsp. paratuberculosis ELISA response by parity and stage of lactation. *Prev. Vet. Med.* **2002**, *54*, 1–10. [CrossRef] [PubMed]
60. Nielsen, S.S.; Toft, N. Effect of days in milk and milk yield on testing positive in milk antibody ELISA to *Mycobacterium avium* subsp. paratuberculosis in dairy cattle. *Vet. Immunol. Immunopathol.* **2012**, *149*, 6–10. [CrossRef] [PubMed]
61. Viazzi, S.; Bahr, C.; Schlageter-Tello, A.; Van Hertem, T.; Romanini, C.; Pluk, A.; Halachmi, I.; Lokhorst, C.; Berckmans, D. Analysis of individual classification of lameness using automatic measurement of back posture in dairy cattle. *J. Dairy Sci.* **2013**, *96*, 257–266. [CrossRef] [PubMed]
62. Wagner, N.; Antoine, V.; Mialon, M.M.; Lardy, R.; Silberberg, M.; Koko, J.; Veissier, I. Machine learning to detect behavioural anomalies in dairy cows under subacute ruminal acidosis. *Comput. Electron. Agric.* **2020**, *170*, 105233. [CrossRef]
63. Naqvi, S.A.; King, M.T.; DeVries, T.J.; Barkema, H.W.; Deardon, R. Data considerations for developing deep learning models for dairy applications: A simulation study on mastitis detection. *Comput. Electron. Agric.* **2022**, *196*, 106895. [CrossRef]

animals

MDPI

Article

Effects of Silirum®-Based Vaccination Programs on Map Fecal Shedding and Serological Response in Seven French Dairy Herds

Fabien Corbiere [1,*], Dorra Guellouz [1], Christian Tasca [1], Laurent Foures [2], Emma Dubaux [2] and Gilles Foucras [1]

1 INRAE, ENVT, UMR 1225 IHAP, 31076 Toulouse, France; gilles.foucras@envt.fr (G.F.)
2 GDS Meuse, 55108 Verdun, France
* Correspondence: fabien.corbiere@envt.fr; Tel.: +33-561-193-234

Simple Summary: Paratuberculosis is an economically important disease in ruminants, and control in affected herds/flocks primarily relies on test and cull strategies and good hygiene and management practices. Vaccination against the causative agent *Mycobacterium avium* subsp. *paratuberculosis* is also used as a tool in some paratuberculosis control programs. The effects of the whole-cell heat-killed Silirum® vaccine on bacterial shedding in the feces and serological response were studied in a controlled field study involving seven French dairy herds, where calves were vaccinated at various ages. A statistically significant reduction of the probability of fecal shedding was observed for cows vaccinated before 4 to 5 months of age compared to non-vaccinated controls. A strong effect of age at vaccination on the serological status investigated in adulthood was also demonstrated, which was also associated with a difference in the protective effect of vaccination.

Abstract: (1) Background: paratuberculosis is an important disease in ruminants, causing worldwide economic losses to the livestock industry. Although vaccination is known not to prevent transmission of the causative agent *Mycobacterium avium* subsp. *paratuberculosis* (Map), it is considered an effective tool for paratuberculosis in infected herds. The objectives of this controlled field study were to evaluate the effects of the whole-cell heat-killed Silirum® vaccine on Map fecal shedding and serological status in dairy herds infected with paratuberculosis. (2) Methods: The serological status (ELISA) and fecal shedding (qPCR) of 358 vaccinated cows were assessed over 3 years in 7 infected dairy herds in the Meuse department, France. Within each herd, cows from the last non-vaccinated birth cohort ($n = 265$) were used as controls. The probability and level of Map fecal shedding and the serological status were modeled using multivariable mixed general linear regression models. (3) Results: Overall, 34.7% of cows tested positive at least once on fecal qPCR, with significant differences between herds, but high shedding levels were observed in only 5.5% of cows. Compared to non-vaccinated seronegative cows, a statistically significant reduction in the probability of Map shedding was found only in cows vaccinated before 4 to 5 months of age that tested negative for Map antibodies throughout the study period (odds ratio = 0.5, 95% confidence interval: 0.3–0.9, $p = 0.008$), but no significant effect of vaccination on the amount of Map shedding could be evidenced. Finally, the younger the cows were when vaccinated, the less they tested positive on the serum ELISA. (4) Conclusions: a beneficial effect of vaccination on Map fecal shedding may exist in cows vaccinated before 4 to 5 months of age. The variability of the serum ELISA response in vaccinated cows remains to be investigated.

Keywords: paratuberculosis; vaccination; serum ELISA; fecal qPCR; longitudinal study; controlled field study

1. Introduction

Paratuberculosis (PTB), also known as Johne's disease, is a severe slow-developing, and incurable proliferative enteritis of cattle, sheep, goats, and other domestic and wild ruminants caused by the intracellular bacterium *Mycobacterium avium* subsp. *paratuberculosis* (Map) [1]. Animals with clinical signs of the disease are generally euthanized or culled from the herd, and sub-clinically infected animals also tend to have reduced milk production [2], a longer calving-to-conception interval [3], and an increased likelihood of culling, leading to large economic losses [4,5]. The control of paratuberculosis is largely hampered by the lack of sensitivity of available diagnostic tests, especially for the detection of sub-clinical infections (i.e., clinically healthy animals) [6], which results in animals shedding Map in their environment without being identified. Moreover, Map remains viable for weeks to months in the environment, facilitating the exposure and contamination of susceptible animals [7–9]. Numerous programs have been developed in many countries to control PTB in domestic animals, and to reduce the within and between herd/flock transmission of Map [10]. At the herd level, control strategies aim to eliminate infected animals from the herd (particularly those affected and infectious—e.g., test-and-cull strategy) to break down the transmission routes of the disease and to reduce the risk of infection, particularly for young animals, by applying strict hygiene and management measures. In some cases, these strategies are successful in reducing the number of subclinically and clinically affected animals [11–13]. However, successful disease control requires the commitment of owners and herd managers over many years. When effective changes are lacking or reduced, clinical cases of PTB continue to occur years and even decades after a control strategy has been put in place [14]. Moreover, the compliance of farmers with recommendations on the control of paratuberculosis can be low, especially when clinical cases have disappeared [15]. Amongst other control strategies, considerable effort has been made to identify candidate genes associated with reduced susceptibility or increased resilience to Map infection, primarily in dairy cattle breeds [16]. Although results are not consistent among studies, these control strategies appear promising [17–20]. It may, however, be years or decades before the benefits of such control strategies can be evaluated, and research studies have only focused on a few breeds.

Vaccination against paratuberculosis was first described as early as 1926 [21], at which time live vaccines were used. Since then, numerous studies have been performed to evaluate the effects of vaccination against Map in cattle, sheep, and goats. It has been demonstrated to be, on average, effective in reducing the clinical incidence of PTB, delaying the onset of the disease and intestinal lesions and reducing fecal shedding of Map [22]. However, vaccination is not widely used due to its possible interference with intradermal testing for bovine tuberculosis [23–25], and is currently allowed in only a few countries around the world [10]. Vaccination is also known to induce a long-lasting antibody response that may interfere with the use of serological tests for the detection of PTB-infected animals [26,27]. In France, after the production of the Neoparasec® vaccine was discontinued in 2001 [28], vaccination against paratuberculosis was not possible, until the Silirum® vaccine (CZ Veterinaria, S.A., Porriño, Spain) was approved in 2014. Its use is, however, only allowed in herds deemed to be free of bovine tuberculosis and where the presence of Map has been confirmed and is subject to authorization by the veterinary authorities. Silirum® is a whole-cell heat-killed Map (strain 316F) in an oil adjuvant. Field studies reporting its effect on the epidemiological dynamic of PTB are scarce and almost all come from the Basque Country, Spain, where it has been widely applied for almost two decades. A significant reduction in the number of culled cows showing lesions associated with PTB and with diffuse or severe PTB lesions was found in vaccinated cows compared to non-vaccinated controls in a single herd where cattle were vaccinated regardless of their age or clinical status [28]. Similarly, a significant attenuation of pre-existing infection in naturally PTB-infected cows vaccinated as adults has been suggested by Alonso-Hearn et al. [29]. Juste et al. (2009) [30] reported a decline in the number of Map shedders and the total amount of Map shed 2 to 4 years after all cattle were vaccinated in 6 dairy herds. This reduction in Map shedding was also demonstrated in animals vaccinated when less than 6 months old or at adult age, with a significant difference depending on the age of vaccination [31]. Based on the analysis of a large cohort from several vaccinated and non-vaccinated herds, the same research group also suggested a specific and non-specific effect of vaccination on cattle lifespan [32]. The Gudair® vaccine (CZ Veterinaria, S.A., Porriño, Spain), related to the Silirum® vaccine and approved for use in sheep and goats, has been intensively used in Australia for almost two decades, and is effective in reducing the incidence of clinical cases, the prevalence of lesion attributable to ovine paratuberculosis, and the prevalence of Map shedding in affected flocks [33,34].

At the initiative of the Groupement de Défense Sanitaire de la Meuse (GDS Meuse), a voluntary vaccination program was launched in 2015 in the Meuse administrative department, France, as a potential aid to control losses associated with paratuberculosis in affected herds already engaged in a control plan. The aim of the present study was, therefore, to evaluate the effects of Silirum®-based vaccination programs on Map fecal shedding and serological response in a subset of these herds. By comparing a cohort of vaccinated and unvaccinated (control) cows reared under the same management conditions within each herd, we sought to evaluate vaccination as a management tool in paratuberculosis control programs.

2. Materials and Methods

2.1. Herd and Cows Selection

Seven dairy herds were selected on a voluntary basis. Herds were infected with Map and involved in the paratuberculosis control plan managed by GDS de la Meuse for between 6 to 18 years at the beginning of the current study. They were amongst the first to join the paratuberculosis vaccination plan launched by GDS de la Meuse after Silirum® was approved in France in 2014. The paratuberculosis control plan relies on hygiene and management measures aiming at reducing calf exposure to Map, including calving in a dedicated paddock, the immediate removal of calves from their dam after birth (before suckling), feeding with colostrum from a dam that tested negative for Map antibodies, and subsequently, with milk replacer or pasteurized waste milk, raising in separate housing with dedicated equipment, and, after weaning, grazing on dedicated pastures with no contact with adult cattle or adult manure. Before vaccination, test-and-cull measures based on the serological testing of all cows over 24 months of age every six months were also implemented. After the introduction of vaccination within each herd, test-and-cull measures were maintained only for non-vaccinated cows. On some occasions, fecal qPCR was performed to evaluate the shedding status of non-vaccinated cows that tested positive for Map antibodies or on vaccinated cows with clinical suspicion of paratuberculosis as a guide for removal from the herd.

All herds were free from tuberculosis, Prim'Holstein herds with free-stall housing for lactating cows, ranging in average size from 81 to 302 cattle over 24 months old and for which vaccination against paratuberculosis had been authorized by the French veterinary services.

In the selected herds, vaccination against paratuberculosis was initiated between April 2015 and April 2016, and was performed only in young animals. For logistic and convenience reasons, at the first vaccination date within each herd, all calves older than 1 month and younger than 12 months were vaccinated. Subsequently, all calves older than 1 month were vaccinated when their number reached a multiple of 5 (i.e., the minimum volume of the Silirum® vaccine bottle (CZ Veterinaria, S.A., Porriño, Spain)). This vaccination schedule explains the variation in age at vaccination within each herd. Each calf received a single dose of 1 mL of Silirum®, administrated subcutaneously by an accredited veterinarian. Animal id and vaccination date were recorded on a database managed by the GDS de la Meuse.

Within each herd, a sample of cows was selected for the present study. The selected cows had to be born on the farm and be in their second lactation in the first year of the study (i.e., between May 2018 and 2019). Because the current study was initiated in April

2018, 2–3 years after the start of vaccination depending on the farm, some cows meeting these criteria had not been vaccinated (last birth cohort of unvaccinated cows) and were included as intra-herd controls. Overall, in this setting, 358 vaccinated and 265 non-vaccinated cows were included. A one-sided power calculation, assuming that 25% of unvaccinated cows would shed Map in their feces, indicated that these numbers allowed for a one-third reduction in shedding frequency in vaccinated cows to be evidenced, with a statistical power $1 - \beta = 80\%$ and a type I error $\alpha = 5\%$.

All farmers participated voluntarily and gave written consent that the test results would not be shared with them until the end of the study to avoid bias due to cows being removed from the herd based on a positive result.

Information related to cows (birth herd, date of birth, and calving dates) was extracted from the French Base de Données National d'Identification (BDNI).

2.2. Sample Collection and Handling

Farms were visited on 5 occasions between May 2018 and January 2021 at 6-month intervals, except for the last sampling date, which was delayed due to the COVID-19 lockdown in France. However, because not all cows had entered their second lactation by the first two sampling dates (i.e., May and October 2018), and considering the high renewal rate in dairy herds, the total number of samples collected per cow was typically between 1 and 3 (see Section 3). A handful of feces was sampled from the rectum of selected animals using single-use gloves without lubricant and was placed in an individually identified sterile plastic bag for transportation. In parallel, a five-milliliter blood sample was also collected from the caudal vein in vacuum tubes without anticoagulant (Vacutainer® System). Feces and serum samples were frozen at −20 °C before analysis, which was carried out within two months from sampling.

Blood and feces samples were collected by accredited veterinarians as part of the routine paratuberculosis control program managed by GDS de La Meuse. All animal owners gave written consent for samples collected from their animals as part of the paratuberculosis control plan managed by GDS de la Meuse to be used in this study. In this context, ethical approval for animal procedures was deemed not necessary by the Ecole Nationale Vétérinaire de Toulouse ethical committee.

2.3. Laboratory Testing

2.3.1. Serological Tests

A commercial ELISA test (ID Screen Paratuberculosis Indirect®, IDVet, Montpellier, France) was applied to serum samples using an overnight incubation protocol following the manufacturer's instructions. All samples were tested in duplicate and the average sample-to-positive (S/P) ratios were used as the final result. Duplicates showing discordant results were systematically retested. Negative and positive controls provided by the manufacturers were included on each ELISA plate, and the manufacturer's guidelines were strictly followed for the interpretation of S/P ratio results: serum samples with S/P values $< 60\%$, between 60 and 70%, and $\geq 70\%$ were considered negative, doubtful, and positive for Map antibodies, respectively.

2.3.2. Fecal Real-Time PCR

First, fecal samples underwent a concentration procedure using the ADIAFILTER system (BioX, Rochefort, Belgium) following the manufacturer's instructions. Ten grams of feces was rehydrated for 1 h in 70 mL of bi-distilled sterile water. The top 10 mL of the supernatant was then filtered and centrifuged using the ADIAFILTER® disposal. Pellets were then resuspended in 500 µL of bi-distilled water and mixed with 300 mg of 100 µm silica beads (PRECELLYS LYSING KITS VK01, Bertin Technologies, Montigny-le-Bretonneux, France) for 30 s at 6800 rpm (i.e., approximately 3200 g) three times in a bead beater (Precellys 24®, Bertin Technologies, Montigny-le-Bretonneux, France). Magnetic bead-based DNA extraction was performed on a Kingfisher Flex® magnetic particle processor (Thermo Fisher Scientific, Courtaboeuf, France), following the NucleoMag 96 tissue protocol (Macherey-Nagel, Hoerdt, France), with the addition of an extraction control (ADIAVET™ PARATB REAL TIME, BioX, Rochefort, Belgium) in each plate well. Samples were subjected to qPCR (ADIAVET™ PARATB REAL TIME, BioX, Rochefort, Belgium), following the manufacturer's instructions. Each sample was also tested for amplification of the internal control. Bi-distilled water and synthetic IS900 DNA provided in the amplification kit were used as negative and positive controls, respectively. In addition, fecal samples from a cow at the clinical stage of paratuberculosis with extensive multibacillary intestinal lesions and from a clinically healthy vaccinated but persistently shedding ewe with only focal paucibacillary intestinal lesions were systematically included in each test plate, following an identical preparation process to that of tested samples. Fifty amplification cycles were performed on a LightCycler 96 (Roche Life Science, Meylan, France), and fluorescent signals were recorded in two channels, with FAM detecting IS900 and VIC detecting the extraction control. Due to the overlapping spectra of the two dyes, a color compensation step was applied. Raw fluorescence data were obtained from the LightCycler 96 and modeled using the qpcR package [35] in R software (version 4.2.2) [36]. Cycle thresholds were determined using the second derivative maximum (CpD2).

All DNA amplifications were ran in duplicate in separate plates, and the average cycle count (Ct values) was used as the final result. Duplicates yielding discordant results (i.e., absolute difference in Ct values > 2.5) were further retested.

Samples that reached fluorescence with a Ct value below 40 were considered positive. A careful examination of late fluorescence curves indicated that they were associated with low but unambiguously positive results up to 40 Ct, while non-specific amplification results could not be ruled out beyond this threshold.

An external quantification curve was constructed using triplicate end-point dilutions of a purified Map culture of the reference K10 strain ATCC 19698 from 10^8 to 10 copies of the IS900 gene per qPCR plate well, using the DNA amplification protocol, as described above. This quantification curve was used to approximate the amount of Map in each qPCR-positive fecal sample.

2.4. *Data Analysis*

2.4.1. Data Description and Univariable Analysis

Frequencies and detection rates were compared using the Fisher exact test. Quantitative variables were compared using the Student *t*-test or the Wilcoxon–Mann–Whitney test for ranks when the distributions were strongly skewed or when the sample size was small. A Benjamini–Hochberg correction for multiple comparisons was applied when needed.

2.4.2. Multivariable Analysis

Logistic mixed-effect models were fitted to data available from all cows to model the effect of vaccination on the probability of Map shedding, indicated by a positive fecal qPCR result. The subset of qPCR-positive fecal samples was used to model the effect of vaccination on Map shedding levels, expressed as log10 estimated number of Map per gram of feces (log10 Map$\cdot$g^{-1}). Because the distribution of the response variable was strongly skewed to the right, a gamma error distribution with a log link was used. Fixed effects included age at vaccination (AGEVACC), age at sampling (AGESAMP), days in milk within the sampling lactation (DIM), and the serological status at the time of sampling (SEROSTAT). The effect of quantitative variables (i.e., AGEVACC, AGESAMP, DIM) was modeled using natural B-splines to ensure smoothness in the relationship between the quantitative variables and the outcome or by creating dummy variables with different cutoff values when appropriate. To account for the clustered structure of the data (several sampling points for each cow, cows from different herds), cow and farm random effects were included.

In both models, the statistical unit was the fecal sample Y_{ijk} from cow *i* at sampling point *j* within the *k* herd.

A final logistic mixed-effect model was fitted to the subset of vaccinated cows to model the effect of age at vaccination and potential confounders, including AGEVAC, DIM, AGESAMP, and fecal qPCR status, on the probability of yielding a positive serum ELISA result at the first sampling point.

When needed, separate analyses were performed according to whether doubtful ELISA results were handled as positive or negative.

Data description and analysis were performed using the R software (version 4.2.2) [36]. Generalized linear mixed-effect models were fitted using the glmer function from the "lme4" package [37] and the glmmTMB function [38]. The marginal effects of continuous variables modeled as splines were computed using the "effects" package [39]. Statistical significance was defined as a *p*-value less than 0.05, and the results meeting this criterion were considered significant.

3. Results

3.1. Data Description

3.1.1. Herds and Sampled Cows

During the study period, 358 vaccinated and 265 non-vaccinated control cows were sampled, with unequal distribution across herds (Table 1). On average, this represented 48.4% of cattle older than 24 months, present in the 7 herds (between 38.0% and 58.6% depending on the herd). A majority of cows (n = 365, 58.6%) were sampled twice, and the proportion of cows sampled only once was significantly higher (p = 0.018) amongst the vaccinated cows (31.3%) than the non-vaccinated ones (22.7%) (Table 2).

Table 1. Herds and sampled cows.

Herd	Herd Size [1]	Control Plan [2]	Vaccination Plan [3]	Serological Incidence (%) [4]	Number of Sampled Cows	Vaccinated Cows	Non-Vaccinated Cows
A	302	February 2003	August 2015	5.7	138	64	74
B	140	February 2009	April 2016	2.5	82	28	54
C	187	February 2007	April 2015	1.8	71	51	20
D	218	September 2006	June 2015	2.2	95	45	50
E	81	October 2012	April 2016	9.1	43	22	21
F	256	August 2002	May 2015	2.1	146	116	30
G	104	November 2000	June 2015	2.1	48	32	16
Overall	1288				623	358	265

[1] Average number of cattle > 24 months old over the study period. [2] Month and year of entrance in the paratuberculosis control plan managed by the GDS de la Meuse. [3] Month and year of entrance in the paratuberculosis vaccination plan managed by the GDS de la Meuse. [4] Serological incidence in the year of vaccination initiation in ≥ 24-month-old cows.

Table 2. Distribution of vaccinated and non-vaccinated cows by sampling frequency.

	Number of Samples during the Study Period				
	1 (% [1])	2 (%)	3 (%)	4 (%)	Overall
Vaccinated cows	112 (31.3)	212 (59.2)	34 (9.5)	0 (0.0)	358
Non-vaccinated cows	60 (22.6)	153 (57.7)	49 (18.5)	3 (1.1)	265
Overall	172 (27.6)	365 (58.6)	83 (13.3)	3 (0.5)	623

[1] Percent are computed row-wise.

The median age at first sampling was 48.0 months (first quartile Q1 = 43.8 months and third quartile Q3 = 53.0 months), and, overall, 254 out of the 1163 samples (21.8%) occurred in cows older than 5 years. On average, vaccinated cows were 7.5 months younger that non-vaccinated controls at the different sampling dates ($p < 10^{-6}$). More details on age at sampling and time since vaccination are provided in Supplementary Table S1.

The selected cows had received a single dose of Silirum® vaccine at a median age of 4.3 months (Q1 = 2.8, Q3 = 6.9). Significant differences were evidenced across herds ($p < 2 \times 10^{-5}$), with cows in Herd A vaccinated at a median age of 6.1 months (Q1 = 4.4, Q3 = 9.2), and those in Herd D at a median age of 3.1 months (Q1 = 2.1, Q3 = 4.3).

3.1.2. Serological Results

Overall, 13.0% ($n = 151$) of serum samples were positive for Map antibodies, and 1.7% ($n = 20$) were doubtful. The proportion of positive or doubtful results was significantly higher among vaccinated animals (23.7%) than among non-vaccinated animals (3.8%) ($p < 10^{-6}$).

In the 246 vaccinated cows sampled at least twice, 167 (67.9%) remained negative for Map antibodies, 35 (14.22%) were always positive, 25 (10.2%) changed from positive to negative, and 19 (7.7%) inversely changed from negative to positive. For these 19 cows, seroconversion was detected 38.6 +/− 4.3 months after vaccination (minimum: 32.8 months; maximum: 48.0 months). On average, in the 60 cows tested at least twice and that yielded a positive ELISA result on their first sampling date, a significant decline of the S/P value was observed between their first and last test (median difference in S/P values: −23.8, Q1 = −70.0, Q3 = −4.6, $p = 0.001$). This decrease was observed in 39 out of the 60 cows, while for 7 other cows an increase in the S/P values was observed, which was greater than 50%. Conversely, among the 186 cows tested at least twice and that yielded a negative ELISA result on their first sampling date, a slight but significant increase in the S/P value was observed, on average, between their first and last test (median difference in S/P values: 2.2, Q1 = −6.8, Q3 = 15.9, $p = 0.004$). An increase in S/P values was observed in almost half of these cows ($n = 88$, 47.3%), and was greater than 50% in 60 of them, while a decrease higher than 50% occurred in 55 other cows.

In the 205 non-vaccinated control cows sampled at least twice, only 10 (4.9%) cows seroconverted and 1 (0.5%) changed from positive to negative; all other cows remained negative on all sampling occasions. On average, the variation in the S/P values between the first and last serological test was significantly positive (median difference in S/P values: 3.8, Q1 = −4.6, Q3 = 16.9, $p = 10^{-4}$).

The proportion of initially negative cows that mounted an antibody response towards Map that could be detected by ELISA did not significantly differ between vaccinated and non-vaccinated cows ($p = 0.122$).

Seropositive cows were observed in all herds. The proportion of vaccinated cows that tested positive or doubtful for Map antibodies at least once differed significantly between herds, from 13.3% (Herd D) to 68.7% (Herd G). These proportions ranged from 0% (Herd C) to 14.3% (Herd E) for non-vaccinated cows, but the overall small number of ELISA-positive cows ($n = 20$) precluded any formal statistical comparisons between herds.

3.1.3. Fecal qPCR Results

The interplate coefficient of variation (cv) of Ct values for the positive control of the qPCR kit and the external positive fecal controls (i.e., feces from the clinically affected cow and the clinically healthy shedding ewe) were 3.2%, 2.7%, and 3.1%, respectively.

Of the 1160 fecal samples tested, 276 (23.8%) were positive by fecal qPCR. However, only 36 of them (12.9%), originating from 34 cows, were considered highly positive (i.e., corresponding to an estimated Map concentration $\geq 1000 \cdot g^{-1}$ of feces) (Table 3, Figure 1). Six samples had an estimated Map concentration higher than the clinically affected cow control sample (i.e., 174,000 Map$\cdot g^{-1}$ of feces), and 26 had an estimated Map concentration lower than the clinically healthy shedding ewe control sample (i.e., 8 Map$\cdot g^{-1}$ of feces). Overall, 216 out of 623 cows (34.7%) (38.3% of vaccinated cows ($n = 137$) and 29.8% of non-vaccinated ones ($n = 79$), $p = 0.03$) tested positive at least once during the study period, but this overall result does not account for the unbalanced distribution of cows between herds.

Table 3. Number (%) of fecal samples and cows according to the estimated Map concentration in the feces (in equivalent number of Map per gram of feces).

	Estimated Map Concentration (Equivalent Number of Map·g^{-1})						Total
	Negative	[5–10^2]	[10^2–10^3]	[10^3–10^4]	[10^4–10^5]	>10^5	
Vaccinated cows							
Number of samples	464 (72.7)	130 (20.4)	20 (3.1)	15 (2.4)	6 (0.9)	3 (0.5)	638
Number of cows [1]	221 (61.7)	97 (27.1)	18 (5.0)	13 (3.6)	6 (1.7)	3 (0.8)	358
Non-vaccinated cows							
Number of samples	420 (80.5)	74 (14.2)	16 (3.1)	4 (0.8)	4 (0.8)	4 (0.8)	522
Number of cows [1]	186 (70.2)	53 (20.0)	14 (5.3)	4 (1.5)	4 (1.5)	4 (1.5)	265
Overall							
Number of samples	884 (76.2)	204 (17.6)	36 (3.1)	19 (1.6)	10 (0.9)	7 (0.6)	1160
Number of cows [1]	407 (65.3)	150 (24.1)	32 (5.1)	17 (2.7)	10 (1.6)	7 (1.1)	623

[1] For a given cow, the highest estimated concentration found across its fecal samples was used.

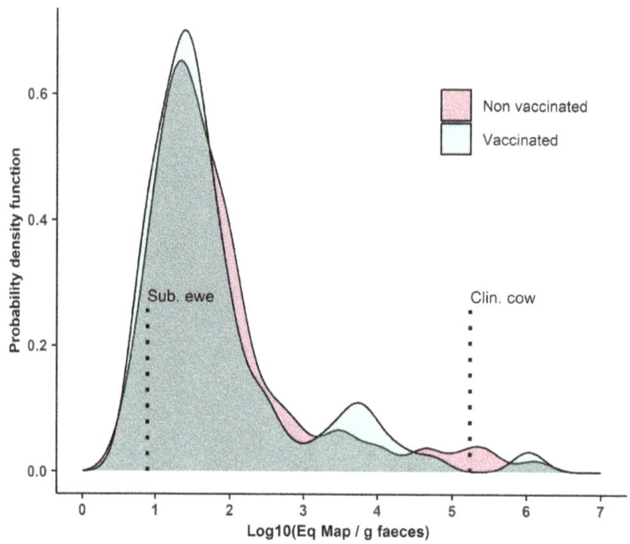

Figure 1. Distribution of the estimated amount of Map shed in the feces of qPCR-positive samples (in log10 equivalent Map·g^{-1}), given the vaccination status. To aid interpretation, non-vaccinated cows are shown in red and vaccinated cows are in green. Vertical dotted lines: values for a clinically affected cow (Clin. cow) and a vaccinated clinically healthy but persistently shedding ewe (Sub. ewe).

Out of the 246 vaccinated cows that were sampled at least twice, 146 (59.3%) consistently tested negative for fecal qPCR and 33 (13.4%) consistently tested positive, while the remaining 82 cows (33.3%) had varying results across the different sampling points, either testing positive or negative. In the 205 non-vaccinated control cows sampled at least twice, 141 (68.8%) were consistently negative and 21 (10.2%) were consistently positive.

Cows that tested positive for qPCR were found in all herds, but in significantly different proportions, from 13.4% (11 out of 82, Herd B) up to 56.2% (27 out of 48, Herd G) and 71.2% (104 out of 146, Herd F). High-shedding cows (i.e., estimated Map concentration $\geq$ 1000·g^{-1} of feces) were found in all herds except Herd C. The detailed distributions of fecal samples and cows according to herd, vaccination status, and estimated Map concentration are provided in Tables S1 and S2, respectively.

3.2. Effect of Silirum® Vaccination on the Probability of Fecal Map Shedding

None of the logistic mixed regression models that were fitted indicated a significant effect on the odds of Map shedding of either days in milk within the lactation ($p = 0.56$) or age at sampling ($p = 0.22$). In all instances, the herd random effect was highly significant ($p < 10^{-6}$), indicating a strong herd effect on the probability of Map shedding.

Overall, the probability of shedding was lower in vaccinated cows compared to non-vaccinated ones, but this result only approached statistical significance (OR = 0.72, 95% confidence interval: 0.50–1.02, $p = 0.07$).

After adjusting for age at vaccination, a significant reduction in the odds of Map shedding was only observed in cows that were vaccinated before 4 months of age ($n = 169$ cows), when compared to non-vaccinated cows (OR = 0.58, 95% CI = 0.42–0.94, $p = 0.025$) (Figure 2A). Distinguishing between cows vaccinated before and after 5 months of age yielded very similar results, but was associated with a slightly lower fit to the data.

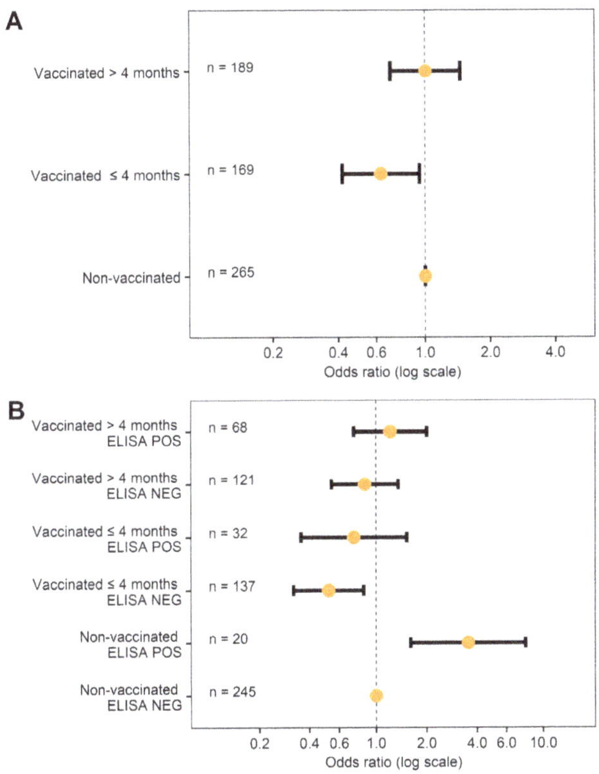

Figure 2. Graphical representation of the effect of vaccination on the odds of a positive result on fecal qPCR. Results are expressed as odds ratio (OR) (orange points) and 95% confidence interval (horizontal black lines). In both panels, the lowest line represents the reference group (OR = 1). The number of cows in each group is indicated within each line. (**A**) Results from the final logistic mixed regression model accounting for vaccination status and age at vaccination. (**B**) Results from final logistic mixed regression model accounting for vaccination status, age at vaccination, and serological status.

Finally, the best fit to the data was provided by models also accounting for the serological status. Indeed, a significant reduction in the odds of shedding was observed in cows vaccinated before 4 months of age and that tested negative for Map antibodies throughout the study period ($n = 137$) (OR = 0.52, CI = 0.32–0.85, $p = 0.008$), but not for those that were vaccinated after 4 months of age and/or tested positive at least once on ELISA ($p > 0.5$) (Figure 2B). Once again, distinguishing between cows vaccinated before and after 5 months of age led to very similar results. Results also indicated that among the non-vaccinated group, cows that tested positive at least once on serum ELISA ($n = 20$) were at higher risk to test positive with fecal qPCR compared to those that always tested negative ($n = 245$) (OR = 3.55, 95% CI: 1.61–7.84, $p = 0.002$).

A strong significant herd effect was found ($p < 10^{-6}$), with Herds A and B associated with the lowest odd of a positive result and Herds F and G with the highest.

Finally, these findings were still found when only samples with an estimated concentration of Map > $100 \cdot g^{-1}$ were considered positive ($n = 72$). Doubtful ELISA results did not influence model outputs and only the results where they were handled as positive are reported.

3.3. *Effect of Silirum® Vaccination on the Level of Fecal Map Shedding*

This analysis was performed on the subset of qPCR-positive fecal samples ($n = 276$). No difference in the estimated amount of Map shedding could be demonstrated between cows that always tested negative on serum ELISA and vaccinated cows, regardless of their age at vaccination or their serological status ($p = 0.58$). Conversely, non-vaccinated cows that tested positive on serum ELISA at least once had significantly higher shedding levels than all other cows ($p < 10^{-4}$). It is noteworthy that high shedders (i.e., estimated Map concentration $\geq 1000 \cdot g^{-1}$ of feces) were observed in all groups (Figure 3). Furthermore, we found that the amount of Map shedding was significantly and positively related to the age at which cows were sampled ($p = 0.006$), whatever their vaccination status (Figure 4). Again, a strong significant effect of herd was observed ($p < 10^{-6}$), with Herds C and D associated with the lowest Map loads and Herds F and G with the highest.

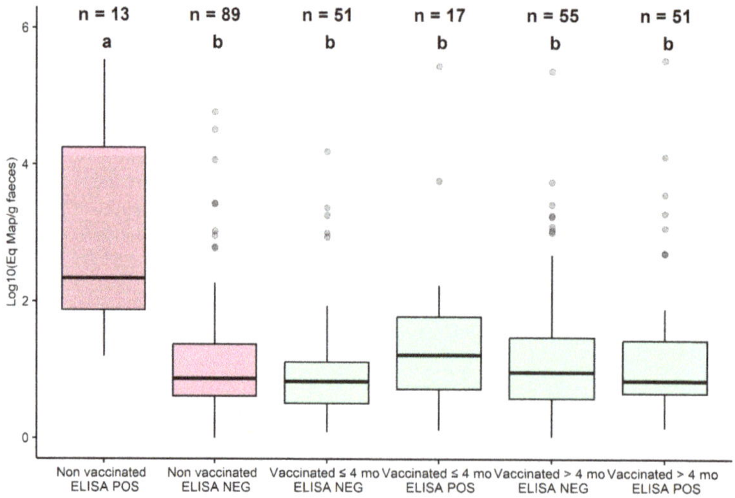

Figure 3. Distribution of the estimated amount of Map shedding in the feces of qPCR-positive samples, given the vaccination status, age at vaccination, and serological status of cows. To aid interpretation, non-vaccinated cows are shown in red and vaccinated cows are in green. The number of fecal samples in each group is reported at the top. Groups not sharing the same letter are statistically different.

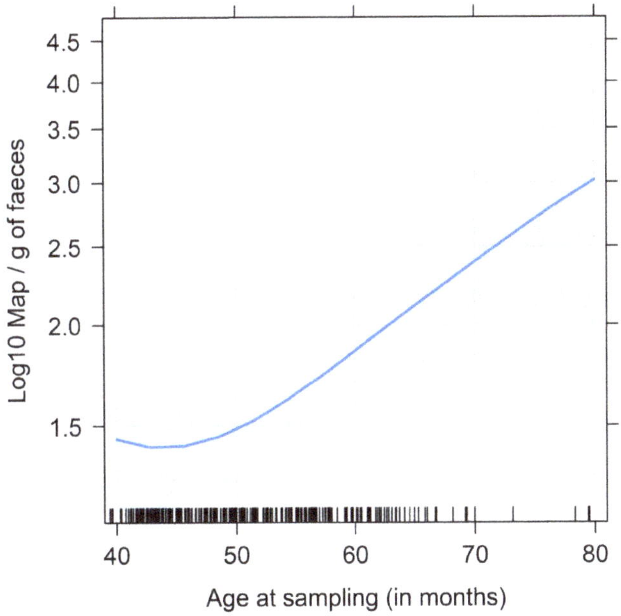

Figure 4. Estimated effect of age at sampling on the amount of Map shedding in the feces. The point estimate is represented by the solid blue line, and its 95% confidence interval is shown by the blue area surrounding the line. The black marks on the bottom represent the observations, giving an idea of their distribution along the *x*-axis. The influence of vaccination status or age at vaccination was not statistically significant. The curve, therefore, represents the marginal effect across all animal categories.

3.4. Effect of Age at Vaccination on Serological Status of Vaccinated Cows

The proportion of vaccinated cows that tested positive or doubtful for Map antibodies at first sampling remained low (81 out of 358, 22.6%). We, therefore, sought to determine the factors influencing the variability of the serological response in vaccinated animals. The analysis was limited to the serological results obtained from the first sampling of each cow and was adjusted on age at vaccination, age at sampling, days in milk within the lactation, and the amount of Map shed in the contemporary fecal sample.

Taking into account the strong herd effect ($p = 10^{-4}$), only age at vaccination had a significant effect ($p = 0.002$), with a higher probability of being positive in Map antibodies for cows that were vaccinated at an older age (Figure 5). As an illustration, a raw description indicated that 34.4% of cows (45 out of 131) vaccinated after 5 months of age tested positive at their first sampling point, whereas only 15.9% (36 out of 227) of cows vaccinated at a younger age did so ($p = 10^{-5}$).

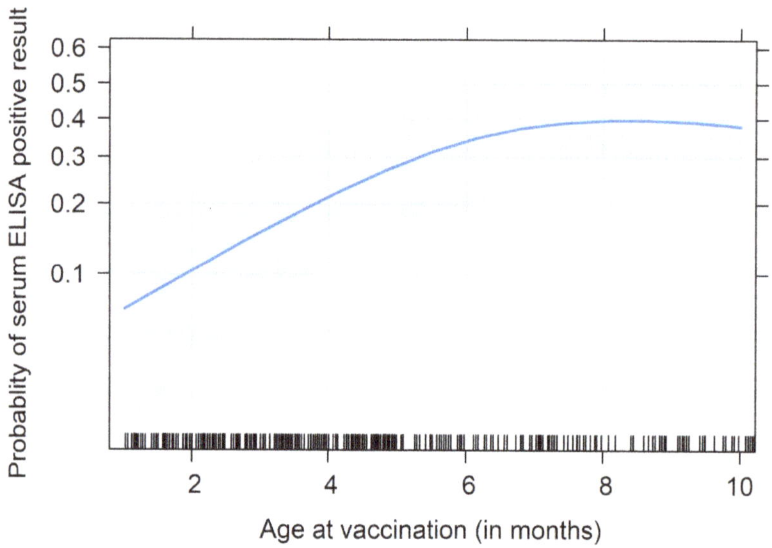

Figure 5. Estimated effect of age at vaccination on the probability of testing positive for Map antibodies at the time of first sampling. The point estimate is represented by the solid blue line, and its 95% confidence interval is shown by the blue area surrounding the line. The black marks on the bottom represent the observations, giving an idea of their distribution along the *x*-axis.

4. Discussion

By comparing a cohort of vaccinated and unvaccinated (control) cows reared under the same management conditions within each herd, this field study sought to evaluate vaccination as a management tool in paratuberculosis control programs. To the best of our knowledge, reports on the effect of Silirum®-based vaccination programs focusing specifically on animals vaccinated when young and including within-herd unvaccinated control individuals are lacking. In many studies, results from bacteriologic culture or fecal qPCR from vaccinated cows were either compared with results for nonvaccinated animals from other farms or with herd prevalence of Map infection before vaccination was implemented [29–31,40,41]. The use of other herds as comparison groups introduces potential bias because the confounding effects of herd management, including practices designed to reduce Map transmission within herd, cannot be entirely controlled without randomization, and comparisons of prevalence within a herd before and after vaccination is initiated ignore the confounding effect of time. Our study was based on a particular vaccination program that targeted only calves under 12 months of age, allowing the use of non-vaccinated animals as controls within each herd. The choice of GDS de la Meuse not to vaccinate animals over 12 months of age at the beginning of the voluntary vaccination program was mainly motivated by financial reasons, but also by the wish to be able to follow the dynamics of infection through serological testing of non-vaccinated cows. Additionally, although a reduction in Map shedding has been described in already infected cows vaccinated as adults [29,30], the control plan managed by the GDS de la Meuse focused mainly on reducing the exposure of young animals to Map. Our choice to focus only on cows within their second lactation at the beginning of the follow-up was motivated by a compromise between the occurrence of Map shedding, which we expected to be higher as cows aged [42], and the high culling rates commonly observed in French dairy herds (average renewal rate in French Prim-Holstein herds 34%, mean lactation number per cow 2.3). In addition, at the initiation of our study in 2018, all first lactation cows had already been vaccinated (in four of the seven herds), precluding the possibility of having a control group of non-vaccinated cows among this lactation number.

Management changes aiming to reduce the transmission of Map were enacted long before the introduction of vaccination in all herds, and these changes have probably reduced the incidence of infection in young animals. In the year before the introduction of vaccination, the serological incidence in non-vaccinated cows was below 3% in 5 of the 7 herds, suggesting that the infection pressure may be low, as also supported by the low frequency of cows that seroconverted during the follow-up in the non-vaccinated control group ($n = 10$, 4.9%) or the vaccinated group ($n = 19$, 7.7%). It was not a primary objective of the study to measure the effect of vaccination alone on the change in the herd prevalence of Map shedding, given that most herd farmers would use vaccination in combination with other disease control actions. This low infection rate may have reduced the possibility of evidencing a strong effect of vaccination, as would be expected in herds with high clinical incidence [31,32,42,43]. This reduction in the infection pressure may have impacted our findings with respect to detecting differences between vaccinates and controls. Similarly, in studies in which all animals except controls are vaccinated [32], an indirect protection of non-vaccinated cows may bias the results towards the null, by reducing the environmental bacterial loads to which they are exposed within the herds. This potential bias can be excluded in our study because adult cows were not vaccinated, and also because the cows used as controls were born a few months before the cohort of calves that comprised the vaccinated group. Conversely, vaccinated cows may not have benefited from herd immunity [44]; with the greater environmental pressure than if all animals in the herd had been vaccinated, the possibility of detecting a protective effect of the vaccine is reduced.

Our findings that vaccinated cattle had detectable fecal shedding less often compared to controls contribute to the growing body of knowledge on the role of vaccination in achieving paratuberculosis control. However, it is well-established that vaccination does not guarantee the complete prevention of actual infection in animals, and that there exists a possibility of high shedding among vaccinated animals [22,29,30,42]. Our results are consistent with these findings, with some cows having fecal samples with high concentrations of Map regardless of their vaccination status or age at vaccination. It is noteworthy that non-vaccinated cows that tested positive on serum ELISA at least once had, on average, significantly higher amounts of Map in their feces than all other cows. Although no individual information was collected regarding clinical status at the time of sampling, we can speculate that some of these unvaccinated ELISA-positive cows were at or near the clinical disease stage [6,45]. If vaccination was able to reduce or delay the onset of clinical paratuberculosis, as reported by others [22], a direct beneficial effect on the overall amounts of Map excreted within a herd could be expected. We also found a significant effect of aging on the amount of Map shedding by infected cows, including vaccinated ones, which raises the question of the duration of the protection offered by the vaccine. Although this issue has been little studied, it has been pointed out in longitudinal studies in sheep or goats vaccinated with the killed whole-cell Gudair® vaccine that it could last over the entire production life of animals [33]. This question, however, still requires further work in cattle for the Silirium® vaccine. The fact that a few vaccinated cows may shed large amounts of Map in their feces, even among those testing negative for Map antibodies, is an argument for maintaining vaccination and good management practices for a very long period of time, even after clinical cases have disappeared, before a beneficial effect on Map transmission dynamics may be evidenced [46,47].

A significant effect of the age at vaccination was observed, with a beneficial effect on the odd of Map shedding demonstrated only in cows vaccinated before 4 to 5 months of age. Our findings are in accordance with other studies [31]; the age at vaccination may play an important role in the mounting of an immune response capable of delaying disease progression at the individual level, or that the exposure to Map during a longer period of time in animals vaccinated at older ages may reduce vaccine efficiency [48, 49]. Experimental and field studies in lambs, goats, and calves have demonstrated that vaccination may be effective when administrated in animals before or around 1 month of age [42,50–52]. However, a decrease in Map shedding has also been shown in cattle vaccinated as adults and presumably already infected [29,30]. Future field and experimental studies focusing on the protective and/or therapeutic effect of vaccination according to the age at which it is administered would be of interest, as this could have practical implications for defining the best vaccination schedule.

In all models, a strong herd effect was evidenced, meaning that conclusions drawn on the overall effect of vaccination may not be valid in all herds. Interestingly Herds F and G were associated with both a higher probability of shedding and a higher amount of Map shedding in feces, even in vaccinated cows, suggesting that paratuberculosis control was less efficient in those herds. Common management practices designed to reduce Map transmission within each herd were monitored, and farmers were interviewed regularly to ensure that all the producers implemented them similarly, but departures from the best practices may have happened occasionally or repeatedly, which may explain different trajectories at the herd level [47]. Others factors that may influence animal immunity, such as nutritional factors, worm burdens, lameness, or the lack of general cleanliness, which favors the maintenance and transmission of Map, could also contribute to differences between farms, but were not investigated.

Vaccination with either attenuated or inactivated paratuberculosis vaccines is known to induce a strong and long-lasting humoral response, but results vary widely according to the vaccine used or age at vaccination. Interindividual differences are also often reported. In a longitudinal field study involving calves vaccinated before 35 days of age with Mycopar® killed whole-cell vaccine, antibody response was still detected, as assessed by serum ELISA, up to 7 years after vaccination in some animals and, conversely, about one third of the vaccinated animals did not appear to develop a serological response [27]. In lambs vaccinated before 3 months of age with the killed whole-cell Gudair® vaccine, only 15% to 50% of sheep still tested positive on the serum ELISA 2 to 4 years post-vaccination, with large differences between flocks [26]. In contrast, in another field study of 5 flocks where lambs had been vaccinated at 3–6 months old with the same vaccine, the vast majority (>95%) still tested positive on the serum ELISA five years post-vaccination, with no decline in S/P values with age [53]. Interindividual heterogeneity in both humoral and cellular response has also been reported in goats during the first months after vaccination with Gudair® [49]. Traditionally, this humoral response has not been associated with protection against Map infection, as opposed to cellular response [54,55], but a recent experimental study in sheep pointed out that a strong initial B cell response soon after vaccination is associated with the clearance of infection [56]. In our study, only a small proportion of cows tested positive on the serum ELISA (22.6% at first sampling time), and no significant protective effect of vaccination on the odd of Map shedding was observed for those cows that tested positive on the ELISA during the study period, although this finding could be due to a lack of statistical power in relation with the small sample size. The individual trajectories of post-vaccination humoral responses were highly diverse, with a decrease in the ELISA S/P values observed in some vaccinated animals, while no notable variation was observed in others. This low proportion of seropositive animals could be related to the fact that the humoral response was only evaluated in adulthood, 4 to 6 years after vaccination, and does not allow inference about the earlier history of this response. Some of the positive results observed in the ELISA test may also correspond to animals that have responded to a natural infection. However, their proportion appears to be too high, particularly compared to non-vaccinated cows, for this phenomenon alone to be invoked. Interestingly, we found that the younger the cows were vaccinated, the less they tested positive on the serum ELISA, in agreement with previous experimental studies in lambs and kids [49,57]. This supports the hypothesis that the antibody response induced by the paratuberculosis vaccine in young animals is lower than in older ones due to incomplete development of their immune system [58]. Vaccination also induces a Th1/IL-17 immune response [25,59], which is known to play a major role in the protective response against intracellular mycobacterial pathogens. Our findings, therefore, do not imply that the protective effect of vaccination in young animals is lower, as also supported by the significant protective effect of vaccination on the odds of Map shedding found in cows vaccinated when younger than 4 to 5 months of age. Additional research is needed, including the investigation of potential genetic factors [60], to more accurately evaluate the inter-individual variability in serological responses following vaccination and to explore possible links with the efficacy of the vaccine.

5. Conclusions

Altogether, our findings contribute to the growing body of knowledge on the role of vaccination in achieving paratuberculosis control, even in herds where good management and hygiene practices are already enacted. The fact that the beneficial effect of vaccination was only demonstrated in cows vaccinated before 4–5 months of age also provides practical information for the management of vaccination schedules in infected herds. Our results also reinforce the idea that even if beneficial in a few years, long-term vaccination is necessary to reduce the risk of new infections.

Finally, non-specific effects of vaccination against paratuberculosis on lifespan have been suggested in cattle and goats [32,49], which could also add to the beneficial effect of vaccination in infected herds. Although still controversial [61], a protective effect against general mortality unrelated to tuberculosis has similarly been suggested by several studies in children vaccinated with BCG or against measles [62–65]. Other specific and non-specific beneficial effect of vaccination, including increased milk yields and carcass weights, were reported [29,30,32]. The primary objectives of the present study did not include the evaluation of the effect of vaccination on cattle lifespan and production performances, but data are currently being collected in all herds involved in the vaccination plan managed by the GDS de la Meuse to investigate these topics.

Supplementary Materials: The following supporting information can be downloaded at: https://www.mdpi.com/2076-2615/13/9/1569/s1; Table S1: Summary statistics of age at sampling (in months) in vaccinated and non-vaccinated cows according to sampling point; Table S2: Number of fecal samples according to the estimated Map concentration in the feces, in each herd, and given the vaccination status of cows; Table S3: Number of cows according to the estimated Map concentration in the feces in each herd, and given their vaccination status; Table S4: Individual data used in the study.

Author Contributions: Conceptualization, F.C., G.F., E.D. and L.F.; methodology, F.C. and G.F.; software, F.C. and D.G.; validation, F.C. and G.F.; formal analysis, C.T., D.G. and F.C.; writing—original draft preparation, F.C.; writing—review and editing, F.C. and G.F.; supervision, F.C., E.D. and L.F.; project administration, F.C.; funding acquisition, not applicable. All authors have read and agreed to the published version of the manuscript.

Funding: This study was funded by "Institut National de Recherche pour l'Agriculture, l'Alimentation et l'Environnement" (INRAE) and the "Groupement de Défense Sanitaire de la Meuse" (GDS 55).

Institutional Review Board Statement: Blood and feces samples were collected by accredited veterinarians as part of the routine paratuberculosis control program managed by GDS de La Meuse. Ethical approval for animal procedures was deemed not necessary by Ecole Nationale Vétérinaire de Toulouse ethical committee.

Informed Consent Statement: All animal owners gave written consent for samples collected from their animals as part of the paratuberculosis control plan managed by GDS de la Meuse to be used in this study.

Data Availability Statement: The data presented in this study are provided as Supplementary Table S4.

Acknowledgments: We wish to acknowledge farmers for their participation in the study. We gratefully acknowledge the veterinarians for collecting samples. We also acknowledge Frank Biet (INRAE Tours) for providing the Map K10 strain.

Conflicts of Interest: The authors declare no conflict of interest. The funders had no role in the design of the study; in the collection, analyses, or interpretation of data; in the writing of the manuscript; or in the decision to publish the results.

References

1. Clarke, C.J. The Pathology and Pathogenesis of Paratuberculosis in Ruminants and Other Species. *J. Comp. Pathol.* **1997**, *116*, 217–261. [CrossRef] [PubMed]
2. McAloon, C.G.; Whyte, P.; More, S.J.; Green, M.J.; O'Grady, L.; Garcia, A.; Doherty, M.L. The Effect of Paratuberculosis on Milk Yield—A Systematic Review and Meta-Analysis. *J. Dairy Sci.* **2016**, *99*, 1449–1460. [CrossRef] [PubMed]
3. Rieger, A.; Meylan, M.; Hauser, C.; Knubben-Schweizer, G. Meta-analysis to estimate the economic losses caused by reduced milk yield and reproductive performance associated with bovine paratuberculosis in Switzerland. *Schweiz. Arch. Tierheilkd.* **2021**, *164*, 737–751. [CrossRef] [PubMed]
4. Garcia, A.B.; Shalloo, L. Invited Review: The Economic Impact and Control of Paratuberculosis in Cattle. *J. Dairy Sci.* **2015**, *98*, 5019–5039. [CrossRef]
5. Rasmussen, P.; Barkema, H.W.; Mason, S.; Beaulieu, E.; Hall, D.C. Economic Losses Due to Johne's Disease (Paratuberculosis) in Dairy Cattle. *J. Dairy Sci.* **2021**, *104*, 3123–3143. [CrossRef]
6. Nielsen, S.S.; Toft, N. Ante Mortem Diagnosis of Paratuberculosis: A Review of Accuracies of ELISA, Interferon-Gamma Assay and Faecal Culture Techniques. *Vet. Microbiol.* **2008**, *129*, 217–235. [CrossRef]
7. Whittington, R.J.; Marshall, D.J.; Nicholls, P.J.; Marsh, I.B.; Reddacliff, L.A. Survival and Dormancy of Mycobacterium Avium Subsp. Paratuberculosis in the Environment. *Appl. Environ. Microbiol.* **2004**, *70*, 2989–3004. [CrossRef]
8. Eppleston, J.; Begg, D.J.; Dhand, N.K.; Watt, B.; Whittington, R.J. Environmental Survival of Mycobacterium Avium Subsp. Paratuberculosis in Different Climatic Zones of Eastern Australia. *Appl. Environ. Microbiol.* **2014**, *80*, 2337–2342. [CrossRef]
9. Salgado, M.; Collins, M.T.; Salazar, F.; Kruze, J.; Bolske, G.; Soderlund, R.; Juste, R.; Sevilla, I.A.; Biet, F.; Troncoso, F.; et al. Fate of Mycobacterium Avium Subsp. Paratuberculosis after Application of Contaminated Dairy Cattle Manure to Agricultural Soils. *Appl. Environ. Microbiol.* **2011**, *77*, 2122–2129. [CrossRef]
10. Whittington, R.; Donat, K.; Weber, M.F.; Kelton, D.; Nielsen, S.S.; Eisenberg, S.; Arrigoni, N.; Juste, R.; Sáez, J.L.; Dhand, N.; et al. Control of Paratuberculosis: Who, Why and How. A Review of 48 Countries. *BMC Vet. Res.* **2019**, *15*, 198. [CrossRef]
11. Ferrouillet, C.; Wells, S.J.; Hartmann, W.L.; Godden, S.M.; Carrier, J. Decrease of Johne's Disease Prevalence and Incidence in Six Minnesota, USA, Dairy Cattle Herds on a Long-Term Management Program. *Prev. Vet. Med.* **2009**, *88*, 128–137. [CrossRef] [PubMed]
12. Collins, M.T.; Eggleston, V.; Manning, E.J.B. Successful Control of Johne's Disease in Nine Dairy Herds: Results of a Six-Year Field Trial. *J. Dairy Sci.* **2010**, *93*, 1638–1643. [CrossRef] [PubMed]
13. Gavin, W.G.; Porter, C.A.; Hawkins, N.; Schofield, M.J.; Pollock, J.M. Johne's Disease: A Successful Eradication Programme in a Dairy Goat Herd. *Vet. Rec.* **2018**, *182*, 483. [CrossRef] [PubMed]
14. Tuberquia-López, B.C.; Correa-Valencia, N.M.; Hernández-Agudelo, M.; Fernández-Silva, J.A.; Ramírez-Vásquez, N.F. Paratuberculosis Control Strategies in Dairy Cattle: A Systematic Review. *Open. Vet. J.* **2022**, *12*, 525. [CrossRef] [PubMed]
15. Wraight, M.D.; McNeil, J.; Beggs, D.S.; Greenall, R.K.; Humphris, T.B.; Irwin, R.J.; Jagoe, S.P.; Jemmeson, A.; Morgan, W.F.; Brightling, P.; et al. Compliance of Victorian Dairy Farmers with Current Calf Rearing Recommendations for Control of Johne's Disease. *Vet. Microbiol.* **2000**, *77*, 429–442. [CrossRef]
16. Kravitz, A.; Pelzer, K.; Sriranganathan, N. The Paratuberculosis Paradigm Examined: A Review of Host Genetic Resistance and Innate Immune Fitness in Mycobacterium Avium Subsp. Paratuberculosis Infection. *Front. Vet. Sci.* **2021**, *8*, 721706. [CrossRef]
17. Alpay, F.; Zare, Y.; Kamalludin, M.H.; Huang, X.; Shi, X.; Shook, G.E.; Collins, M.T.; Kirkpatrick, B.W. Genome-Wide Association Study of Susceptibility to Infection by Mycobacterium Avium Subspecies Paratuberculosis in Holstein Cattle. *PLoS ONE* **2014**, *9*, e111704. [CrossRef]
18. Canive, M.; González-Recio, O.; Fernández, A.; Vázquez, P.; Badia-Bringué, G.; Lavín, J.L.; Garrido, J.M.; Juste, R.A.; Alonso-Hearn, M. Identification of Loci Associated with Susceptibility to Mycobacterium Avium Subsp. Paratuberculosis Infection in Holstein Cattle Using Combinations of Diagnostic Tests and Imputed Whole-Genome Sequence Data. *PLoS ONE* **2021**, *16*, e0256091. [CrossRef]
19. Sanchez, M.-P.; Guatteo, R.; Davergne, A.; Saout, J.; Grohs, C.; Deloche, M.-C.; Taussat, S.; Fritz, S.; Boussaha, M.; Blanquefort, P.; et al. Identification of the ABCC4, IER3, and CBFA2T2 Candidate Genes for Resistance to Paratuberculosis from Sequence-Based GWAS in Holstein and Normande Dairy Cattle. *Genet. Sel. Evol.* **2020**, *52*, 14. [CrossRef]
20. Sanchez, M.-P.; Tribout, T.; Fritz, S.; Guatteo, R.; Fourichon, C.; Schibler, L.; Delafosse, A.; Boichard, D. New Insights into the Genetic Resistance to Paratuberculosis in Holstein Cattle via Single-Step Genomic Evaluation. *Genet. Sel. Evol.* **2022**, *54*, 67. [CrossRef]
21. Vallée, H.; Rinjard, P. Etudes Sur l'entérite Paratuberculeuse des Bovidés. *Rev. Gen. Med. Vet.* **1926**, *35*, 1–9.
22. Bastida, F.; Juste, R.A. Paratuberculosis Control: A Review with a Focus on Vaccination. *J. Immune Based Ther. Vaccines* **2011**, *9*, 8. [CrossRef] [PubMed]

23. Serrano, M.; Elguezabal, N.; Sevilla, I.A.; Geijo, M.V.; Molina, E.; Arrazuria, R.; Urkitza, A.; Jones, G.J.; Vordermeier, M.; Garrido, J.M.; et al. Tuberculosis Detection in Paratuberculosis Vaccinated Calves: New Alternatives against Interference. *PLoS ONE* **2017**, *12*, e0169735. [CrossRef]
24. Garrido, J.M.; Vazquez, P.; Molina, E.; Plazaola, J.M.; Sevilla, I.A.; Geijo, M.V.; Alonso-Hearn, M.; Juste, R.A. Paratuberculosis Vaccination Causes Only Limited Cross-Reactivity in the Skin Test for Diagnosis of Bovine Tuberculosis. *PLoS ONE* **2013**, *8*, e80985. [CrossRef] [PubMed]
25. Gupta, S.K.; Wilson, T.; Maclean, P.H.; Rehm, B.H.A.; Heiser, A.; Buddle, B.M.; Wedlock, D.N. Mycobacterium Avium Subsp. Paratuberculosis Antigens Induce Cellular Immune Responses in Cattle without Causing Reactivity to Tuberculin in the Tuberculosis Skin Test. *Front. Immunol.* **2023**, *13*, 1087015. [CrossRef] [PubMed]
26. Reddacliff, L.; Eppleston, J.; Windsor, P.; Whittington, R.; Jones, S. Efficacy of a Killed Vaccine for the Control of Paratuberculosis in Australian Sheep Flocks. *Vet. Microbiol.* **2006**, *115*, 77–90. [CrossRef]
27. Tewari, D.; Hovingh, E.; Linscott, R.; Martel, E.; Lawrence, J.; Wolfgang, D.; Griswold, D. Mycobacterium Avium Subsp. Paratuberculosis Antibody Response, Fecal Shedding, and Antibody Cross-Reactivity to Mycobacterium Bovis in M. Avium Subsp. Paratuberculosis-Infected Cattle Herds Vaccinated against Johne's Disease. *Clin. Vaccine Immunol.* **2014**, *21*, 698–703. [CrossRef] [PubMed]
28. García-Pariente, C.; Pérez, V.; Geijo, M.V.; Moreno, O.; Muñoz, M.; Fuertes, M.; Puentes, E.; Doce, J.; Ferreras, M.C.; Garcia Marin, J.F. The Efficacy of a Killed Vaccine against Paratuberculosis (SILIRUM®) in Cattle. In A Field Study. In Proceedings of the 8th International Colloquium on Paratuberculosis, Copenhagen, Denmark, 14–17 August 2005; p. 282.
29. Alonso-Hearn, M.; Molina, E.; Geijo, M.; Vazquez, P.; Sevilla, I.A.; Garrido, J.M.; Juste, R.A. Immunization of Adult Dairy Cattle with a New Heat-Killed Vaccine Is Associated with Longer Productive Life Prior to Cows Being Sent to Slaughter with Suspected Paratuberculosis. *J. Dairy Sci.* **2012**, *95*, 618–629. [CrossRef]
30. Juste, R.A.; Alonso-Hearn, M.; Molina, E.; Geijo, M.; Vazquez, P.; Sevilla, I.A.; Garrido, J.M. Significant Reduction in Bacterial Shedding and Improvement in Milk Production in Dairy Farms after the Use of a New Inactivated Paratuberculosis Vaccine in a Field Trial. *BMC Res. Notes* **2009**, *2*, 233. [CrossRef]
31. Serrano, M.; Elguezabal, N.; Plazaola, J.M.; Lauzurika, J.; Ocabo, B.; Geijo, M.V.; Molina, E.; Sevilla, I.A.; Arrazuria, R.; Alonso-Hearn, M.; et al. Effect of Age at Vaccination on Map Shedding: An Ten-Year Follow-Up. In Proceedings of the 13th International Colloquium on Paratuberculosis, Nantes, France, 20–24 June 2016; p. 137.
32. Juste, R.A.; Geijo, M.V.; Elguezabal, N.; Sevilla, I.A.; Alonso-Hearn, M.; Garrido, J.M. Paratuberculosis Vaccination Specific and Non-Specific Effects on Cattle Lifespan. *Vaccine* **2021**, *39*, 1631–1641. [CrossRef]
33. Dhand, N.K.; Eppleston, J.; Whittington, R.J.; Windsor, P.A. Changes in Prevalence of Ovine Paratuberculosis Following Vaccination with Gudair(R): Results of a Longitudinal Study Conducted over a Decade. *Vaccine* **2016**, *34*, 5107–5113. [CrossRef] [PubMed]
34. Links, I.J.; Denholm, L.J.; Evers, M.; Kingham, L.J.; Greenstein, R.J. Is Vaccination a Viable Method to Control Johne's Disease Caused by Mycobacterium Avium Subsp. Paratuberculosis? Data from 12 Million Ovine Vaccinations and 7.6 Million Carcass Examinations in New South Wales, Australia from 1999–2009. *PLoS ONE* **2021**, *16*, e0246411. [CrossRef] [PubMed]
35. Ritz, C.; Spiess, A.-N. QpcR: An R Package for Sigmoidal Model Selection in Quantitative Real-Time Polymerase Chain Reaction Analysis. *Bioinformatics* **2008**, *24*, 1549–1551. [CrossRef] [PubMed]
36. R Foundation for Statistical Computing R: A Language and Environment for Statistical Computing. Available online: https://www.R-project.org/ (accessed on 26 March 2023).
37. Bates, D.; Mächler, M.; Bolker, B.; Walker, S. Fitting Linear Mixed-Effects Models Using Lme4. *J. Stat. Softw.* **2015**, *67*, 1–48. [CrossRef]
38. Brooks, M.E.; Kristensen, K.; van Benthem, K.J.; Magnusson, A.; Berg, C.W.; Nielsen, A.; Skaug, H.J.; Mächler, M.; Bolker, B.M. GlmmTMB Balances Speed and Flexibility Among Packages for Zero-Inflated Generalized Linear Mixed Modeling. *R J.* **2017**, *9*, 378–400. [CrossRef]
39. Fox, J. Effect Displays in R for Generalised Linear Models. *J. Stat. Softw.* **2003**, *8*, 1–27. [CrossRef]
40. Körmendy, B. The Effect of Vaccination on the Prevalence of Paratuberculosis in Large Dairy Herds. *Vet. Microbiol.* **1994**, *41*, 117–125. [CrossRef]
41. Kalis, C.H.; Hesselink, J.W.; Barkema, H.W.; Collins, M.T. Use of Long-Term Vaccination with a Killed Vaccine to Prevent Fecal Shedding of Mycobacterium Avium Subsp Paratuberculosis in Dairy Herds. *Am. J. Vet. Res.* **2001**, *62*, 270–274. [CrossRef]
42. Knust, B.; Patton, E.; Ribeiro-Lima, J.; Bohn, J.J.; Wells, S.J. Evaluation of the Effects of a Killed Whole-Cell Vaccine against Mycobacterium Avium Subsp Paratuberculosis in 3 Herds of Dairy Cattle with Natural Exposure to the Organism. *J. Am. Vet. Med. Assoc.* **2013**, *242*, 663–669. [CrossRef]
43. Koets, A.; Ravesloot, L.; Ruuls, R.; Dinkla, A.; Eisenberg, S.; Lievaart-Peterson, K. Effects of Age and Environment on Adaptive Immune Responses to Mycobacterium Avium Subsp. Paratuberculosis (MAP) Vaccination in Dairy Goats in Relation to Paratuberculosis Control Strategies. *Vet. Sci.* **2019**, *6*, 62. [CrossRef]
44. Carpenter, T.E. Evaluation of Effectiveness of a Vaccination Program against an Infectious Disease at the Population Level. *Am. J. Vet. Res.* **2001**, *62*, 202–205. [CrossRef] [PubMed]

45. Navarro-Gonzalez, N.; Fourichon, C.; Blanquefort, P.; Delafosse, A.; Joly, A.; Ngwa-Mbot, D.; Biet, F.; Boichard, D.; Schibler, L.; Journaux, L.; et al. Longitudinal Study of Mycobacterium Avium Ssp. Paratuberculosis Fecal Shedding Patterns and Concurrent Serological Patterns in Naturally Infected Dairy Cattle. *J. Dairy Sci.* **2019**, *102*, 9117–9137. [CrossRef] [PubMed]
46. Cho, J.; Tauer, L.W.; Schukken, Y.H.; Gómez, M.I.; Smith, R.L.; Lu, Z.; Grohn, Y.T. Economic Analysis of Mycobacterium Avium Subspecies Paratuberculosis Vaccines in Dairy Herds. *J. Dairy Sci.* **2012**, *95*, 1855–1872. [CrossRef] [PubMed]
47. Dhand, N.K.; Plain, K.M.; Green, A.C.; Martinez, E.; Eppleston, J.; Ly, A.; Arif, S.; Emery, D. Factors Influencing the Effectiveness of the Gudair Vaccine for Controlling Johne's Disease in Sheep Flocks in Australia. *Prev. Vet. Med.* **2021**, *193*, 105394. [CrossRef] [PubMed]
48. Hasvold, H.J.; Valheim, M.; Berntsen, G.; Storset, A.K. In Vitro Responses to Purified Protein Derivate of Caprine T Lymphocytes Following Vaccination with Live Strains of Mycobacterium Avium Subsp. Paratuberculosis. *Vet. Immunol. Immunopathol.* **2002**, *90*, 79–89. [CrossRef]
49. Fernández, M.; Royo, M.; Fuertes, M.; Arteche-Villasol, N.; Ferreras, M.C.; Benavides, J.; Pérez, V. Effects of Paratuberculosis Vaccination at Different Ages in a Dairy Goat Herd: A 2-Year Follow-Up. *Animals* **2022**, *12*, 3135. [CrossRef]
50. Mercier, P.; Brémaud, I.; Gautier, M.-P. Vaccination of Kids under One Month of Age with a Killed Vaccine and Reduction in the Frequency of Faecal Shedding of Mycobacterium Avium Subspecies Paratuberculosis. *Small Rumin. Res.* **2014**, *121*, 425–433. [CrossRef]
51. Gwozdz, J.; Thompson, K.; Manktelow, B.; Murray, A.; West, D. Vaccination against Paratuberculosis of Lambs Already Infected Experimentally with Mycobacterium Avium Subspecies Paratuberculosis. *Aust. Vet. J.* **2000**, *78*, 560–566. [CrossRef]
52. Muñoz, M.; Garcia Marin, J.; Garcia-Pariente, C.; Reyes, L.; Verna, A.; Moreno, O.; Fuertes, M.; Doce, J.; Puentes, E.; Garrido, J.M.; et al. IAP: 2005: Efficacy of a Killed Vaccine (SILIRUM®) in Calves Challenged with Map. In Proceedings of the 8th International Colloquium on Paratuberculosis, Copenhagen, Denmark, 14–17 August 2005; pp. 208–217.
53. Mathevon, Y.; Lafort, M.-P.; Tasca, C.; Galan, E.; Falguières, R.; Corbiere, F.; Foucras, G. Vaccination against Paratuberculosis in sheep: Methodological developments and evaluation of efficacy. In Proceedings of the 13th International Colloquium on Paratuberculosis, Nantes, France, 20–24 June 2016; p. 143.
54. Koets, A.P.; Eda, S.; Sreevatsan, S. The within Host Dynamics of Mycobacterium Avium Ssp. Paratuberculosis Infection in Cattle: Where Time and Place Matter. *Vet. Res.* **2015**, *46*, 61. [CrossRef]
55. DeKuiper, J.L.; Cooperider, H.E.; Lubben, N.; Ancel, C.M.; Coussens, P.M. Mycobacterium Avium Subspecies Paratuberculosis Drives an Innate Th17-Like T Cell Response Regardless of the Presence of Antigen-Presenting Cells. *Front. Vet. Sci.* **2020**, *7*, 108. [CrossRef]
56. Pooley, H.B.; Begg, D.J.; Plain, K.M.; Whittington, R.J.; Purdie, A.C.; de Silva, K. The Humoral Immune Response Is Essential for Successful Vaccine Protection against Paratuberculosis in Sheep. *BMC Vet. Res.* **2019**, *15*, 223. [CrossRef] [PubMed]
57. Corpa, J.M.; Pérez, V.; García Marín, J.F. Differences in the Immune Responses in Lambs and Kids Vaccinated against Paratuberculosis, According to the Age of Vaccination. *Vet. Microbiol.* **2000**, *77*, 475–485. [CrossRef] [PubMed]
58. Chase, C.C.L.; Hurley, D.J.; Reber, A.J. Neonatal Immune Development in the Calf and Its Impact on Vaccine Response. *Vet. Clin. North. Am. Food Anim. Pract.* **2008**, *24*, 87–104. [CrossRef]
59. Arteche-Villasol, N.; Gutiérrez-Expósito, D.; Vallejo, R.; Espinosa, J.; Elguezabal, N.; Ladero-Auñon, I.; Royo, M.; del Carmen Ferreras, M.; Benavides, J.; Pérez, V. Early Response of Monocyte-Derived Macrophages from Vaccinated and Non-Vaccinated Goats against in Vitro Infection with Mycobacterium Avium Subsp. Paratuberculosis. *Vet. Res.* **2021**, *52*, 69. [CrossRef]
60. Purdie, A.C.; Plain, K.M.; Pooley, H.; Begg, D.J.; de Silva, K.; Whittington, R.J. Correlates of Vaccine Protection against Mycobacterium Avium Sub-Species Paratuberculosis Infection Revealed in a Transcriptomic Study of Responses in Gudair® Vaccinated Sheep. *Front. Vet. Sci.* **2022**, *9*, 1004237. [CrossRef] [PubMed]
61. Kandasamy, R.; Voysey, M.; McQuaid, F.; de Nie, K.; Ryan, R.; Orr, O.; Uhlig, U.; Sande, C.; O'Connor, D.; Pollard, A.J. Non-Specific Immunological Effects of Selected Routine Childhood Immunisations: Systematic Review. *BMJ* **2016**, *355*, i5225. [CrossRef]
62. Vaugelade, J.; Pinchinat, S.; Guiella, G.; Elguero, E.; Simondon, F. Non-Specific Effects of Vaccination on Child Survival: Prospective Cohort Study in Burkina Faso. *BMJ* **2004**, *329*, 1309. [CrossRef]
63. Higgins, J.P.T.; Soares-Weiser, K.; López-López, J.A.; Kakourou, A.; Chaplin, K.; Christensen, H.; Martin, N.K.; Sterne, J.A.C.; Reingold, A.L. Association of BCG, DTP, and Measles Containing Vaccines with Childhood Mortality: Systematic Review. *BMJ* **2016**, *355*, i5170. [CrossRef]
64. Pfeiffer, G.; Fisker, A.B.; Nebié, E.; Hengelbrock, J.; Sié, A.; Becher, H.; Müller, O. Non-Specific Effects of Childhood Vaccinations—A Case Control Study Nested into a Health and Demographic Surveillance System in Rural Burkina Faso. *Vaccine* **2017**, *35*, 7114–7120. [CrossRef]
65. Quinn, M.; Edmond, K.M.; Fawzi, W.W.; Hurt, L.; Kirkwood, B.R.; Masanja, H.; Muhihi, A.J.; Newton, S.; Noor, R.A.; Williams, P.L.; et al. Non-Specific Effects of BCG and DTP Vaccination on Infant Mortality: An Analysis of Birth Cohorts in Ghana and Tanzania. *Vaccine* **2022**, *40*, 3737–3745. [CrossRef]

animals MDPI

Article

The Upregulation of *Cathepsin G* Is Associated with Resistance to Bovine Paratuberculosis

Maria Canive [1], Gerard Badia-Bringué [1,2] and Marta Alonso-Hearn [1,*]

1 NEIKER-Basque Research and Technology Alliance (BRTA), 20850 Derio, Spain
2 Doctoral Program in Molecular Biology and Biomedicine, Universidad del País Vasco/Euskal Herriko Unibertsitatea (UPV/EHU), 48940 Leioa, Spain
* Correspondence: malonso@neiker.eus

Simple Summary: *Cathepsin G* (CTSG) is a serine protease that participates in the killing of pathogens and tissue remodeling at sites of inflammation. We previously found that the presence of the minor allele in the cis-eQTLs-rs41976219 (AC) was associated with increased CTSG mRNA levels in blood samples from Holstein cows. In this study, we test whether genetic variation in the cis-eQTL-rs41976219 is associated with changes in CTSG protein expression and control of *Mycobacterium avium* subsp. *paratuberculosis* (MAP) infection. We demonstrate that the heterozygous genotype for the rs41976219 (AC) results in higher CTSG protein levels in the supernatants of infected CD14+-monocyte-derived macrophages (MDMs) after 2 h of infection and a significantly lower intracellular MAP load at 7 d p.i. than in MDMs from cattle with the AA genotype. Furthermore, the rs41976219 genotype CC is more frequent in healthy cows than in cows with PTB-associated lesions in gut tissues. Higher CTSG levels are detected in plasmas from cows with the CC genotype when compared with cows with the AA + AC genotypes for the rs41976219. We conclude that the presence of the minor allele in the rs41976219 increases the *CTSG* expression in plasma samples and MAP-infected macrophages, and can influence PTB pathogenesis by modulating MAP load within infected macrophages.

Abstract: An in silico genomic–transcriptomic combined approach allowed the identification of a polymorphism (cis-eQTL-rs41976219) in the *Bos taurus* genome associated with the CTSG mRNA expression in bovine blood samples, which suggests that individual genetic variation might modulate the *CTSG* transcriptional response. In the current study, a sandwich ELISA is used to measure the CTSG protein levels in supernatants of monocyte-derived macrophages (MDMs) isolated from cows with the AA (n = 5) and AC (n = 11) genotypes for the rs41976219 and infected ex vivo with MAP. Cows with the AC genotype have significantly higher CTSG protein levels (1.85 ng/mL) in the supernatants of enriched CD14+-MDMs after 2 h of infection when compared with infected CD14+-MDMs from cows with the AA genotype (1.68 ng/mL). Statistically significant differences in the intracellular MAP load at 7 d p.i. are observed between animals with the AA (2.16 log CFUs) and AC (1.44 log CFUs) genotypes. Finally, the association between the rs41976219 allelic variants and resistance to PTB is tested in a larger cattle population (n = 943) classified according to the presence (n = 442) or absence (n = 501) of PTB-associated lesions. The presence of the two minor alleles in the rs41976219 (CC) is more frequent among healthy cows than in cows with PTB-associated lesions in gut tissues (2.2% vs. 1.4%, OR = 0.61). In agreement with this, the CTSG levels in plasma samples of cows without lesions in gut tissues and with the CC (n = 8) genotype are significantly higher than in the plasmas of cows with the AA + AC (n = 36) genotypes.

Keywords: paratuberculosis; *Cathepsin G*; monocyte-derived macrophages; mycobacterial growth-inhibition assay; modulation of immune function; polymorphisms; functional variants; cis-eQTL; therapeutics; immunomodulators

1. Introduction

Paratuberculosis (PTB) is a chronic granulomatous enteritis of ruminants caused by *Mycobacterium avium* susbp. *paratuberculosis* (MAP). PTB is a major problem for animal health and the dairy industry. Therefore, the World Organization for Animal Health (WOAH) requires member countries to maintain epidemiological surveillance with the notification of disease cases. The economic impact of PTB on the US dairy industry has been estimated between USD 250 million per year to USD 1.5 billion annually [1]. In Europe, it has been estimated at USD 364.31 million per year. According to their extension in the intestine, cellular infiltrate, and amount of MAP, PTB-associated lesions were classified into focal, multifocal, and diffuse (diffuse paucibacillary or lymphoplasmacytic, diffuse intermediate, and diffuse multibacillary or histiocytic) [2,3]. In addition, MAP has been related to human gastrointestinal inflammatory diseases, such as Crohn's disease (CD), inflammatory bowel disease (IBD), ulcerative colitis, and IBD-associated colorectal cancer [4,5]. MAP has also been associated with multiple sclerosis, rheumatoid arthritis, and diabetes type 1 [6–9]. There is no effective treatment for MAP infection and vaccination with heat-inactivated vaccines interferes with the diagnosis of *Mycobacterium bovis* [10–12].

Several studies have demonstrated that host genetics governs variations in susceptibility to MAP infection, and therefore selective breeding could decrease the susceptibility to MAP infection [13]. In recent years, genome-wide association studies (GWAS) have identified single nucleotide polymorphisms (SNPs) associated with susceptibility to MAP infection [14–17]. However, limited progress has been made in identifying SNPs associated with resistance to MAP infection. Moreover, how the identified genetic variants exert their effect is unknown and only a few functional mutations associated with susceptibility to MAP infection have been identified [18,19]. In 2001, the concept of "genetical genomics" where genomic variants are associated with transcript abundance was introduced by Jansen and Nap [20]. This concept supports the idea that integrating genotypes and gene-expression data could contribute to a better understanding of the susceptibility or resistance to microbial infections. Recently, a combined genomic–transcriptomic approach allowed for the identification of several cis-eQTLs, SNPs associated with gene expression and usually located within 1 Mb upstream of the transcription start site of a gene locus, significantly associated with PTB susceptibility [21]. More specifically, 192 cis-eQTLs were significantly associated (FDR ≤ 0.05) with the expression of 145 genes in peripheral blood (PB) samples obtained from Holstein cows with different types of PTB-associated lesions or without lesions in gut tissues [21]. One cis-eQTL, the rs41976219, located on Bos Taurus chromosome 21, was associated with the *Cathepsin G* (CTSG) mRNA expression in PB samples. The minor allele in the cis-eQTLs-rs41976219 (AC) was associated with higher CTSG mRNA expression levels (mean = 114.75 fragments per kilobase per million, FPKMs) while the most frequent homozygous genotype (AA) correlated with very low levels of CTSG mRNA (mean = 0.23 FPKMs).

Neutrophil and macrophage function depends on, among other factors, the presence of *Cathepsins* (*CTS*) that are involved in the innate (pathogen-recognizing and killing) and adaptive (antigen-processing and presentation) responses facilitating rapid pathogen elimination and the induction of long-term immunity [22]. The name "*Cathepsin*" refers to two serine proteases (*CTS A* and *G*), two aspartic proteases (*CTS D* and *E*), and eleven lysosomal cysteine proteases (*CTS B*, *C*, *F*, *H*, *K*, *L*, *O*, *S*, *V*, *X*, and *W*) [23,24]. Following synthesis, *CTSs* are transported to the endosomal/lysosomal compartment where they are activated following the removal of the N-terminal propeptide [25,26]. The presence of *CTSs* in the endo-lysosomal compartment of infected monocytes directs the interaction with and killing of bacteria and contributes to the processing of bacterial antigens for the induction of adaptive immune response [27]. It is not surprising that bacteria manipulate the expression of *CTSs* to favor their intracellular survival in monocytes. On the other hand, we hypothesize that the presence of SNPs in the host genome might also influence *CTSs'* expression and subsequent proteolytic activity. Understanding how the individual regulation of *CTS* expression participates in the pathogenesis of mycobacterial infections is needed for the development of novel therapeutical alternatives to antibiotics. In this study, we test whether individual genetic variation in the cis-eQTL-rs41976219 is associated with changes in CTSG protein expression, the control of MAP infection, and PTB progression.

2. Materials and Methods

2.1. Animals

For the mycobacterial growth-inhibition assays (MGIAs) and the CTSG quantification assay, PB samples were collected from 16 Holstein Friesian cows from a commercial dairy farm in the Basque Country (Spain). Only adult cows (2 years or older)were included in the study. Animals were not diagnosed with any disease in the period during which samples were collected, and all of them were ELISA negative for the detection of anti-MAP antibodies in the serum at the sampling time (IDEXX Map Ab test, Montpellier, France).

The case–control population consisted of 940 culled Holstein cattle from eight Spanish regions: Basque Country (n = 401, 42.52%), Catalonia (n = 211, 22.38%), Navarre (n = 193, 20.47%), Cantabria (n = 65, 6.89%), Aragon (n = 35, 3.71%), Castile and Leon (n = 24, 2.55%), La Rioja (n = 7, 0.71%), and Asturias (n = 4, 0.41%). Three slaughtered cows of unknown origin were also included in the study. Only cows (2 years or older, 5.6 years mean age) were included in the analysis. The infection status of the 943 culled animals was determined by the histopathological analysis of gut tissues, as previously described [28]. The stained tissue sections were examined by light microscopy and classified into four groups: those with no lesions, and those with focal, multifocal, or diffuse lesions [2].

2.2. MAP Strain, Bacterial Culture, and Preparation of Bacterial Suspensions

MAP reference strain K10 was obtained from the American Type Culture Collection (ATCC) (Manassas, VA, USA). The bovine K10 isolate of MAP was grown in T25 tissue culture flasks at 37 ± 1 °C in 8 mL of Middlebrook 7H9 broth (Difco Laboratories, Detroit, MI, USA) supplemented with 10% (v/v) oleic acid-albumin-dextrose-catalase (Becton, Dickinson and Company, Franklin Lakes, NJ, USA), 0.05% (v/v) Tween-80 (Sigma-Aldrich, St. Louis, MO, USA), and 2 mg/L of mycobactin J (Allied Monitor Inc., Fayette, MO, USA) for 20 days at 37 °C. Bacterial cells were harvested by centrifugation at 3000× g for 20 min in a Beckman Coulter Allegra X-12 centrifuge. Then, bacterial pellets were resuspended in 2 mL of Hank's balanced salt solution (HBSS), and the resultant suspension was passed 20 times through a 27-gauge needle. The turbidity of the bacterial suspension was adjusted to a McFarland standard of 1 with a Densimat (bioMerieux, Marcy l'Etoile, France). Only the top fraction of the suspension containing dispersed bacteria was used for the infection assays.

2.3. Peripheral Blood Mononuclear Cells' (PBMCs) Purification, CD14+ Monocyte Selection, Ex Vivo Differentiation to Monocyte-Derived Macrophages (MDMs), and MAP Infection

PBMCs were isolated from PB using Ficoll-Paque density gradient centrifugation. Briefly, fifteen milliliters of PB were drawn from the tail vein of healthy Holstein cows into heparinized Vacutainer tubes (Becton, Dickinson and Company, Sparks, MD, USA) and diluted at 1:2 in Hanks balanced salt solution (HBSS). Leucosep tubes (Greiner, Kremsmünster, Austria) were filled with 15 mL of Ficoll-Paque gradient (1.084 g/cm^3) (GE HealthCare, Uppsala, Sweden) and centrifuged at 1000× g for 30 s at room temperature. Subsequently, the diluted blood was overlaid on the top of the Ficoll-Paque and centrifuged at 800× g for 15 min. The plasma layer was removed and the cell interphase containing PBMCs was collected and transferred to a clean tube. PBMCs were washed twice in HBSS and centrifuged at 400 g for 10 min to remove platelets. Additionally, CD14+ monocytes were magnetically enriched from PBMCs using CD14+ microbeads and magnetically activated cell-sorting (MACS) technology (Miltenyi Biotech, Bergisch Galdbach, Germany), as previously described [29]. CD14+ monocytes were enriched from PBMCs to maximize monocyte purification and subsequent differentiation to MDMs. PBMCs and CD14+-selected monocytes were resuspended in RPMI-1640 supplemented with 20 mM L-glutamine, 10% heat-inactivated bovine serum, 100 U ml-1 penicillin G, and 100 mg ml-1 streptomycin sulfate (Lonza, Spain), and seeded into 24-well plates at a density of 1×10^6 cells/mL. The cells were incubated overnight at 37 °C in a humidified 5% CO_2 incubator. The following day, non-adherent cells were washed with PBS, and adherent cells were incubated in medium for 7 days at 37 °C to allow for differentiation into MDMs before infection. MDMs and CD14+-MDMs were inoculated in duplicate with a single-cell suspension of MAP K10 strain at a multiplicity of infection (MOI) of 10:1 (bacteria:cell). At 2 h p.i., the supernatants were removed and the cells were washed twice with HBSS to remove extracellular bacteria. The infected MDMs were lysed at this time (2 h p.i.) or cultured at 37 °C for 7 d in medium. At each time point, the infected MDMs were lysed with 0.5 mL of 0.1% Triton X-100 (Sigma-Aldrich) in sterile water for 10 min.

2.4. Mycobacterial Growth-Inhibition Assay (MGIA) Assessment Using the BACTEC MGIT 960 System

The MGIA assessment using the BACTEC MGIT 960 system (Becton, Dickinson and Company, Sparks, MD, USA) was performed, as previously described [21,29,30]. Each supplemented mycobacteria growth-indicator tube (MGIT) contained 7 mL of modified Middlebrook 7H9 broth base with casein peptone and an oxygen-sensitive fluorescent compound (tris-4,7-diphenyl-1,10-phenathroline ruthenium chloride pentahydrate) embedded in silicone on the bottom of the tube. The fluorescent compound fluoresced upon exposure to CO_2 produced from metabolically active mycobacteria. Each tube was supplemented with 800 µL of an enrichment supplement (BBL MGIT OADC growth supplement) and an antibiotic mixture (BBL MGIT PANTA Antibiotic Mixture) (Becton, Dickinson, and Company). The tubes were also supplemented with 2 $\mu g\ ml^{-1}$ of mycobactin. The MGIT tubes were inoculated with 0.1 mL of the initial MAP suspension and with 0.5 mL of the 2 h and 7 d p.i. cell lysates. The tubes were incubated at 37 $\pm$ 2 °C for up to 42 days in a Bactec MGIT 960 instrument (Becton, Dickinson, and Company). The earliest instrumental indicator of positivity (time to detection (TTD)) for each tube was recorded. The predicted number of bacteria in each positive tube was calculated using standard curves that related TTD to the estimated log CFUs [30]. For the standard curves, a 10-fold dilution series of the MAP cellular suspension (McFarland = 1) was prepared and the TTDs of 100 µL of each dilution in the BACTEC MGIT 960 system were recorded. MAP (log CFUs) was plotted versus TTDs (in days) and the obtained mathematical equation was used to determine the estimated log CFUs for each sample. The log CFU ratios were calculated by dividing the estimated log CFUs at day 7 by that at 2 h p.i. Lower log-CFU ratios reflected greater resistance to the infection. A Student's *t*-test was performed for a comparison of the mean log CFUs between two groups (GraphPad Prism 8, San Diego, CA, USA). Differences were considered significant when *p*-values were $\leq$0.05.

2.5. Bovine CTSG ELISA

The CTSG levels in plasma samples and supernatants of MAP-infected MDMs were assessed by a quantitative sandwich ELISA according to the manufacturer's instructions (MyBioSource, San Diego, CA, USA). The sensitivity of this kit was 0.1 ng/mL and the detection range was 0.25–8 ng/mL. Statistical analysis was performed using an unpaired *t*-test for the comparison of the mean CTSG concentration between the two groups (GraphPad Prism 8, San Diego, CA, USA). Differences were considered significant when *p*-values were $\leq$0.05.

2.6. Genotyping, Case–Control Population, and Odds Ratio (OR)

The blood samples of all the animals included in the study were collected from the coccygeal (tail) vein of each animal into EDTA Vacutainer tubes (BD Vacutainer System, Becton, Dickinson, and Company, Sparks, MD, USA). Genomic DNA was extracted from the blood buffy coat using the QIAmp DNA Blood Mini Kit, according to the manufacturer's instructions (Qiagen, Hilden, Germany). Purified genomic DNA was quantified spectrophotometrically and subsequently genotyped with the EuroG10K MD

BeadChip at the molecular genetic laboratory service of the Spanish Federation of Holstein Cattle (CONAFE) using the Infinium iScan software for allele assignation (Illumina, San Diego, CA, USA). Genotypes obtained from this case–control population have been used previously for several GWAS studies [16,17,31].

To quantify the strength of the association between the rs41976219 genotypes and the presence or absence of PTB-associated lesions, ORs were calculated using logistic regression analysis under five different genetic models (co-dominant, dominant, recessive, over-dominant, and log-additive) with SNPassoc 2.0-11 (Victor Moreno, Juan R Gonzalez, Dolors Pelegri, Barcelona, Spain). Age was included as a covariate in the analysis. If the OR was less than 1, the presence of one event (a specific genotype) reduced the odds of the other event (presence of PTB-associated lesions). For each genetic model, SNPassoc provided genotype frequencies, ORs, and 95% CI with the major homozygous genotype deemed as the baseline.

3. Results

3.1. The Presence of the Minor Allele in the rs41976219 (AC) Resulted in Increased CTSG Protein Levels in Supernatants of MAP-Infected CD14+-MDMs after 2 h of Infection

The heterozygous genotype (AC) in the cis-eQTL-rs41976219 was previously associated with higher CTSG mRNA expression (21). Consequently, the rs41976219 genotypes may directly impact CTSG protein expression. To address this, MDMs and CD14+-MDMs purified from 16 cows with the homozygous (AA; n = 5) and heterozygous (AC; n = 11) genotypes for the rs41976219 were infected ex vivo with MAP for 2 h and 7 days. Using a bovine CTSG quantitative sandwich ELISA, the CTSG protein levels were measured in plasma and supernatants obtained from MAP-infected MDMs and CD14+-MDMs purified from cows with the AA and AC genotypes for the rs41976219. As presented in Figure 1, we can observe that the CTSG levels are significantly higher in the supernatants of MAP-infected CD14+-MDMs obtained from cows carrying the C allele (1.85 ± 0.11 ng/mL) after 2 h of infection than in infected CD14+-MDMs from cows with the AA genotype (1.68 ± 0.06 ng/mL) (p = 0.008). No significant differences in the CTSG protein levels in the supernatants of MAP-infected MDMs between cows with the AA and AC genotypes were observed. Although higher CTSG protein levels were also observed in the plasmas of cows with the AC genotype (2.72 ± 1.64 ng/mL) when compared with the AA genotype (1.67 ± 0.65 ng/mL), this difference was not statistically significant (p = 0.19). Our results confirm that an increase in CTSG mRNA expression correlates with higher levels of CTSG protein expression in the supernatants of MAP-infected CD14+-MDMs obtained from cows with the AC genotype following 2 h of infection. The use of selected CD14+-MDMs instead of MDMs presented better variation in the genotype-dependent CTSG expression after MAP infection.

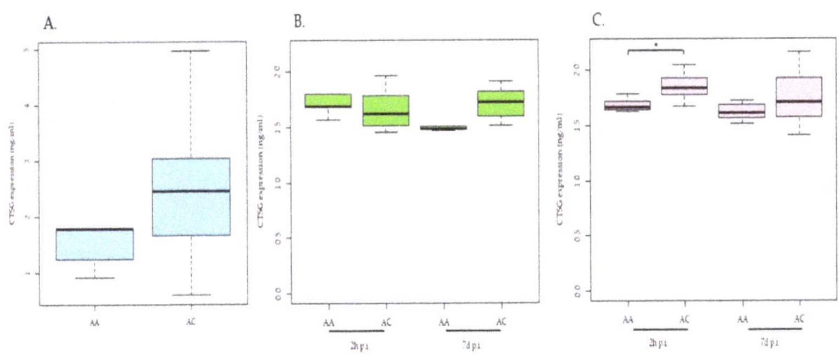

Figure 1. CTSG levels in carriers of different genotypes of cis-eQTL-rs41976219. MDMs and CD14+-MDMs purified from 16 cows with the homozygous (AA; n = 5) and heterozygous (AC; n = 11) genotypes for rs41976219 infected ex vivo with MAP for 2 h and 7 days. CTSG levels measured in plasmas (**A**) and

supernatants of MDMs (**B**) and CD14+-MDMs (**C**) after 2 h and 7 d of MAP infection by ELISA. Statistically significant differences are observed in the mean CTSG levels of expression in the supernatants of MAP-infected CD14+-MDMs after 2 h of infection depending on the genotype of the cis-eQTL-rs41976219. The median and upper and lower quartiles of each group are represented by horizontal and vertical lines, respectively. The lowest point is the minimum of the dataset and the highest point is the maximum of the dataset. *p*-values are calculated using an unpaired *t*-test. * *p*-value $\leq$ 0.05.

3.2. Association between the Inhibition of MAP Growth within MDMs and the cis-eQTL-rs41976219 AC Genotype Using an MGIA Assay

The association between MDMs' resistance to MAP infection and the cis-eQTL-rs41976219 genotypes was tested using an ex vivo MGIA assay (Figure 2). MDMs were purified from the PB of cows with homozygous (AA, n = 5) and heterozygous (AC, n = 11) genotypes for the rs41976219 and infected ex vivo with MAP. The bacterial load was estimated at 2 h and 7 d p.i. using the BACTEC MGIT system. MDMs obtained from cows with the AA genotype presented a greater MAP uptake at 2 h p.i. when compared with cows with the AC genotype; 3.74 $\pm$ 0.19 and 3.69 $\pm$ 0.17 log CFUs, respectively. Although the presence of the AC genotype reduced MAP internalization within MDMs, this difference was not statistically significant (*p*-value = 0.409). Statistically significant differences in the intracellular MAP load at 7 days p.i. were observed between animals with the AA (2.32 $\pm$ 1.06 log CFUs) and AC (1.44 $\pm$ 0.97 log CFUs) genotypes (p = 0.028). Moreover, statistically significant differences in the ratios (log CFU 7 d/log CFUs 2 h) of the animals with the AA (0.58) and AC (0.33) genotypes were observed (p = 0.04). These results suggest a significant effect of the minor allele in the rs41976219 on limiting/controlling intracellular MAP and confirms that cis-regulatory variation modulates resistance to MAP infection.

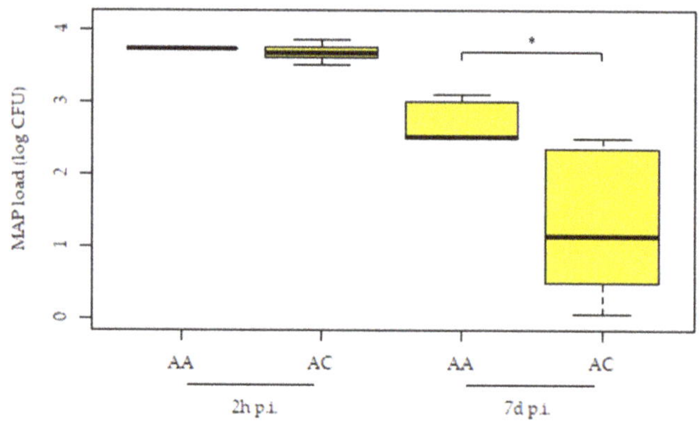

Figure 2. Host MDMs expressing the cis-eQTL-rs41976219 AC genotype show an increased ability to control MAP infection. MDMs purified from PB of cows with the homozygous (AA, n = 5) and heterozygous (AC, n = 11) genotypes for the rs41976219 and infected ex vivo with MAP. Intracellular bacterial load estimated at 2 h and 7 days p.i. The median and upper and lower quartiles of each group are represented by horizontal and vertical lines, respectively. The lowest point is the minimum of the dataset and the highest point is the maximum of the dataset. *p*-values are calculated using an unpaired t-test. * *p*-value $\leq$ 0.05.

3.3. The cis-eQTL-rs41976219 T/T Genotype Correlated with PTB Control

The infectious status of the animals included in the case–control study (n = 943) was previously determined by the histopathological analysis of gut tissues [28]. To quantify the strength of the association between the rs41976219 genotypes and the presence or absence of PTB-associated lesions, ORs were calculated using logistic regression analysis. Table 1 presents the allelic frequencies of the rs41976219 in cases and controls and the calculated ORs. As can be observed with the recessive model, the CC genotype in the cis-eQTL-rs41976219 was more frequent among healthy cows (2.2%) compared to cows with PTB-associated lesions in gut tissues (1.4%); OR = 0.61 (CI: 0.22–1.67).

Table 1. Frequencies of the rs41976219 genotypes in healthy cows and cows with PTB-associated lesions and odds ratios (ORs) for each genetic model.

	No Lesion (N)	Frequency (%)	Lesion (N)	Frequency (%)	OR (95% CI)
Codominant					
AA	422	84.2	369	83.5	1
AC	68	13.6	67	15.2	1.13 (0.78–1.62)
CC	11	2.2	6	1.4	0.62 (0.23–1.7)
Dominant					
AA	422	84.2	369	83.5	1
AC-CC	79	15.8	73	16.5	1.06 (0.75–1.5)
Recessive					
AA-AC	490	97.8	436	98.6	1
C/C	11	2.2	6	1.4	0.61 (0.22–1.67)
Overdominant					
AA-CC	433	86.4	375	84.8	1
AC	68	13.6	67	15.2	1.14 (0.79–1.64)
log-Additive					
0, 1, 2	501	53.1	442	46.9	1

3.4. The Presence of Two Minor Alleles in the cis-eQTL-rs41976219 (CC) Increased CTSG Protein Levels in the Plasmas of Healthy Cows

The CTSG levels were measured by ELISA in plasma samples of cows without lesions in their gut tissues and with the CC (n = 8), AA (n = 17) and/or AC (n = 19) genotypes. As observed in Figure 3 the CTSG levels in plasmas of cows carrying the CC genotype are significantly higher (2.15 ng/mL) than in the plasmas of cows displaying the AA and AC genotypes (0.82 ng/mL) for rs41976219 (p = 0.039).

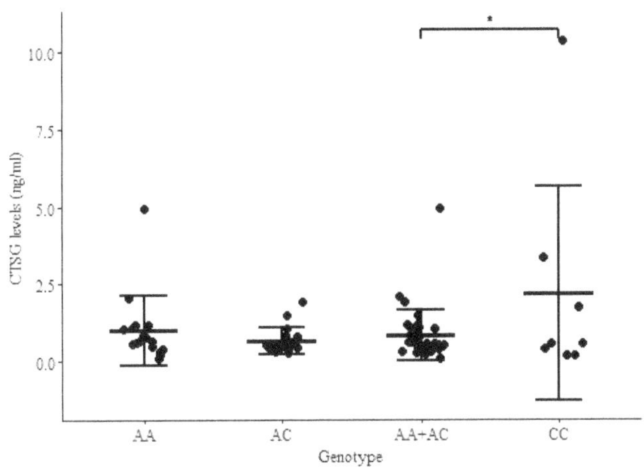

Figure 3. Plasma CTSG levels in carriers of different genotypes of cis-eQTL-rs41976219. CTSG levels measured in the plasmas of cows without lesions in gut tissues and with the AA, AC, and CC genotypes for cis-eQTL-rs41976219 by ELISA. Data are shown as the mean ± SD of 44 cattle, 8 carriers of the CC, 17 of the AA, and 19 of the AC genotypes. Statistically significant differences are observed in the mean CTSG levels in the plasmas of cows with the CC genotype when compared with cows with the AA + AC genotypes. p-values are calculated using an unpaired t-test. * p-value $\leq$ 0.05.

4. Discussion

To control MAP infection, we first need to understand the host–pathogen interactions [32–35]. Second, developing new diagnostic tools or improving the sensitivity of the ones already existing would help diagnose the infection earlier and avoid the late stages of the disease [36,37]. This should be followed by developing new and affordable therapeutic approaches for the treatment of infected animals and the prevention of MAP spread. At present, it is widely accepted that host antimicrobial proteins and peptides able to target the intracellular bacteria without damaging the host play important roles in the mammalian innate defense and often have immunomodulatory effects [38]. Although therapeutic interventions with these immunomodulatory agents could control intracellular pathogens, including MAP, research on host antimicrobial proteins and peptides against MAP is still scarce. A previous study demonstrated that synthetic *Cathelicidin* LL-37 reduces MAP internalization and pro-inflammatory cytokines in macrophages [39].

The CTSG was the first human PBMC protein proven to have antibacterial activity in vitro against Gram+ and Gram− bacteria [40–42]. CTSG is known for its ability to kill pathogens and also for its role in tissue remodeling and inflammation resolution. CTSG is a 26 kDa serine protease expressed in azurophilic granules of neutrophils, B cells, myeloid dendritic cells (DCs), plasmacytoid DCs, and cells of the monocyte/macrophage lineage [43,44]. Although CTSG is mainly located in the endo-lysosomal compartment, it is also active in the cytosol, extracellular space, and cell surface. CTSG is released after cell infection enhancing *Platelet-activating Factor*, *Tumor Necrosis Factor α*, *Interleukin 1β*, and *Interleukin-8 (CXCL8/IL8)* production. *CXCL8/IL8* acts as a strong neutrophil chemoattractant to the site of infection and as a major proinflammatory cytokine. Since the presence of CTSG in the endo-lysosomal compartment of infected cells permits direct interaction with and killing of bacteria, it is not surprising that mycobacteria can control CTSG expression and activity to favor their intracellular survival in monocytes [27]. A comparative study using a pathogenic strain *M. tuberculosis* and a non-pathogenic strain *M. smegmatis* revealed that following infection with *M. tuberculosis*, mRNAs for the majority of *CTSs* were downregulated in human primary macrophages (*CTSB*, *C*, *D*, *E*, *G*, *K*, *O*, *S*, *V*, and *W*), in contrast to the non-pathogenic *M. smegmatis* that induced the upregulation of most *CTSs*. Such a global downregulation of *CTSs* expression decreased pathogen killing and improved intracellular bacterial survival [45]. More specifically, the mechanism that *M. tuberculosis* uses to control the expression and proteolytic activity of CTSG was proposed by Danelishvili et al. [46]. They proposed that the bacterial protein Rv3364c, secreted by infected macrophages, binds to and inhibits the activity of membrane CTSG leading apoptosis inhibition. Walter et al. conducted a survival study using mice deficient in *CTSs G*, *K*, *L*, *B*, and *Z* and infected with *M. tuberculosis* [47]. Their results showed that only mice deficient in serine protease CTSG succumbed to *M. tuberculosis* infection more rapidly than wild-type mice, whereas mice deficient in the other *CTSs* did not demonstrate such an effect. In the context of MAP infection, our previous RNA-sequencing analysis revealed the downregulation of *CXCL8/IL8* in the PB and ileocecal valve (ICV) of MAP-infected cattle, regardless of the stage of the infection [35]. In addition, the downregulation of the CTSG mRNA expression was also observed in PB samples of MAP-infected animals suggesting the inhibition of neutrophil recruitment and CTSG-dependent killing response upon MAP infection [35].

Genomic-transcriptomic approaches allow for the identification of SNPs that affect the expression levels of genes associated with complex traits. We previously observed that the minor allele in cis-eQTLs-rs41976219 was associated with individual increased CTSG mRNA levels in blood samples obtained from Holstein cows [21]. In contrast, the CTSG mRNA was practically absent in the blood samples of cattle with the AA genotype. In the current study, we demonstrated that the heterozygous genotype for rs41976219 (AC) resulted in higher CTSG protein levels in the supernatants of infected MDMs after 2 h of infection and a significant lower intracellular MAP load at 7 d p.i. when compared with cattle with the AA genotype. Since it was difficult to find animals with the two minor alleles (CC) for the rs41976219 due to its low frequency, only animals with the genotypes AA and AC were included in the first part of the study. Although the number of animals was low (n = 16), we demonstrated that the heterozygous genotype for the rs41976219 (AC) resulted in higher CTSG protein levels in the supernatants of infected CD14+-MDMs following 2 h of infection and significantly lower intracellular MAP load at 7 d p.i. than in MDMs from cattle with the AA genotype. The association between the CC allelic variant and resistance to PTB was subsequently demonstrated in a larger cattle population (n = 943) using logistic regression analysis. In addition, the CTSG levels in plasma samples of cows without lesions in gut tissues and with the CC (n = 8) genotype were significantly higher than in the plasmas of cows with the AA + AC (n = 36) genotypes (p = 0.039). Since some of the animals without lesions might not have been exposed to MAP, variability was observed in the levels of CTSG in the plasmas of the animals with the CC genotype. Consequently, more samples are needed to support our data. On the other hand, MAP infection blocks CTSG expression in the blood of infected animals (35). Therefore, plasma samples obtained from animals with lesions were not tested by ELISA.

In agreement with our results, Rivera-Marrero et al. demonstrated that in THP-1 cells, the expression and activity of CTSG were downregulated upon exposure to a pathogenic strain of *M. tuberculosis*, and that the addition of CTSG to THP-1 cells before infection decreased *M. tuberculosis* viability [48]. In the alveolar macrophages of mice infected with the attenuated *M. bovis* bacillus Calmette–Guérin (BCG) strain, CTSG is strongly upregulated and, together with *neutrophil elastase* (*NE*), participates in the effective elimination of pathogens [49]. Although engulfed pathogenic mycobacteria induce the arrest of phagosome maturation and acidification in macrophages, CTSG and *NE* are neutral serine proteases and so can optimally digest the engulfed bacteria at a neutral pH level. Moreover, Steinwede et al. demonstrated that human CTSG encapsulated in liposomes reduced the mycobacterial load in the lungs of mice infected with pathogenic *M. bovis* [50].

5. Conclusions

CTSG is relevant for the control of MAP infection, but pathogenic mycobacteria, including MAP, downregulate the levels of CTSG in infected macrophages to facilitate their survival. On the other hand, individual genetic variation affects CTSG expression levels in MAP-infected MDMs. Our study revealed that the CTSG levels were significantly higher in the supernatants of MAP-infected CD14+-MDMs obtained from cows carrying the AC genotype for rs41976219 compared to those detected in supernatants obtained from cows with the most abundant AA genotype. Interestingly, MDMs obtained from hosts with the AC genotype for rs41976219 exhibited an enhanced ability to clear MAP infection. Furthermore, the rs41976219 genotype CC was more frequent in healthy cows than in cows with PTB-associated lesions in their gut tissues. Higher CTSG levels were measured in the plasmas of healthy cows with the CC genotype when compared with cows with the AA and AC genotypes. We can conclude that the presence of the minor allele in ciseQTL-rs41976219 increases CTSG expression in plasma samples and MAP-infected macrophages, and contributes to the control bacterial load within infected macrophages.

Author Contributions: Conceptualization, M.A.-H.; methodology, M.C.; validation, M.C. and G.B.-B.; formal analysis, M.A.-H., M.C. and G.B.-B.; writing—original draft preparation, M.C.; writing—review and editing, M.A.-H.; visualization, G.B.-B.; supervision, M.A.-H.; project administration, M.A.-H.; funding acquisition, M.A.-H. All authors have read and agreed to the published version of the manuscript.

Funding: This research was funded by grants RTI2018-094192-R-CO1 and PID2021-122197OR-C21, from the Spanish Ministry of Science and Innovation (MCIN/AEI/10.13039/501100011033), and by European Regional Funds (FEDER) "Una manera de hacer Europa". G.B.-B. was awarded a fellowship grant from MCIN/AEI/10.13039/501100011033 and "FSE Invierte en tu futuro"; grant PRE2019-090562.

Institutional Review Board Statement: Animals included in this study were not submitted to any in vivo experimentation. For the MGIA assay, blood samples were drawn from healthy Holstein Friesian cows from a commercial dairy farm located in the Basque Country. Animals used in the MGIA assay were submitted only to procedures that according to the European (Directive 2010/63/EU of the European Parliament and of the Council of 22 September 2010 on the protection of animals used for scientific purposes. Chapter 1, Article 1, Section 5, paragraphs b and f) and Spanish (Real Decreto 53/2013, de 1 de febrero, por el que se establecen las normas básicas aplicables para la protección de los animales utilizados en experimentación y otros fines científícos, incluyendo la docencia, Article 2, Section 5, Paragraphs b and f) legislation on experimental animals are exempt from its application. The animals, belonging to a registered commercial farm supervised by the local livestock authority (Servicio de Ganaderıa de la Diputación Foral de Alava) were not killed in relation to this study. Slaughtered animals used in the case–control study were sampled under pertinent legislation for safeguarding animal welfare (Basque Government Decree 454/1994, Spanish Government Law 32/2007, Royal decree 731/2007, and European council regulation number 1099/2009) and were analyzed by histopathology in a previous research project (28).

Informed Consent Statement: Not applicable.

Data Availability Statement: The original contributions presented in this study are included in the article. Further inquiries can be directed to the corresponding authors.

Acknowledgments: We would like to acknowledge the manager of our collaborating farm in Alava (Spain) for his support in the sampling work. We gratefully acknowledge Kyle P. Hearn for the careful editing of the manuscript.

Conflicts of Interest: The authors declare no conflict of interest. The funders had no role in the design of the study; in the collection, analyses, or interpretation of data; in the writing of the manuscript; or in the decision to publish the results.

References

1. Rasmussen, P.; Barkema, H.W.; Mason, S.; Beaulieu, E.; Hall, D.C. Economic losses due to Johne's disease (paratuberculosis) in dairy cattle. *J. Dairy Sci.* **2021**, *104*, 3123–3143. [CrossRef]
2. González, J.; Geijo, M.V.; García-Pariente, C.; Verna, A.; Corpa, J.M.; Reyes, L.E.; Ferreras, M.C.; Juste, R.A.; García Marín, J.F. Pérez, V. Histopathological classification of lesions associated with natural paratuberculosis infection in cattle. *J. Comp. Patho* **2005**, *133*, 184–196. [CrossRef]
3. Balseiro, A.; Perez, V.; Juste, R.A. Chronic regional intestinal inflammatory disease: A trans-species slow infection? *Comp Immunol. Microbiol. Infect. Dis.* **2019**, *62*, 88–100. [CrossRef]
4. Juste, R.A.; Elguezabal, N.; Pavón, A.; Garrido, J.M.; Geijo, M.; Sevilla, I.; Cabriada, J.L.; Tejada, A.; García-Campos, F.; Casado, R. et al. Association between *Mycobacterium avium* subsp. *paratuberculosis* DNA in blood and cellular and humoral immune respons in inflammatory bowel disease patients and controls. *Int. J. Infect. Dis.* **2009**, *13*, 247–254. [CrossRef]
5. Pierce, E.S. Could *Mycobacterium avium* subspecies *paratuberculosis* cause Crohn's disease, ulcerative colitis and colorectal cancer *Infect. Agents Cancer* **2018**, *13*, 1. [CrossRef]
6. Mameli, G.; Cocco, E.; Frau, J.; Marrosu, M.G.; Sechi, L.A. *Epstein Barr* Virus and *Mycobacterium avium* subsp. *paratuberculosi* peptides are recognized in sera and cerebrospinal fluid of MS patients. *Sci. Rep.* **2016**, *6*, 22401. [CrossRef]
7. Niegowska, M.; Rapini, N.; Piccinini, S.; Mameli, G.; Caggiu, E.; Bitti, M.L.M.; Sechi, L.A. Type 1 Diabetes at-risk children highl recognize *Mycobacterium avium* subspecies *paratuberculosis* epitopes homologous to human Znt8 and Proinsulin. *Sci. Rep.* **2016**, *6*, 2226 [CrossRef]
8. Bo, M.; Erre, G.L.; Niegowska, M.; Piras, M.; Taras, L.; Longu, M.G.; Passiu, G.; Sechi, L.A. Interferon regulatory factor 5 is potential target of autoimmune response triggered by *Epstein-Barr* virus and *Mycobacterium avium* subsp. *paratuberculosis* in rheumatoid arthritis: Investigating a mechanism of molecular mimicry. *Clin. Exp. Rheumatol.* **2018**, *36*, 376–381.
9. Naito, T.; Yokoyama, N.; Kakuta, Y.; Ueno, K.; Kawai, Y.; Onodera, M.; Moroi, R.; Kuroha, M.; Kanazawa, Y.; Kimura, T.; et a Clinical and genetic risk factors for decreased bone mineral density in Japanese patients with inflammatory bowel disease. *Gastroenterol. Hepatol.* **2018**, *33*, 1873–1881. [CrossRef]
10. Garrido, J.M.; Vazquez, P.; Molina, E.; Plazaola, J.M.; Sevilla, I.A.; Geijo, M.V.; Alonso-Hearn, M.; Juste, R.A. Paratuberculosi vaccination causes only limited cross-reactivity in the skin test for diagnosis of bovine tuberculosis. *PLoS ONE* **2013**, *8*, e8098 [CrossRef]
11. Garry, F. Control of Paratuberculosis in Dairy Herds. *Vet. Clin. Food Anim. Pract.* **2011**, *27*, 599–607. [CrossRef]
12. Groenendaal, H.; Nielen, M.; Jalvingh, A.W.; Horst, S.H.; Galligan, D.T.; Hesselink, J.W. A simulation of Johne's disease contro *Prev. Vet. Med.* **2002**, *54*, 225–245. [CrossRef]
13. Alonso-Hearn, M.; Badia-Bringué, G.; Canive, M. Genome-wide association studies for the identification of cattle susceptible an resilient to paratuberculosis. *Front. Vet. Sci.* **2022**, *9*, 935133. [CrossRef]

4. Kiser, J.N.; White, S.N.; Johnson, K.A.; Hoff, J.L.; Taylor, J.F.; Neibergs, H.L. Identification of loci associated with susceptibility to *Mycobacterium avium* subspecies *paratuberculosis* (Map) tissue infection in cattle. *J. Anim. Sci.* **2017**, *95*, 1080–1091. [CrossRef]
5. McGovern, S.P.; Purfield, D.C.; Ring, S.C.; Carthy, T.R.; Graham, D.A.; Berry, D.P. Candidate genes associated with the heritable humoral response to *Mycobacterium avium* ssp. *paratuberculosis* in dairy cows have factors in common with gastrointestinal diseases in humans. *J. Dairy Sci.* **2019**, *102*, 4249–4263. [CrossRef]
6. Canive, M.; Badia-Bringué, G.; Vázquez, P.; González-Recio, O.; Fernández, A.; Garrido, J.M.; Juste, R.A.; Alonso-Hearn, M. Identification of loci associated with pathological outcomes in Holstein cattle infected with *Mycobacterium avium* subsp. *paratuberculosis* using whole-genome sequence data. *Sci. Rep.* **2021**, *11*, 20177. [CrossRef]
7. Canive, M.; González-Recio, O.; Fernández, A.; Vázquez, P.; Badia-Bringué, G.; Lavín, J.L.; Garrido, J.M.; Juste, R.A.; Alonso-Hearn, M. dentification of loci associated with susceptibility to *Mycobacterium avium* subsp. *paratuberculosis* infection in Holstein cattle using combinations of diagnostic tests and imputed whole-genome sequence data. *PLoS ONE* **2021**, *16*, e0256091. [CrossRef]
8. Aryngaziyev, B.; Beltramo, C.; Dondo, A.; Karymsakov, T.; Varello, K.; Goria, M.; Di Blasio, A.; Nodari, S.; Colussi, S.; Modesto, P.; et al. Polymorphisms associated to bovine paratuberculosis: Investigation of their role in dna-protein interactions and transcriptional regulation. *Vet. Ital.* **2020**, *56*, 109–114. [CrossRef]
9. Kiser, J.N.; Wang, Z.; Zanella, R.; Scraggs, E.; Neupane, M.; Cantrell, B.; Van Tassell, C.P.; White, S.N.; Taylor, J.F.; Neibergs, H.L. Functional Variants Surrounding Endothelin 2 Are Associated with *Mycobacterium avium* Subspecies *paratuberculosis* Infection. *Front. Vet. Sci.* **2021**, *8*, 280. [CrossRef]
0. Jansen, R.C.; Nap, J.P. Genetical genomics: The added value from segregation. *Trends Genet.* **2001**, *17*, 388–391. [CrossRef]
1. Canive, M.; Fernandez-Jimenez, N.; Casais, R.; Vázquez, P.; Lavín, J.L.; Bilbao, J.R.; Blanco-Vázquez, C.; Garrido, J.M.; Juste, R.A.; Alonso-Hearn, M. Identification of loci associated with susceptibility to bovine paratuberculosis and with the dysregulation of the MECOM, eEF1A2, and U1 spliceosomal RNA expression. *Sci. Rep.* **2021**, *11*, 313. [CrossRef] [PubMed]
2. Conus, S.; Simon, H.U. Cathepsins and their involvement in immune responses. *Swiss Med. Wkly* **2010**, *140*, w13042. [CrossRef] [PubMed]
3. Turk, V. NEW EMBO MEMBERS' REVIEW: Lysosomal cysteine proteases: Facts and opportunities. *EMBO J.* **2001**, *20*, 4629–4633. [CrossRef] [PubMed]
4. Rossi, A.; Deveraux, Q.; Turk, B.; Sali, A. Comprehensive search for cysteine cathepsins in the human genome. *Biol. Chem.* **2004**, *385*, 363–372. [CrossRef] [PubMed]
5. Chwieralski, C.E.; Welte, T.; Bühling, F. Cathepsin-regulated apoptosis. *Apoptosis* **2006**, *11*, 143–149. [CrossRef]
6. Turk, V.; Stoka, V.; Vasiljeva, O.; Renko, M.; Sun, T.; Turk, B.; Turk, D. Cysteine cathepsins: From structure, function and regulation to new frontiers. *Biochim. Biophys. Acta Proteins Proteom.* **2012**, *1824*, 68–88. [CrossRef]
7. Szulc-Dąbrowska, L.; Bossowska-Nowicka, M.; Struzik, J.; Toka, F.N. Cathepsins in Bacteria-Macrophage Interaction: Defenders or Victims of Circumstance? *Front. Cell. Infect. Microbiol.* **2020**, *10*, 601072. [CrossRef]
8. Vázquez, P.; Ruiz-Larrañaga, O.; Garrido, J.M.; Iriondo, M.; Manzano, C.; Agirre, M.; Estonba, A.; Juste, R.A. Genetic association analysis of paratuberculosis forms in Holstein-Friesian cattle. *Vet. Med. Int.* **2014**, *2014*, 321327. [CrossRef]
9. Juste, R.A.; Alonso-Hearn, M.; Garrido, J.M.; Abendaño, N.; Sevilla, I.A.; Gortazar, C.; De La Fuente, J.; Dominguez, L. Increased lytic efficiency of bovine macrophages trained with killed mycobacteria. *PLoS ONE* **2016**, *11*, e0165607. [CrossRef]
0. Abendaño, N.; Sevilla, I.; Prieto, J.M.; Garrido, J.M.; Juste, R.A.; Alonso-Hearn, M. Quantification of *Mycobacterium avium* subsp. *paratuberculosis* strains representing distinct genotypes and isolated from domestic and wildlife animal species by use of an automatic liquid culture system. *J. Clin. Microbiol.* **2012**, *50*, 2609–2617. [CrossRef]
1. Canive, M.; Badia-Bringué, G.; Vazquez, P.; Garrido, J.M.; Juste, R.A.; Fernandez, A.; Gonzalez-Recio, Ó.; Alonso-Hearn, M. A Genome-Wide Association Study for Tolerance to Paratuberculosis Identifies Candidate Genes Involved in DNA Packaging, DNA Damage Repair, Innate Immunity, and Pathogen Persistence. *Front. Immunol.* **2022**, *13*, 820965. [CrossRef]
2. Abendaño, N.; Juste, R.A.; Alonso-Hearn, M. Anti-inflammatory and antiapoptotic responses to infection: A common denominator of human and bovine macrophages infected with *Mycobacterium avium* subsp. *paratuberculosis*. *BioMed Res. Int.* **2013**, *2013*, 908348. [CrossRef] [PubMed]
3. Abendaño, N.; Tyukalova, L.; Barandika, J.F.; Balseiro, A.; Sevilla, I.A.; Garrido, J.M.; Juste, R.A.; Alonso-Hearn, M. *Mycobacterium avium* subsp. *paratuberculosis* isolates induce in vitro granuloma formation and show successful survival phenotype, common anti-inflammatory and antiapoptotic responses within ovine macrophages regardless of genotype or host of origin. *PLoS ONE* **2014**, *9*, e104238. [CrossRef] [PubMed]
4. Fitzgerald, L.E.; Abendaño, N.; Juste, R.A.; Alonso-Hearn, M. Three-dimensional in vitro models of granuloma to study bacteria-host interactions, drug-susceptibility, and resuscitation of dormant mycobacteria. *BioMed Res. Int.* **2014**, *2014*, 623856. [CrossRef] [PubMed]
5. Alonso-Hearn, M.; Canive, M.; Blanco-Vazquez, C.; Torremocha, R.; Balseiro, A.; Amado, J.; Varela-Martinez, E.; Ramos, R.; Jugo, B.M.; Casais, R. RNA-Seq analysis of ileocecal valve and peripheral blood from Holstein cattle infected with *Mycobacterium avium* subsp. *paratuberculosis* revealed dysregulation of the CXCL8/IL8 signaling pathway. *Sci. Rep.* **2019**, *9*, 14845. [CrossRef]
5. Badia-Bringué, G.; Canive, M.; Casais-Goyos, R.; Blanco-Vázquez, C.; Amado, J.; Iglesias, N.; González, A.; Bascones, M.; Juste, R.A.; Alonso-Hearn, M. Evaluation of a droplet digital PCR (ddPCR) assay for quantification of *Mycobacterium avium* subsp. *paratuberculosis* (MAP) DNA in whole-blood and fecal samples from MAP-infected Holstein cattle. *Front. Vet. Sci.* **2022**, *9*, 944189. [CrossRef]

37. Blanco-Vázquez, C.; Alonso-Hearn, M.; Iglesias, N.; Vázquez, P.; Juste, R.A.; Garrido, J.M.; Balseiro, A.; Canive, M.; Amado, J Queipo, M.A.; et al. Use of ATP-binding cassette subfamily A member 13 (ABCA13) for sensitive detection of focal pathologica forms of subclinical bovine paratuberculosis. *Front. Vet. Sci.* **2022**, *9*, 816135. [CrossRef]
38. Arranz-Trullén, J.; Lu, L.; Pulido, D.; Bhakta, S.; Boix, E. Host antimicrobial peptides: The promise of new treatment strategies against tuberculosis. *Front. Immunol.* **2017**, *8*, 1499. [CrossRef]
39. Cirone, K.M.; Lahiri, P.; Holani, R.; Tan, Y.L.; Arrazuria, R.; De Buck, J.; Barkema, H.W.; Cobo, E.R. Synthetic cathelicidin LL-37 reduces *Mycobacterium avium* subsp. *paratuberculosis* internalization and pro-inflammatory cytokines in macrophages. *Cell Tissue Res.* **2020**, *379*, 207–217. [CrossRef]
40. Odeberg, H.; Olsson, I. Antibacterial activity of cationic proteins from human granulocytes. *J. Clin. Investig.* **1975**, *56*, 1118–1124 [CrossRef]
41. Bangalore, N.; Travis, J.; Onunka, V.C.; Pohl, J.; Shafer, W.M. Identification of the primary antimicrobial domains in human neutrophil *Cathepsin G*. *J. Biol. Chem.* **1990**, *265*, 13584–13588. [CrossRef]
42. Kretschmer, D.; Breitmeyer, R.; Gekeler, C.; Lebtig, M.; Schlatterer, K.; Nega, M.; Stahl, M.; Stapels, D.; Rooijakkers, S.; Peschel, A *Staphylococcus aureus* Depends on Eap Proteins for Preventing Degradation of Its Phenol-Soluble Modulin Toxins by Neutrophi Serine Proteases. *Front. Immunol.* **2021**, *12*, 701093. [CrossRef] [PubMed]
43. Reeves, E.P.; Lu, H.; Jacobs, H.L.; Messina, C.G.M.; Bolsover, S.; Gabellall, G.; Potma, E.O.; Warley, A.; Roes, J.; Segal, A.W. Killing activity of neutrophils is mediated through activation of proteases by K+ flux. *Nature* **2002**, *416*, 291–297. [CrossRef] [PubMed]
44. Burster, T.; Macmillan, H.; Hou, T.; Boehm, B.O.; Mellins, E.D. Cathepsin G: Roles in antigen presentation and beyond. *Mol Immunol.* **2010**, *47*, 658–665. [CrossRef]
45. Pires, D.; Marques, J.; Pombo, J.P.; Carmo, N.; Bettencourt, P.; Neyrolles, O.; Lugo-Villarino, G.; Anes, E. Role of *Cathepsins* in *Mycobacterium tuberculosis* Survival in Human Macrophages. *Sci. Rep.* **2016**, *6*, 32247. [CrossRef]
46. Danelishvili, L.; Everman, J.L.; McNamara, M.J.; Bermudez, L.E. Inhibition of the plasma-membrane-associated serine protease Cathepsin G by *Mycobacterium tuberculosis* Rv3364c suppresses caspase-1 and pyroptosis in macrophages. *Front. Microbiol.* **2012**, *2*, 281. [CrossRef
47. Walter, K.; Steinwede, K.; Aly, S.; Reinheckel, T.; Bohling, J.; Maus, U.A.; Ehlers, S. *Cathepsin G* in Experimental Tuberculosis Relevance for Antibacterial Protection and Potential for Immunotherapy. *J. Immunol.* **2015**, *195*, 3325–3333. [CrossRef]
48. Rivera-Marrero, C.A.; Stewart, J.; Shafer, W.M.; Roman, J. The down-regulation of *Cathepsin G* in THP-1 monocytes after infection with *Mycobacterium tuberculosis* is associated with increased intracellular survival of bacilli. *Infect. Immun.* **2004**, *72*, 5712–5721 [CrossRef]
49. Srivastava, M.; Meinders, A.; Steinwede, K.; Maus, R.; Lucke, N.; Bühling, F.; Ehlers, S.; Welte, T.; Maus, U.A. Mediator responses of alveolar macrophages and kinetics of mononuclear phagocyte subset recruitment during acute primary and secondary mycobacterial infections in the lungs of mice. *Cell. Microbiol.* **2007**, *9*, 738–752. [CrossRef]
50. Steinwede, K.; Maus, R.; Bohling, J.; Voedisch, S.; Braun, A.; Ochs, M.; Schmiedl, A.; Länger, F.; Gauthier, F.; Roes, J.; et al *Cathepsin G* and *Neutrophil Elastase* Contribute to Lung-Protective Immunity against Mycobacterial Infections in Mice. *J. Immunol.* **2012**, *188*, 4476–4487. [CrossRef]

animals

MDPI

Article

Differences in Faecal Microbiome Taxonomy, Diversity and Functional Potential in a Bovine Cohort Experimentally Challenged with *Mycobacterium avium* subsp. *paratuberculosis* (MAP)

Chloe Matthews [1,2,*], Aaron M. Walsh [2], Stephen V. Gordon [3], Bryan Markey [3], Paul D. Cotter [2,4] and Jim O' Mahony [1,*]

1 Department of Biological Sciences, Munster Technological University, Bishopstown, T12 P928 Cork, Ireland
2 Teagasc Food Research Centre, Moorepark, P61 C996 Fermoy, Ireland; paul.cotter@teagasc.ie (P.D.C.)
3 School of Veterinary Medicine, University College Dublin, D04 V1W8 Dublin, Ireland
4 APC Microbiome Ireland, University College Cork, T12 R229 Cork, Ireland
* Correspondence: chloe.mmatthews093@gmail.com (C.M.); jim.omahony@mtu.ie (J.O'.M.)

Simple Summary: Johne's disease is a global economic burden. The disease contributes to reduced milk yield, reduced fertility, higher rates of susceptibility to other diseases and eventual death. Unhealthy animals are less efficient, producing higher kg of greenhouse gases per kg of output (milk and meat). Therefore, prognosis and diagnosis are important in terms of animal welfare and future climate change policies by reducing the impact of livestock on climate change. In the past decade, our knowledge around the microbiome in health and disease has increased. Next generation sequencing technologies have provided a new understanding of the interactions between the host microbiome and pathogens. The microbiome is a target for treatment, prevention of disease and a source of novel biomarkers of disease risk. The potential to improve animal health is highly dependent on our ever growing knowledge of the gut microbiome and their interactions with the host. With the aim to gain knowledge in the early months of exposure to MAP in calves, we investigated the dynamics of the gut microbiome (using faecal sample) of calves 3, 6 and 9 months post inoculation with MAP relative to unexposed controls.

Abstract: Mycobacterium avium subspecies paratuberculosis (MAP) is the causative agent of Johne's disease in ruminants, a chronic enteritis which results in emaciation and eventual loss of the animal. Recent advances in metagenomics have allowed a more in-depth study of complex microbiomes, including that of gastrointestinal tracts, and have the potential to provide insights into consequences of the exposure of an animal to MAP or other pathogens. This study aimed to investigate taxonomic diversity and compositional changes of the faecal microbiome of cattle experimentally challenged with MAP compared to an unexposed control group. Faecal swab samples were collected from a total of 55 animals [exposed group (n = 35) and a control group (n = 20)], across three time points (months 3, 6 and 9 post-inoculation). The composition and functional potential of the faecal microbiota differed across time and between the groups ($p < 0.05$), with the primary differences, from both a taxonomic and functional perspective, occurring at 3 months post inoculation. These included significant differences in the relative abundance of the genera Methanobrevibacter and Bifidobacterium and also of 11 other species (4 at a higher relative abundance in the exposed group and 7 at a higher relative abundance in the control group). Correlations were made between microbiome data and immunopathology measurements and it was noted that changes in the microbial composition correlated with miRNA-155, miR-146b and IFN-γ. In summary, this study illustrates the impact of exposure to MAP on the ruminant faecal microbiome with a number of species that may have relevance in veterinary medicine for tracking exposure to MAP.

Keywords: Johne's disease; MAP; microbiome; experimental model; immune function

1. Introduction

Mycobacterium avium subspecies *paratuberculosis* (MAP) is the causative agent of Johne's disease, a chronic enteritis, principally affecting the distal ileum. The pathogen induces thickening and corrugation of the intestinal wall, causing emaciation and eventual animal loss. MAP is found worldwide, and its spread is of major concern. In particular, globalisation has led to an increase in the movement of goods and commodities, including livestock and livestock products [1], with associated risk of the spread of infectious diseases in animals. Therefore, rapid diagnosis using minimal quantities of easily accessible biological samples may pave the way for protection of animal production systems in herds (at a local level) and countries with a disease-free status. While direct detection of pathogens is ideal in situations where this is challenging, indirect markers of infection or exposure merit consideration. In this regard identifying changes in the gastrointestinal (GIT) microbiome that occur in response to pathogen infection or exposure to a pathogen has the potential to be of value. This reflects the fact that next generation sequencing has transformed microbiological research by enabling high throughput metagenomic analysis of complex microbial communities from a number of different environments and hosts [2–5]. Metagenomic analysis has already revealed changes in the microbiome associated with exposure to pathogens in cattle [6,7], pigs [8,9], mice [10], and humans [11]. Such an approach, if developed, could be particularly valuable for the detection of difficult isolates, slow-growing pathogens such as MAP. These microorganisms require specialist culture media and laboratory protocols, while growth can take place from approximately 8–12 weeks for Type C strains and longer for Type S strains, while in some instances culture is not always successful.

The microbiome in early life plays an important role in health across mammalian species. In calves, the development of the microbiome is influenced by a number of different factors including diet (colostrum, milk replacer or calf starter), exposure to the dam, antibiotic use, the age of the animal and the sampling location [12]. The establishment of an optimal microbiome has been positively associated with calf health and growth, the prevention of neonatal diarrhoea [13] and pneumonia [14] as well as a rapid increase in weight [12]. Microbial colonisation also contributes to the development of the intestinal epithelial, mucosal layer and lymphoid structures [15]. In contrast, exposure to pathogens early in life may be detrimental to the host, impacting on short- and/or long-term health and productivity [12]. Although calves may be infected with MAP through intra-uterine transmission in clinically affected cows, it is more common via the faecal-oral route when they are exposed to faecal contaminated teats, pasture, water, supplements, bedding, milk or colostrum [16].

There have previously been a small number of 16S rRNA-based analyses of the faecal microbiome of animals naturally infected with MAP relative to controls. In one instance, an over-representation of the families Planococcaceae and Paraprevotellaceae and an under-representation of the genera *Faecalibacterium* and *Akkermansia* were noted in infected cattle [17]. Enrichment of lysine and histidine metabolism pathways and an underrepresentation of glutathione metabolism and leucine and isoleucine degradation pathways within the ileal mucosa-associated microbiome of MAP-infected cattle was also shown using the predictive software, PICRUSt (https://github.com/picrust/picrust, accessed on 25 February 2023). In a separate study, an increase in the genus *Psychrobacter* and decrease in the genera *Oscillospira*, *Ruminococcus* and *Bifidobacterium* was reported in cows infected with MAP [18]. Fecteau et al., 2016 [19] showed differences in the faecal microbiome between MAP-positive and MAP negative cattle using 16S rRNA sequencing technologies, with the microbiome of MAP-positive animals having a higher abundance of Actinobacteria in comparison to controls.

In addition to studies of naturally infected animals, there have been a number of studies whereby cattle have been experimentally challenged with MAP [20,21]. Experimental challenge models have the potential to provide useful information including host immune response following exposure and corresponding interactions with the normal gut microbiome. Such studies were carried out by Rankin and colleagues when MAP was known as *Mycobacterium johnei*. The group used large, single oral doses of MAP in their studies to investigate the portal of entry [22]. Other experiments, such as those conducted by [23], used smaller doses but inoculated weekly. More recently, a long-term MAP experimental study was conducted in Holstein Friesian cattle over a 33 month period [24,25]. It was found that there was significant differential expression of miR-155 in animals exposed to MAP. MiRNAs have been used to detect changes in animals associated with exposure to pathogens, with several studies using them as biomarkers for disease. Differential miRNA profiles in local anatomical sites may facilitate better understanding of the immunopathology of MAP infection [25].

The main aim of this study is to explore the taxonomy and metabolic activity of the microbiome of calves experimentally challenged with MAP relative to controls to determine if MAP exposure impacts the associated faecal microbiome. Diversity measures between exposed and control groups will be examined to determine if exposure to MAP influences the diversity profile of the microbiome. Associations between gut microbiome composition and immunomodulatory measures including miRNAs miR-155 and miR-146b, ELISA sero-positivity and IFN-γ will also be determined.

2. Materials and Methods

2.1. Animals and Experimental Design

This study employed samples collected from a long-term MAP experimental challenge. Specifically, rectal faecal swabs were collected from a group of animals that were exposed to MAP (n = 35) and a control group (n = 20), across three time points (months 3, 6 and 9 post-inoculation) [24].

Detailed information relating to animal selection criteria and husbandry can be found in the thesis of Britton 2017 [25]. Briefly, 55 Holstein-Friesian calves were sourced from two Autumn calving herds (herd A and herd B). Both farms had MAP seropositivity rates of $\leq$10% and neither farm had a history of Johne's disease. All dams were MAP faecal culture negative and seronegative. In terms of diet in early life, calves received 2–4 L of colostrum from their own dam shortly after birth and were then removed from the mothers. Calves were housed indoors in small groups and received Gain® Easi-Mix calf milk replacer (Gain Feeds, Portlaoise, Ireland) until weaning at approximately 3 months of age. Animals were then maintained on an ad libitum silage-based diet. Thirty five calves aged between 3 and 6 weeks were orally inoculated on two consecutive days with 10 mL (3.8×10^9 CFU) of MAP strain CIT003. A control group of 20 calves received a placebo. The animals were housed indoors in group housing on a semi-slatted unit with access to a solid-floor, straw-bedded area, for the duration of the experiment with the groups housed separately (approximately 0.5 km apart). Blood, faecal and faecal rectal swab samples were collected from all animals throughout the trial at regular intervals (premmune and 2, 3, 6, 10, 12, 16, 20, 24, 28, 31 and 33 months post inoculation and premmune and 3, 6, 9, 12, 16, 20, 24, 28, 31 and 33 months post inoculation, respectively) in order to determine MAP infection status. Cell-mediated immunity was measured using an IFN-γ assay (Bovigam®) test and serum anti-MAP antibodies were measured using a commercially available kit (Idexx Johne's Identification and Verification ELISA, IDEXX Laboratories, Inc. One IDEXX Drive, Westbrook, ME, USA). Faecal samples were cultured for 42 days using the TREK ESP para-JEM system (Thermo Fisher Scientific, Waltham, MA, USA). Briefly, faecal samples were mixed thoroughly before 2 g was suspended in 35 mL sterile distilled water. The samples were then vortexed and allowed to stand for 30 min. About 5 mL of the resuspended faecal material was added to 25 mL 0.9% hexadecylpyridinium chloride in 50% brain heart infusion broth and was incubated overnight at 37 °C. Replicates, taken from the same samples were transported on foam swabs, before being frozen at −80 °C until DNA extraction. Storage of swabs at −80 °C allowed for faecal rectal samples to be processed at a later date for DNA extraction and microbiome analysis. Confirmatory MAP specific histological changes (including tissue from distal ileum and ileocecal lymph nodes) were not detected during the 33 month trial, and it is unknown whether any of the challenged animals would have progressed to faecal shedding of MAP and disease. These tissue samples were collected at post-mortem examination at months 12, 24 and 33 post inoculation. The number of animals euthanized at each timepoint is as follows; 8 exposed and 5 controls were euthanized at timepoints 12 and 24, while 19 exposed and 9 controls were euthanized at month 33 (1 animal from the control group was euthanized at month 16 due to issues unrelated to the trial). ELISA seropositivity and IFN-ɣ was also be determined via commercial Bovigam® (Prionics AG, Schlieren, Switzerland).

2.2. DNA Extractions and Library Preparations

DNA extractions were carried out on 165 faecal swabs from months 3, 6 and 9 post inoculation using the Qiagen PowerSoil Pro kit (Qiagen Inc., Germantown, MD, USA). One sample (from the control group) was lost during DNA extraction (n = 164). The foam tip of each faecal swab was cut under sterile conditions into initial bead beating tubes supplied in the extraction kit. Extractions were performed as per protocol with an additional heat step, post bead beating, at 70 °C for 15 min which was adapted from [26]. The heat step was performed in bead-beating microfuge tubes as per [26]. Extracted DNA was quantified using the Qubit High Sensitivity DNA assay (Biosciences, Dublin, Ireland). Whole-metagenome shotgun libraries were prepared using the Illumina Nextera XT DNA Library Preparation Guide except that the tagmentation time was increased from 5 to 7 min. Samples were sequenced on the Illumina NextSeq platform at the Teagasc sequencing facility (Teagasc, Food Research Centre, Moorepark, Co. Cork, Ireland), using the high output reagent cartridge V2 300 cycle kit, following standard Illumina sequencing protocols.

2.3. Bioinformatics Analysis of Exposed vs. Control Animals across Time

Raw whole-genome shotgun sequencing reads were filtered on the basis of quality and quantity using the KneadData package using default parameters (http://huttenhower.sph.harvard.edu/kneaddata, accessed on 25 February 2023). Bovine genome reads (*Bos taurus*) were filtered and excluded from further analysis. Compositional assignment was carried out using MetaPhlAn2 using default settings [27]. The functional potential of the microbiome was determined using SUPER-FOCUS [28].

2.4. Statistical Analysis Comparing Exposed to Controls and across Time, and Correlating Immunopathology with Taxonomic Output

Statistical analysis was carried out in R studio, R version 3.5.3 (11 March 2019) (RStudio, Inc. PBC, Boston, MA, USA). Analysis was carried out to compare the microbiome of animals in exposed versus control groups and across time (be-tween months 3, 6 and 9). Data were checked for normality using the Shapiro-Wilks normality test in R. Each test for each dataset being used returned a p-value of <0.0001, indicating the data were non-normal. Alpha diversity analysis was carried out using the "vegan" package using Shannon, Simpson and Observed Species indices, in addition to multidimensional scaling (MDS). The Wilcoxon rank sum test was used to determine statistical differences in alpha diversity between groups. Permutational analysis of variance (PERMANOVA, was performed using the "adonis" function from vegan (version 2.5-6). Results from significance testing using adonis were confirmed using multivariate analysis using the betadisper function from the "vegan" package. This measures the variance of a group of samples by calculating the average distance of group members to the group centroid (https://www.rdocumentation.org/packages/vegan, accessed on 25 February 2023). PPD-j induced IFN-ɣ and sero-positive results were correlated (Spearman Rank correlation using the "corrplot" package) with microbial abundances. The same approach was applied to identify correlations with the expression of microRNAs; miR-16, miR-21, miR-29a, miR-146b, miR-155 and miR-223, and cell-mediated immunity, measured using an IFN-ɣ assay using purified protein derivative of MAP (PPD-J) was carried out by [25]. For this test, 22.5 ug/mL MAP

PPDj (kindly provided by Dr. Douwe Bakker, Central Veterinary Institute, Lelystad, The Netherlands) was used.

The linear discriminant analysis (LDA) effect size was determined using LEfSe [27]. LEfSe uses the two-tailed non-parametric Kruskal–Wallis test to examine the significance of differences of species and functional potential in the two groups. The groups were analysed separately across time. A set of pair-wise tests among the two groups was performed using the Wilcoxon test. Finally, LDA was performed to estimate the effect size of each differently abundant species. The microbiome of each of the groups was thought to be significantly different if their differences had a p-value < 0.05 and an LDA score (log10) > 2. Data visualisation was performed using the "ggplot2" package in R.

3. Results

3.1. The Taxonomic Profile of Calves Exposed to MAP Differs from Controls

Metagenomic sequencing was used to investigate how exposure to MAP at 3–6 weeks of age impacts the composition and functional potential of the faecal microbiome of calves across three time points post inoculation. In total, the study consisted of 164 samples from 35 exposed and 20 control animals across three time-points (months 3, 6 and 9 post inoculation). Rectal faecal samples and tissue samples remained culture negative throughout the 33 month challenge. However, taxonomic profiling using MetaPhlAn2 revealed a significant difference ($p < 0.05$) between groups across time with respect to diversity and composition. Phylum level analysis revealed an increase in the relative abundance of members of Euryarchaeaota and Firmicutes in the exposed group with a decreased relative abundance of Actinobacteria and Bacteriodetes in the exposed group (Figure 1). The relative abundance of the top 25 most dominant species in each individual animal was examined. Overall, the faecal microbiome, from both the exposed and control groups, was dominated by a combination of four species; *Methanobrevibacter unclassified, Bifidobacterium pseudolongum, Butyrivib-rio unclassified*, and *Peptostreptococcaceae unclassified* (Figure 2). Discriminatory species were identified using linear discriminant analysis (LDA) effect size (LEfSE), which allows for the identification of species or pathways which explain differences between groups (Figure 3). Abundances of Bifidobacterium species decreased over time, particularly *B. pseudolongum, B. angulatum* and *B. adolescentis* (Figure 3). LEfSE also showed a number of species that significantly differed (p-value < 0.05) in relative abundance between the groups (Figure 3). Supporting data can be found in Supplementary Table S1. These included an unclassified Methanobrevibacter species that was present at significantly greater relative abundance in the exposed group at months 3 and 6 and an unclassified species of Butyrivibrio that was significantly more abundant in the exposed group across all time points. *B. angulatum* and *B. adolescentis* were significantly more abundant in the exposed group at month 3, whereas *B. pseudo-longum* was significantly more abundant in the control group at months 3 and 6. Overall, month 3 showed the greatest number (11) of significant differences, with 7 at higher relative abundance in the control group and 4 at higher relative abundance in the exposed group. Other taxa that were more abundant in the control group at month 3 were *Alistipes shahii, Parabacteriodes distasonis, Prevotella stercorcea, Bacteroides vulgatus* and *Subdoligranulum unclassified*. It is important to note that MAP sequences were not detected at any time point.

3.2. Differences in Faecal Microbiome Alpha and Beta Diversity between MAP Exposed and Control Animals

Alpha diversity analysis showed differences over time; alpha diversity decreased over time specifically between month 3 and month 9. More specifically, differences in the species-level alpha diversity (observed species, Shannon and Simpson index) of the faecal microbial communities within the groups (exposed versus control) were determined for all samples (Figure 4).

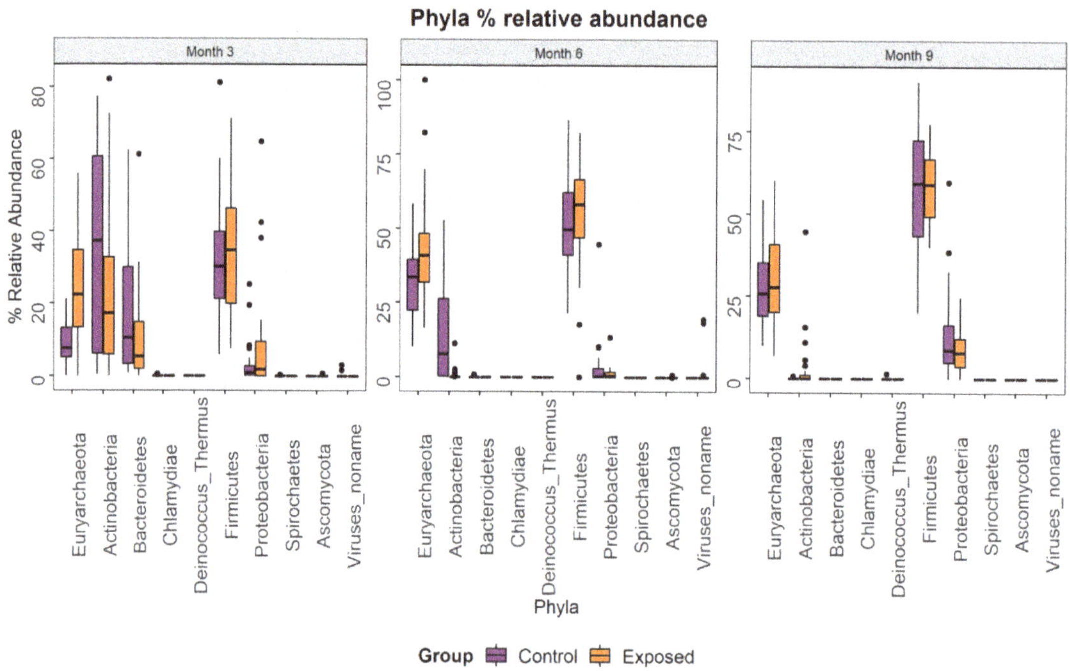

Figure 1. Phylum level analysis revealed an increase in the relative abundance of members of Euryarchaeaota and Firmicutes in the exposed group with a decreased relative abundance of Actinobacteria and Bacteriodetes in the exposed group, with the greatest differences observed in month 3.

With respect to beta diversity, multidimensional scaling was used to visualise the level of similarity between samples, as calculated using the Bray-Curtis dissimilarity metric. PERMANOVA revealed significant differences between groups ($p < 0.001$) and between sero-positive and sero-negative animals ($p < 0.05$). Time-related changes can be observed in the Bray-Curtis index values as overall community structure changed across time (Supplementary Table S2), and progressed towards a more similar microbiome (Figure 5). Microbiomes from samples collected at month 3 had a significantly higher variability than samples collected at months 6 and 9. Betadisper analysis confirmed significant differences between both the exposed and control animals and in month 3 vs. month 6 and month 3 vs. month 9. Betadisper analysis also confirmed that exposed animals were less similar to each other whereas control animals showed more similarity ($p < 0.01$) (Supplementary Table S2). Differences related to herd of origin can be found in Supplementary Figure S2.

3.3. *The Functional Potential of the Microbiome of Animals Exposed to MAP Differs from That of Controls*

Analysis of the functional potential of the faecal microbiomes revealed that genes related to sulphur metabolism, respiration, stress response, RNA metabolism, clustering based subsystems, co-factors and vitamins were significantly ($p < 0.05$) enriched in the exposed group 3 months' post-challenge, relative to month 3 controls. Genes associated with membrane transport, potassium metabolism, virulence, iron acquisition and metabolism, regulation and cell signalling, and phages, prophages and transposable elements were more abundant in the control group. A similar number of significantly different pathways was noted at month 6 but a smaller number of differences was evident at month 9 (Figure 6). A more in-depth analysis of functional data established that genes associated with a number of hydrogenase pathways, including formate hydrogenase, were present at significantly higher relative abundance in the exposed group overall. Maltose and maltodextrin utilisation, glycogen metabolism, glycolysis and gluconeogenesis and genes associated with the resistance of fluoroquinolones were also more abundant in the exposed group. Genes associated with Ton and Tol transport systems (a major group of pattern recognition receptors), cellulosomes, mannose metabolism, histidine degradation and ammonia assimilation pathways were more abundant in the control group (Supplementary Figure S1, Supplementary Tables S3 and S4).

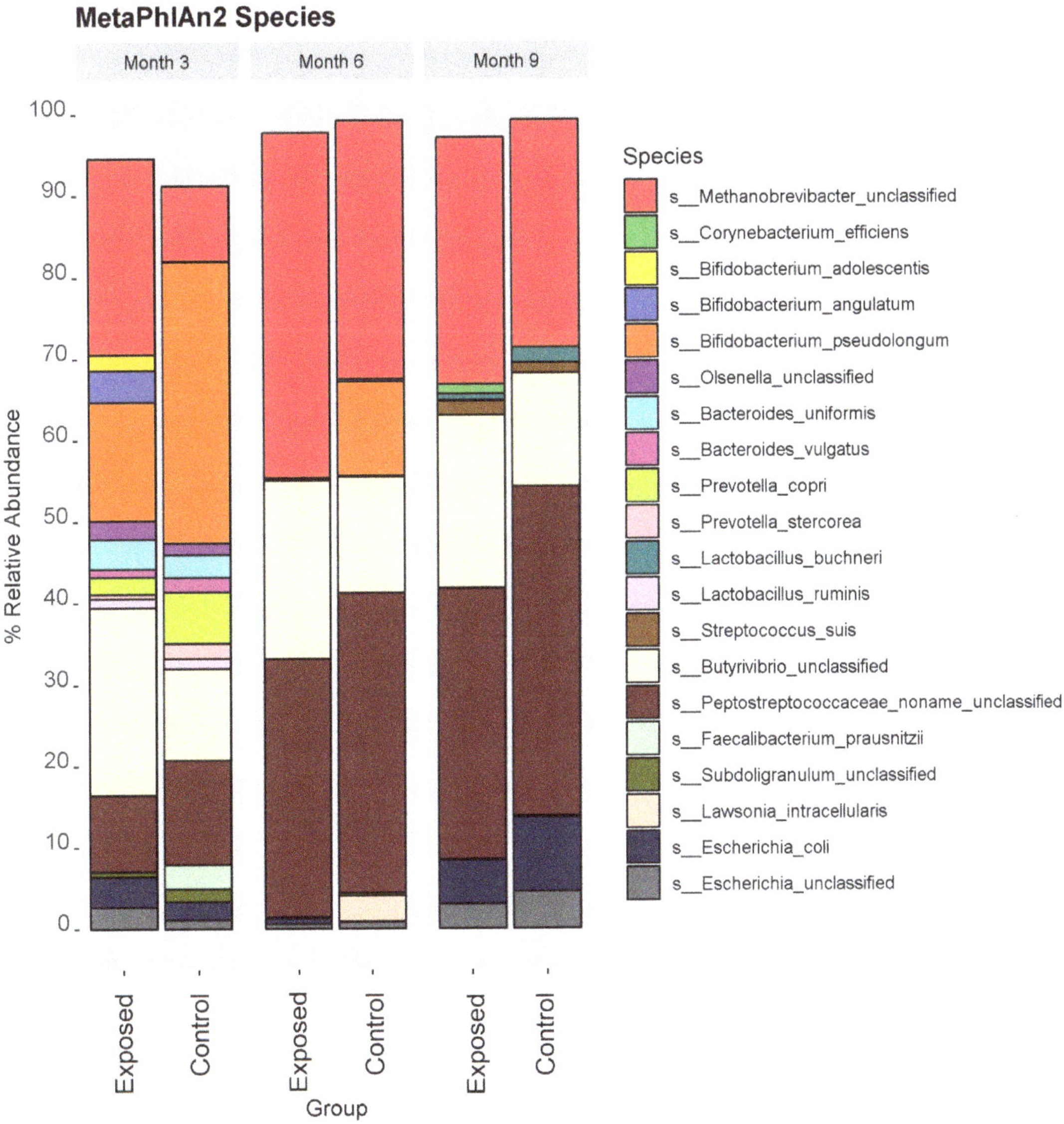

Figure 2. The microbiome was dominated by a combination of four species; Methanobrevibacter unclassified, Bifidobacterium pseudolongum, Butyrivibrio unclassified and Peptostreptococcaceae unclassified.

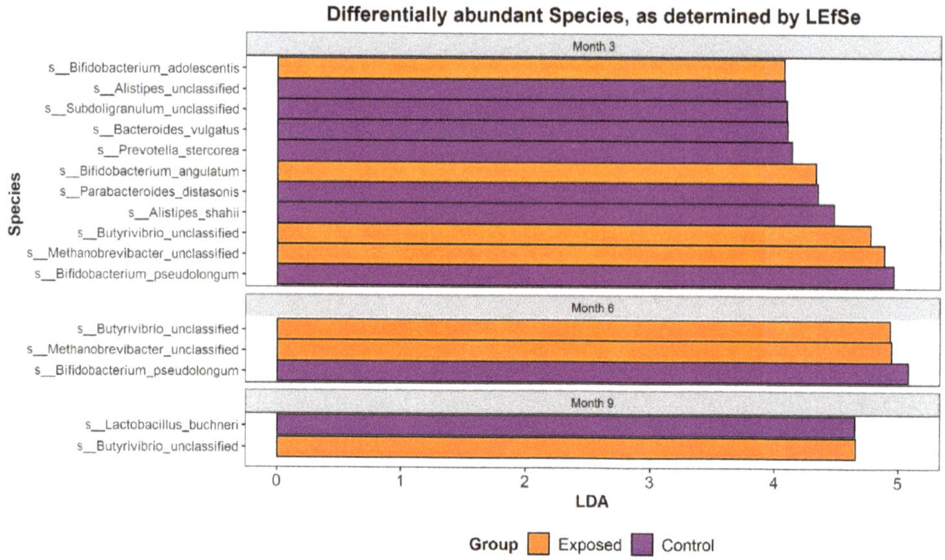

Figure 3. Discriminatory species were identified using linear discriminant analysis (LDA) effect size (LEfSE), which allows for the identification of species which explain differences between groups. Abundances of Bifidobacterium species decreased over time, particularly *B. pseudolongum*, *B. angulatum* and *B. adolescentis*.

3.4. Correlations between the Gut Microbiome and Immunopathology Measurements Reveals Important Associations between the Microbiome and Host Immune Response

In total, 12 animals were seropositive at one or more timepoints across the 33 month trial from month 16–33. Spearman Rank Correlations were used to examine the correlations between faecal microbiome species and seropositive animals. A sero-positive result reflects an instance where an animal had one ELISA positive result during the duration of the trial. Seven taxa were found to have significantly positive ($p < 0.05$) correlations with sero-positive results (ρ ranged between 0.16 and 0.22) (Table 1). *Enterococcus hirae* appears in 4 animals from the exposed group that had sero-positive results specifically at 3 months post inoculation in comparison to animals from the same timepoint in the control group. This species did not appear at other timepoints. ELISA results can be found in Supplementary Table S5.

Table 1. Correlation coefficient and *p*-values of species associated with an ELISA seropositive result using Spearman rank correlation.

Species	*p*-Value	Correlation Co-Efficient
Methanobrevibacter ruminantium	0.039	0.162
Corynebacterium efficiens	0.007	0.211
Bacteroides thetaiotaomicron	0.045	0.157
Enterococcus hirae	0.007	0.209
Butyrivibrio unclassified	0.005	0.220
Erysipelotrichaceae bacterium 21_3	0.006	0.213
Penicillium chrysogenum	0.006	0.213

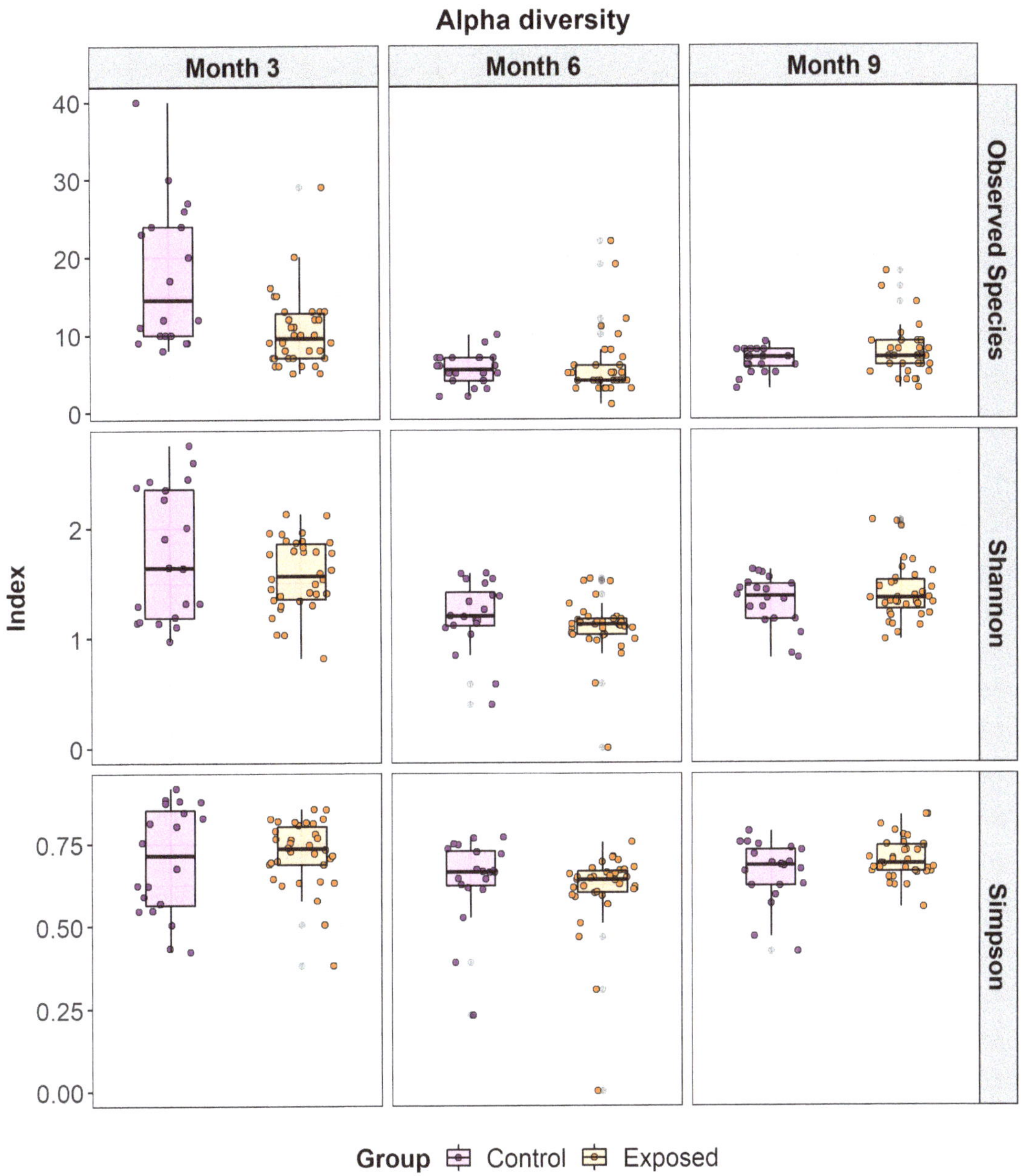

Figure 4. Diversity of the microbiome of both the exposed and control animals across time. Alpha diversity within subjects using Shannon, Simpson and Observed species measures of species-level output from MetaPhlAn2. The control group shows a higher within animal diversity; however, this was non-significant.

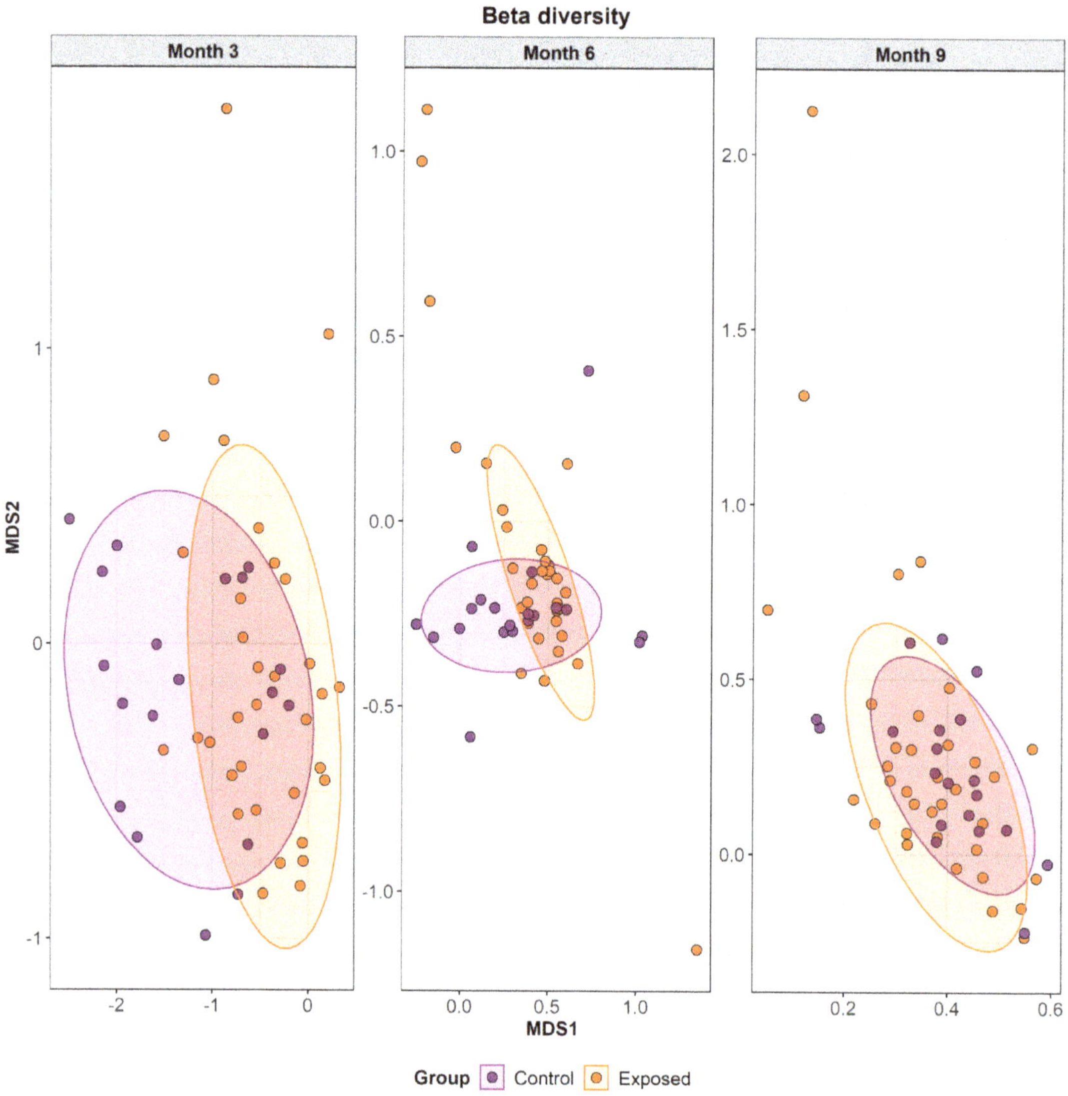

Figure 5. Bray-Curtis beta diversity among animals in both the exposed and control group, across time. Month 3 shows the most variation between the groups. The microbiome becomes more stable as animals age or potentially as a result of MAP leaving the system as can be observed at the month 6 and 9 time points. MDS1 and MDS2 represent the two dimensional space in which the points are arranged.

Circulating miRNAs have been shown to have significant potential as biomarkers for a range of human diseases. Notably, miR-155, an important regulator of gene expression that has the ability to control infection and inflammation in the gut, was significantly decreased in the exposed group in comparison to the control group [25]. The further analysis of the newly generated faecal microbiome data revealed that the relative abundances of *C. efficiens* and *B. adolescentis* were negatively correlated, and *B. pseudolongum*, unclassified *Alistipes* and *Desulfovibrio piger* were positively correlated, with the calibrated normalised relative quantity of miR-155 (Table 2). The only other significant correlation was a positive correlation between *B. adolescentis* and expression of miR-146b, an miRNA which, when overexpressed activates and upregulates the NF-κB pathway, thereby inhibiting autophagy, improving intestinal epithelial function and reducing intestinal inflammation [28].

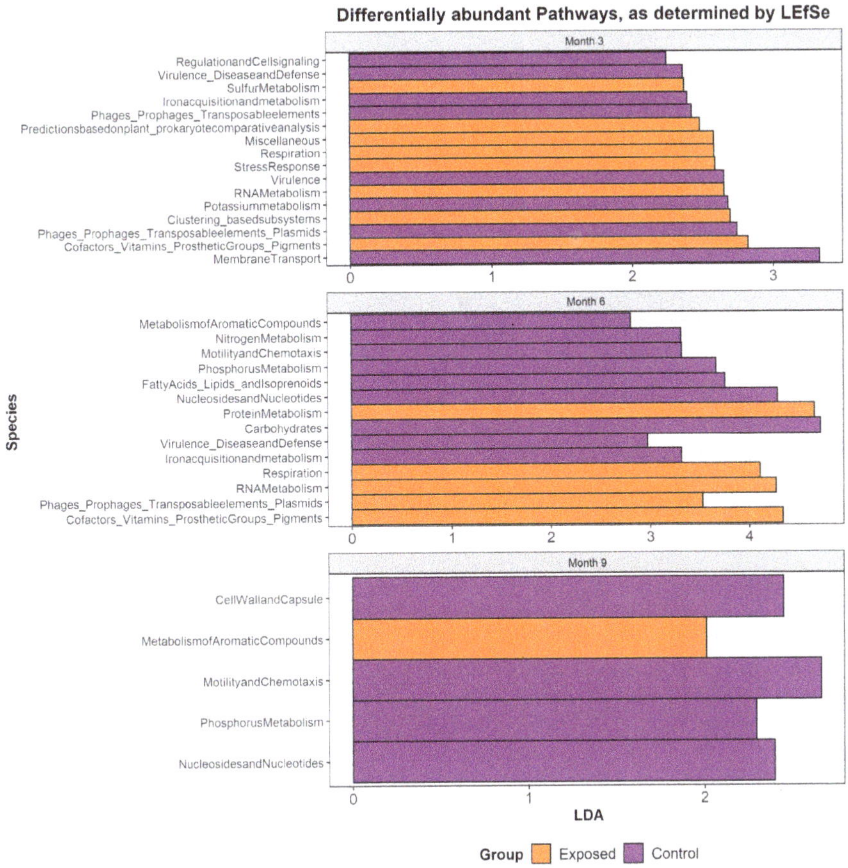

Figure 6. Functional potential of the microbiome at level 1 subsystem from the SUPER-FOCUS output. SUPER-FOCUS uses the SEED database, which is a subsystem database, assigning function at four different levels. Subsystems are divided into hierarchies. Number s on the x axis refers to the linear discriminant analysis (LDA) score. The linear discriminant analysis (LDA) effect size was determined using LEfSe. LEfSe uses the two-tailed non-parametric Kruskal–Wallis test to examine the significance of differences of species and functional potential in the two groups. A set of pairwise tests among the two groups was performed using the Wilcoxon test. LDA was performed to estimate the effect size of each differently abundant species. The functional potential of each of the groups was thought to be significantly different if their differences had a p-value < 0.05 and an LDA score (log10) > 2.

Finally, PPD-j induced IFN-gamma levels from 3 months post inoculation positively correlated with Bacteroides thetaiotaomicron, Bacteroidales bacterium_ph8, Dorea formicigenerans, Faecalibacterium prausnitzii and Escherichia unclassified (Table 3). PPDj results for month 3 post inoculation can be found in Supplementary Table S6.

Table 2. Correlation coefficient and *p*-values of species associated with miR-155 expression using Spearman rank correlation.

Species	*p*-Value	Correlation Co-Efficient
Corynebacterium efficiens	0.048	−0.319
Bifidobacterium adolescentis	0.014	−0.39
Bifidobacterium pseudolongum	0.025	0.357
Alistipes unclassified	0.033	0.342
Desulfovibrio piger	0.044	0.325

Table 3. Correlation coefficient and *p*-values of species associated with IFN-ɣ using PPDj month 3 post inoculation using Spearman rank correlation.

Species	*p*-Values	Correlation Co-Efficient
Bacteroides thetaiotaomicron	0.031	0.293
Bacteroidales bacterium_ph8	0.049	0.27
Dorea formicigenerans	0.012	0.34
Faecalibacterium prausnitzii	0.024	0.308
Escherichia unclassified	0.037	0.285

4. Discussion

Several studies have shown the host microbiome can change in response to both pathogen exposure and infection [19,29,30]. However, relatively little is known about how host microbiomes may respond to MAP exposure. Our study identifies changes in the microbiome of calves inoculated with a high dose of MAP relative to controls. These differences are most apparent at 3 months' post inoculation. We also identify associations between gut microbiome composition and immunomodulatory measures including miRNAs miR-155 and miR-146b, ELISA sero-positivity and IFN-ɣ. Further differences were highlighted through diversity analysis between the exposed and control groups.

There have been some previous studies of the GIT microbiomes of animals exposed to MAP. Authors of [19] showed differences in the faecal microbiome between MAP-positive and MAP negative cattle using 16S rRNA amplicon-based sequencing, with the microbiome of MAP-positive animals having a higher abundance of Actinobacteria in comparison to controls. Our results also show a higher abundance of members of the phylum Actinobacteria, including *Bifidobacterium* spp., in the MAP exposed group. In addition, we also identified an increase in the relative abundance of members of Euryarchaeaota and Firmicutes, with a decreased relative abundance of Actinobacteria and Bacteriodetes, in the MAP-exposed group. Generally, differences were most evident 3 months post inoculation. Our analysis highlighted a number of species that may have relevance with respect to identifying MAP exposure, including a higher abundance of *Methanobrevibacter* in challenged animals. Archaea have not yet been reported in the pathogenesis of MAP and may warrant further investigation. The higher abundance of methane producing archaea in the exposed group may result in higher methane production in the lower gastrointestinal tract. Non-invasive methods to detect this include the measurement of methane emissions through open-circuit respiration chambers may be of value from the perspective of the detection of methane emissions in flatus and faeces. Given that other methane emission units are designed for eructed methane (to estimate methane from rumen fermentation), the open circuit chamber may be the most appropriate method given that methane under examination is methane generated in the lower gastrointestinal tract.

Furthermore, a number of *Bifidobacterium*, including *B. angulatum*, were present at higher relative abundance in the MAP-exposed group. This species is considered a specialised member of the *Bifidobacterium* genus and had been thought to be exclusively associated with animal faeces [31,32]. However, the species has also been reported in a number of Crohn's disease and ulcerative colitis studies [33]. Notably, [33] examined antibody-mediated immune responsiveness to the cell surface of mucosal bacteria for IgG and IgA serum antibodies and noted that *B. angulatum* IgG responses were significantly higher in the ulcerative colitis cohort. The other *Bifidobacterium* species enriched in the exposed group was *B. adolescentis*. This species is known to produce lactate and a small quantity of acetate that may drive butyrate formation in other bacteria. Other taxa enriched among exposed animals included a variety of representatives of the genus *Butyrivibrio*. *Butyrivibrio* are butyrate producers with significant importance as they promote T-cell differentiation that can ultimately suppress pro-inflammatory responses [34]. Butyrate supplementation of the diet of calves has also been shown to have a beneficial effect in terms of growth and performance [35], potentially due to the regulation of the immune system and maintenance of the epithelial barrier. From the other extreme, *B. pseudolongum* was found at a lower relative abundance in the exposed group. This may be a result of competition for an ecological niche with other *Bifidobacterium* species. *Prevotella stercorea* was also present at lower relative abundance in the exposed group in month 3 in comparison to the exposed group. According to authors of [36], this species has a weak negative correlation with *E. coli* O157:H7 prevalence and enumeration in the faecal microbiome of cattle. Although MAP sequences were not detected, alteration in the gut microbiome suggests exposure to MAP. MAP was inoculated at a high dose but may have been passed from the animal in the days and weeks following inoculation. The changes found between groups at month 3 and weaning at months 6 and 9 may be a result of MAP passing through the GIT. This in itself is interesting as it indicates that if an animal is resistant to infection, it may take up to 6 months for the microbiome to fully recover from exposure. To fully identify if this was the reason, samples from months 1 and 2 post-inoculation should have been taken. These samples were not available for this study but should be noted for future investigations. It is also recommended to take samples in the days that follow inoculation to identify how long it takes for MAP to be cleared for the GIT when infection does not occur. The presence of potentially beneficial microbes including species of Bifidobacterium may also indicate that these microbes are aiding in the recovery of the microbiome. However, the influence of the environment on the microbiome cannot be ignored. *Enterococcus hirea* was detected at month 3 but was not detected in months 6 or 9. This may be a result of housing conditions and not experimental conditions.

Changes across time are highly likely to be a result of age and dietary modifications. Gradual weaning of calves from milk or commercial milk replacer typically takes place after the 12th week of life, where the animals are gradually introduced to calf starter feed which usually consist of concentrate (18–20% crude protein and less than 7% fibre) and forage. Overall, the community was highly variable with increasing alpha diversity but decreasing inter-animal variation (beta diversity) as the animals aged, a pattern that is consistent with the previous studies of the calf microbiome [37]. As discussed above, factors which may influence the microbiome during these timepoints may include dietary changes, changes in the environment and maturing of the intestinal microbiome. Alpha diversity was lower in the exposed group in comparison to the control group. Low diversity is a common feature of a disturbed microbiome, which in this case was due to the exposure to MAP.

The functional potential of the faecal microbiome was examined showing significant differences in pathways related to sulphur metabolism, respiration, stress response, RNA metabolism, clustering based subsystems, co-factors and vitamins which were enriched in the MAP exposed relative to the control group at month 3 post exposure. Genes associated with membrane transport, potassium metabolism, virulence, iron acquisition and metabolism, regulation and cell signalling, and phages, prophages and transposable elements were more abundant in the control group. Virulence factors found in MAP may have downregulated these pathways found in the microbiome, contributing to the higher relative abundance found in the control group. Bacterial derived hydrogen may contribute to the proliferation of methanogens in months 3 and 6, which may be a consequence of MAP exposure. Effect of methanogens is interpreted as a shift in the flow of electrons away from the formation of electron sink products lactate and ethanol to methane via hydrogen, favouring an increase in acetate which is in turn converted to methane and carbon dioxide. MAP needs to convert superoxide to hydrogen peroxide. Therefore, hydrogen peroxide may be important in methane production, as there is an increase in available hydrogen. A higher abundance of genes associated with hydrogenase pathways in the exposed group may contribute to the higher number of methanogens seen in the exposed group in the early days of exposure. Hydrogen peroxide is thought to have a positive effect on methane production and is used in a number of industrial systems to increase methane production.

Pathways involving sulphur metabolism pathways within the GIT microbiome of humans can have important roles; however, less is known about their role in cattle. When it is not assimilated, the end product of the anaerobic microbial degradation of sulphur-compounds is predominantly hydrogen sulphide (H_2S). Gastrointestinal H_2S is a neuromodulator and plays a critical role in controlling physiological responses such as motility and epithelial cell health. It has also been suggested that H_2S has a potential pathogenic role, such as with respect to inflammatory bowel disease in humans. A number of bacterial taxa have been associated with sulphur metabolism, including *M. tuberculosis* in the GIT. Genes associated with stress response were enriched in the microbiome of the exposed group, which may reflect a stressful environment arising as a consequence of MAP exposure.

The microbiome has long been implicated in the activation or suppression of the immune system in response to pathogens. Correlations with IFN-ɣ and seropositivity were made for a number of species with immunogenic properties. Commensal microbes are important modulators of host physiology, metabolism and immunity. Immunomodulatory effects of commensal microbes have been noted; for example, studies have examined *Enterococcus hirae* as a therapeutic immunomodulatory microbe, with a role in memory Th1 immune response. Subclinical Johne's disease is characterised by a Th1 response effective at controlling and limiting the spread of intracellular infections [38]. Although found in only 2 samples at a mean abundance of 0.14, it was noted that *Erysipelotrichaceae bacterium* 21_3 was significantly positively correlated with sero-positive animals. Members of this family are thought to be highly immunogenic [39].

The potential of microbiome-activated immunity was explored through identifying correlations between immunomodulatory measures and relative abundances of microorganisms present in the gut of exposed and control animals. Weak correlations were made with a number of taxa, namely *B. thetaiotaomicron*, *C. efficiens*, *Butyrivibrio unclassified* and *Bifidobacterium* spp., some of which have been discussed in detail above, were noted. These were found in only a small number of animals but may warrant further investigation in studies where animals progress to an infected state. Of the others, *B. thetaiotaomicron* is notable as this species encodes a number of glycoside hydrolases and polysaccharide lyases, degrading complex polysaccharides to monosaccharides that can be readily used by non-glycophagic species, overall, enriching the availability of nutrients in the gut [40,41] and inhibiting the effects of IFN-ɣ on epithelial function [42]. As miRNAs are also a promising target for Johne's disease prognosis, correlating miRNAs with the microbiome may provide a robust method for novel prognostics. New targets are regularly under investigation. Recently, authors of [43] found two miRNAs of interest associated with immune response to MAP infection.

There are a number of ways through which the investigations described in this study can be further advanced. Although, faecal samples in this study were collected via rectal swabs, collection of samples through ileal cannulation can provide a better indication of the regional microbiome associated with MAP exposure. Although this type of study could better inform underlying immunopathology associated with MAP exposure, as an invasive procedure it would require robust justification on welfare grounds. Ingestible sensors are among a wave of new technologies that may improve the way in which we sample the microbiome. At present, faecal samples provide the best proxy for GIT microbiome representation. In light of this, a more rigorous sampling period in the hours and days post-inoculation in experimental models may be needed to eliminate the possibility of excretion of the microorganism during this timeframe.

Despite its limitations, this study provides new insights into the molecular mechanisms underpinning MAP exposure in calves, particularly at 3 months post inoculation. With this being said, results should be used cautiously. Extensive work on naturally infected animals would be needed to further verify the results achieved in this study. This study adds to the growing amount of data relating to the impact of exposure to a pathogen on the microbiome. We show that the microbiome changes over time in response to exposure and have identified a number of potential biomarkers that merit further investigation. In parallel, the merits of carrying out immunological and pathological analyses in conjunction with microbiome studies are highlighted.

Supplementary Materials: The following supporting information can be downloaded at: https://www.mdpi.com/2076-2615/13/10/1652/s1, Table S1: LDA scores and p-vales associated for species in each group across time; Table S2: PERMANOVA and betadisper analysis for microbial community; Table S3: Superfocus level 1; Table S4: Level 3 Pathways; Table S5: A summary of the positive and inconclusive IDEX ELISA results during the course of the experimental challenge study, with S/P% provided. Animals which remained negative are not included. This table has been adapted from Britton, 2017; Table S6: IGRA PPDj ΔOD readings from animals at 3 months post inoculation; Figure S1: LEfSE analysis of Level 3 SUPER-FOCUS pathways in month 3; Figure S2: Bray-Curtis beta diversity among animals in both the exposed and control groups, across both time and herd. Month 6 shows the microbiome becoming more stable in herd B, where samples in the exposed group are most similar. However, results were non-significant.

Author Contributions: C.M. performed laboratory work, bioinformatics, statistics, data analysis and wrote the manuscript. P.D.C. and J.O'.M. critically evaluated the manuscript. A.M.W. assisted with bioin-formatics. B.M. and S.V.G. critically evaluated the manuscript. All authors have read and agreed to the published version of the manuscript.

Funding: This research was supported by the Department of Agriculture, Food and the Marine under Grant Awards No. 15/S/651 (NexusMAP) and 11/S/141 (IconMAP).

Institutional Review Board Statement: The experimental challenge study was conducted in accordance with the Irish Heath Products Regulatory Authority (Earlsfort Terrace, Dublin, Ireland) requirements and approved by the UCD Animal Review Ethics Committee (UCD, Dublin, Ireland). Ethics approval code AREC-P-12-61-Markey. No further ethics were required for this study.

Informed Consent Statement: Not applicable.

Data Availability Statement: All data generated or analysed during this study are included in this published article [and its supplementary informationsupplementary files].

Acknowledgments: The authors would like to thank Louise Britton and colleagues for their significant contribution to sample collection and original trial work that made samples available for the microbiome study. The authors would also like to thank the Teagasc Moorepark Sequencing platform, in particular Fiona Crispie and Laura Finnegan, for sequencing and providing expertise in the area.

Conflicts of Interest: The authors declare that there are no competing interests.

Abbreviations

Abbreviations	
DNA	Deoxyribonucleic acid
GIT	Gastrointestinal tract
IFN-ɣ	Interferon gamma
JD	Johne's Disease
LDA	Linear discriminant analysis
MAP	*Mycobacterium avium* subsp. paratuberculosis
miRNA	microRNA
PERMANOVA	Permutational multivariate analysis of variance
qPCR	real-time PCR

1. Bate, A.M.; Jones, G.; Kleczkowski, A.; Naylor, R.; Timmis, J.; White, P.C.L.; Touza, J. Livestock Disease Management for Trading Across Different Regulatory Regimes. *EcoHealth* **2018**, *15*, 302–316. [CrossRef] [PubMed]
2. Pitta, D.W.; Pinchak, W.E.; Indugu, N.; Vecchiarelli, B.; Sinha, R.; Fulford, J.D. Metagenomic Analysis of the Rumen Microbiome of Steers with Wheat-Induced Frothy Bloat. *Front. Microbiol.* **2016**, *7*, 689. [CrossRef] [PubMed]
3. Walsh, A.M.; Crispie, F.; O'Sullivan, O.; Finnegan, L.; Claesson, M.J.; Cotter, P.D. Species Clas-sifier Choice Is a Key Consideration When Analysing Low-Complexity Food Microbiome Data. *Microbiome* **2018**, *6*, 50. [CrossRef] [PubMed]
4. Rosa, B.A.; Supali, T.; Gankpala, L.; Djuardi, Y.; Sartono, E.; Zhou, Y.; Fischer, K.; Martin, J.; Tyagi, R.; Bolay, F.K.; et al. Differential Human Gut Microbiome Assemblages during Soil-Transmitted Helminth Infections in Indonesia and Liberia. *Microbiome* **2018**, *6*, 33. [CrossRef]
5. Patra, A.; Park, T.; Kim, M.; Yu, Z. Rumen Methanogens and Mitigation of Methane Emission by Anti-Methanogenic Compounds and Substances. *J. Anim. Sci. Biotechnol.* **2017**, *8*, 13. [CrossRef]
6. Muñoz-Vargas, L.; Opiyo, S.O.; Digianantonio, R.; Williams, M.L.; Wijeratne, A.; Habing, G. Fecal Microbiome of Periparturient Dairy Cattle and Associations with the Onset of Salmonella Shedding. *PLoS ONE* **2018**, *13*, e0196171. [CrossRef]
7. Li, R.W.; Wu, S.; Li, W.; Huang, Y.; Gasbarre, L.C. Metagenome Plasticity of the Bovine Abomasal Microbiota in Immune Animals in Response to Ostertagia Ostertagi Infection. *PLoS ONE* **2011**, *6*, e24417. [CrossRef]
8. Naqid, I.A.; Owen, J.P.; Maddison, B.C.; Gardner, D.S.; Foster, N.; Tchórzewska, M.A.; La Ragione, R.M.; Gough, K.C. Prebiotic and Probiotic Agents Enhance Antibody-Based Immune Responses to Salmonella Typhimurium Infection in Pigs. *Anim. Feed. Sci. Technol.* **2015**, *201*, 57–65. [CrossRef]
9. Argüello, H.; Estellé, J.; Leonard, F.C.; Crispie, F.; Cotter, P.D.; O'Sullivan, O.; Lynch, H.; Walia, K.; Duffy, G.; Lawlor, P.G.; et al. Influence of the Intestinal Microbiota on Colonization Resistance to Salmonella and the Shedding Pattern of Naturally Exposed Pigs. *mSystems* **2019**, *4*, e00021-19. [CrossRef]
10. Heras, V.L.; Clooney, A.G.; Ryan, F.J.; Cabrera-Rubio, R.; Casey, P.G.; Hueston, C.M.; Pinheiro, J.; Rudkin, J.K.; Melgar, S.; Cotter, P.D.; et al. Short-Term Consumption of a High-Fat Diet In-creases Host Susceptibility to Listeria Monocytogenes Infection. *Microbiome* **2019**, *7*, 7. [CrossRef]
11. Hryckowian, A.J.; Van Treuren, W.; Smits, S.A.; Davis, N.M.; Gardner, J.O.; Bouley, D.M.; Sonnenburg, J.L. Microbiota-Accessible Carbohydrates Suppress Clostridium Difficile Infection in a Murine Model. *Nat. Microbiol.* **2018**, *3*, 662–669. [CrossRef]
12. Malmuthuge, N.; Griebel, P.J.; Guan, L.L. The Gut Microbiome and Its Potential Role in the De-velopment and Function of Newborn Calf Gastrointestinal Tract. *Front. Vet. Sci.* **2015**, *2*, 36. [CrossRef]
13. Ma, T.; Villot, C.; Renaud, D.; Skidmore, A.; Chevaux, E.; Steele, M.; Guan, L.L. Linking Perturba-tions to Temporal Changes in Diversity, Stability, and Compositions of Neonatal Calf Gut Microbiota: Prediction of Diarrhea. *ISME J.* **2020**, *14*, 2223–2235. [CrossRef]
14. Kargar, S.; Roshan, M.; Ghoreishi, S.M.; Akhlaghi, A.; Kanani, M.; Abedi Shams-Abadi, A.R.; Ghaffari, M.H. Extended Colostrum Feeding for 2 Weeks Improves Growth Performance and Re-duces the Susceptibility to Diarrhea and Pneumonia in Neonatal Holstein Dairy Calves. *J. Dairy Sci.* **2020**, *103*, 8130–8142. [CrossRef]
15. Malmuthuge, N.; Griebel, P.J.; Guan, L.L. Taxonomic Identification of Commensal Bacteria Asso-ciated with the Mucosa and Digesta throughout the Gastrointestinal Tracts of Preweaned Calves. *Appl. Environ. Microbiol.* **2014**, *80*, 2021–2028. [CrossRef]
16. Windsor, P.A.; Whittington, R.J. Evidence for Age Susceptibility of Cattle to Johne's Disease. *Vet. J.* **2010**, *184*, 37–44. [CrossRef]
17. Derakhshani, H.; De Buck, J.; Mortier, R.; Barkema, H.W.; Krause, D.O.; Khafipour, E. The Features of Fecal and Ileal Mucosa-Associated Microbiota in Dairy Calves during Early Infection with My-cobacterium Avium Subspecies Paratuberculosis. *Front. Microbiol.* **2016**, *7*, 426. [CrossRef]
18. Kaevska, M.; Videnska, P.; Sedlar, K.; Bartejsova, I.; Kralova, A.; Slana, I. Faecal Bacterial Composition in Dairy Cows Shedding Mycobacterium Avium Subsp. Paratuberculosis in Faeces in Comparison with Nonshedding Cows. *Can. J. Microbiol.* **2016**, *62*, 538–541. [CrossRef]
19. Fecteau, M.-E.; Pitta, D.W.; Vecchiarelli, B.; Indugu, N.; Kumar, S.; Gallagher, S.C.; Fyock, T.L.; Sweeney, R.W. Dysbiosis of the Fecal Microbiota in Cattle Infected with Mycobacterium Avium Subsp. Paratuberculosis. *PLoS ONE* **2016**, *11*, e0160353. [CrossRef]

20. Fernández, B.; Colavecchia, S.B.; Ingratta, G.G.; Jolly, A.; Stempler, A.; Fortuny, M.L.; Paolicchi, F.A.; Mundo, S.L. Early IgG2 in Calves Experimentally Infected with Mycobacterium Avium Subsp. Paratuberculosis. *Vet. Immunol. Immunopathol.* **2019**, *213*, 109886. [CrossRef] [PubMed]
21. Begg, D.J.; Plain, K.M.; de Silva, K.; Gurung, R.; Gunn, A.; Purdie, A.C.; Whittington, R.J. Immunopathological Changes and Apparent Recovery from Infection Revealed in Cattle in an Experimental Model of Johne's Disease Using a Lyophilised Culture of Mycobacterium Avium Subspecies Paratuberculosis. *Vet. Microbiol.* **2018**, *219*, 53–62. [CrossRef]
22. Payne, J.M.; Rankin, J.D. The Pathogenesis of Experimental Johne's Disease in Calves. *Res. Vet. Sci.* **1961**, *2*, 167–176. [CrossRef]
23. Gilmour, N.J.L.; Nisbet, D.I.; Brotherston, J.G. Experimental Oral Infection of Calves with Mycobacterium Johnei. *J. Comp. Pathol.* **1965**, *75*, 281–286. [CrossRef]
24. Farrell, D.; Shaughnessy, R.G.; Britton, L.; MacHugh, D.E.; Markey, B.; Gordon, S.V. The Identification of Circulating MiRNA in Bovine Serum and Their Potential as Novel Biomarkers of Early Mycobacterium Avium Subsp Paratuberculosis Infection. *PLoS ONE* **2015**, *10*, e0134310. [CrossRef] [PubMed]
25. Britton, L.E. The Immunopathology of Mycobacterium Avium Subspecies Paratuberculosis (MAP) Infection in Cattle as a Basis for Novel Diagnostics. Ph.D. Thesis, University College Dublin, Dublin, Ireland, 2017.
26. Yu, Z.; Morrison, M. Improved Extraction of PCR-Quality Community DNA from Digesta and Fecal Samples. *Biotechniques* **2004**, *36*, 808–812. [CrossRef]
27. Truong, D.T.; Franzosa, E.A.; Tickle, T.L.; Scholz, M.; Weingart, G.; Pasolli, E.; Tett, A.; Hut-tenhower, C.; Segata, N. MetaPhlAn2 for Enhanced Metagenomic Taxonomic Profiling. *Nat. Methods* **2015**, *12*, 902–903. [CrossRef]
28. Silva, G.G.Z.; Green, K.T.; Dutilh, B.E.; Edwards, R.A. SUPER-FOCUS: A Tool for Agile Func-tional Analysis of Shotgun Metagenomic Data. *Bioinformatics* **2015**, *32*, 354–361. [CrossRef]
29. Wang, S.; Huang, Y.; Zhou, C.; Wu, H.; Zhao, J.; Wu, L.; Zhao, M.; Zhang, F.; Liu, H. The Role of Autophagy and Related MicroRNAs in Inflammatory Bowel Disease. *Gastroenterol. Res. Pract.* **2018**, *2018*, 7565076. [CrossRef]
30. Salaheen, S.; Kim, S.W.; Karns, J.S.; Hovingh, E.; Haley, B.J.; Van Kessel, J.A.S. Metagenomic Analysis of the Fecal Microbiomes from Escherichia Coli O157:H7-Shedding and Non-Shedding Cows on a Single Dairy Farm. *Food Control.* **2019**, *102*, 76–80. [CrossRef]
31. Arrazuria, R.; Elguezabal, N.; Juste, R.A.; Derakhshani, H.; Khafipour, E. Mycobacterium Avium Subspecies Paratuberculosis Infection Modifies Gut Microbiota under Different Dietary Conditions in a Rabbit Model. *Front. Microbiol.* **2016**, *7*, 446. [CrossRef]
32. Turroni, F.; Foroni, E.; Pizzetti, P.; Giubellini, V.; Ribbera, A.; Merusi, P.; Cagnasso, P.; Bizzarri, B.; De'Angelis, G.L.; Shanahan, F.; et al. Exploring the Diversity of the Bifidobacterial Population in the Human Intestinal Tract. *Appl. Environ. Microbiol.* **2009**, *75*, 1534–1545. [CrossRef] [PubMed]
33. Lamendella, R.; Santo Domingo, J.W.; Kelty, C.; Oerther, D.B. Bifidobacteria in Feces and Envi-ronmental Waters. *Appl. Environ. Microbiol.* **2008**, *74*, 575–584. [CrossRef]
34. Furrie, E.; Macfarlane, S.; Cummings, J.H.; Macfarlane, G.T. Systemic Antibodies towards Mucosal Bacteria in Ulcerative Colitis and Crohn's Disease Differentially Activate the Innate Immune Response. *Gut* **2004**, *53*, 91–98. [CrossRef]
35. Singh, N.; Gurav, A.; Sivaprakasam, S.; Brady, E.; Padia, R.; Shi, H.; Thangaraju, M.; Prasad, P.D.; Manicassamy, S.; Munn, D.H.; et al. Activation of Gpr109a, Receptor for Niacin and the Commensal Metabolite Butyrate, Suppresses Colonic Inflammation and Carcinogenesis. *Immunity* **2014**, *40*, 128–139. [CrossRef]
36. O'Hara, E.; Kelly, A.; McCabe, M.S.; Kenny, D.A.; Guan, L.L.; Waters, S.M. Effect of a Butyr-ate-Fortified Milk Replacer on Gastrointestinal Microbiota and Products of Fermentation in Artifi-cially Reared Dairy Calves at Weaning. *Sci. Rep.* **2018**, *8*, 14901. [CrossRef]
37. Kim, M.; Kuehn, L.A.; Bono, J.L.; Berry, E.D.; Kalchayanand, N.; Freetly, H.C.; Benson, A.K. The Impact of the Bovine Faecal Microbiome on Escherichia Coli O157: H7 Prevalence and Enumeration in Naturally Infected Cattle. *J. Appl. Microbiol.* **2017**, *123*, 1027–1042. [CrossRef]
38. Dill-Mcfarland, K.A.; Breaker, J.D.; Suen, G. Microbial Succession in the Gastrointestinal Tract of Dairy Cows from 2 Weeks to First Lactation. *Sci. Rep.* **2017**, *7*, 40864. [CrossRef]
39. Roussey, J.A.; Steibel, J.P.; Coussens, P.M. Regulatory T Cell Activity and Signs of T Cell Unresponsiveness in Bovine Paratuberculosis. *Front. Vet. Sci.* **2014**, *1*, 20. [CrossRef]
40. Kaakoush, N.O. Insights into the Role of Erysipelotrichaceae in the Human Host. *Front. Cell. Infect. Microbiol.* **2015**, *5*, 84. [CrossRef]
41. Sonnenburg, J.L.; Xu, J.; Leip, D.D.; Chen, C.H.; Westover, B.P.; Weatherford, J.; Buhler, J.D.; Gordon, J.I. Glycan Foraging in Vivo by an Intestine-Adapted Bacterial Symbiont. *Science* **2005**, *307*, 1955–1959. [CrossRef]
42. Curtis, M.M.; Hu, Z.; Klimko, C.; Narayanan, S.; Deberardinis, R.; Sperandio, V. The Gut Com-mensal Bacteroides Thetaiotaomicron Exacerbates Enteric Infection through Modification of the Metabolic Landscape. *Cell Host Microbe* **2014**, *16*, 759–769. [CrossRef] [PubMed]
43. Resta-Lenert, S.; Barrett, K.E. Probiotics and Commensals Reverse TNF-α- and IFN-γ-Induced Dysfunction in Human Intestinal Epithelial Cells. *Gastroenterology* **2006**, *130*, 731–746. [CrossRef] [PubMed]

animals

MDPI

Review

Modeling Paratuberculosis in Laboratory Animals, Cells, or Tissues: A Focus on Their Applications for Pathogenesis, Diagnosis, Vaccines, and Therapy Studies

Ana Jolly [1,*], Bárbara Fernández [1,2,3], Silvia Leonor Mundo [1,2,3] and Natalia Elguezabal [4,*]

1 Cátedra de Inmunología, Facultad de Ciencias Veterinarias, Universidad de Buenos Aires, Av. Chorroarín 280, Buenos Aires C1427CWO, Argentina; bfernandez@fvet.uba.ar (B.F.); smundo@fvet.uba.ar (S.L.M.)
2 Instituto de Investigaciones en Producción Animal (INPA), CONICET-Universidad de Buenos Aires, Av. Chorroarín 280, Buenos Aires C1427CWO, Argentina
3 Instituto de Investigación y Tecnología en Reproducción Animal (INITRA), Facultad de Ciencias Veterinarias, Universidad de Buenos Aires, Av. Chorroarín 280, Buenos Aires C1427CWO, Argentina
4 Departamento de Sanidad Animal, NEIKER-Instituto Vasco de Investigación y Desarrollo Agrario-Basque Research and Technology Alliance (BRTA), 48160 Derio, Spain
* Correspondence: ajolly@fvet.uba.ar (A.J.); nelguezabal@neiker.eus (N.E.)

Simple Summary: Paratuberculosis is a chronic infectious disease that affects a wide variety of domestic and wild animals. It is considered one of the diseases with the highest economic impact on the ruminant industry. Despite many efforts and intensive research, the existing measures for paratuberculosis control are still insufficient. Experimental models are a fundamental tool for making progress in the knowledge and control of this disease. Here, we review the potential and limitations of the different experimental approaches currently used in paratuberculosis research, focusing on laboratory animals and cell-based models.

Abstract: Paratuberculosis is a chronic granulomatous enteritis caused by *Mycobacterium avium* subsp. *Paratuberculosis* that affects a wide variety of domestic and wild animals. It is considered one of the diseases with the highest economic impact on the ruminant industry. Despite many efforts and intensive research, paratuberculosis control still remains controversial, and the existing diagnostic and immunoprophylactic tools have great limitations. Thus, models play a crucial role in understanding the pathogenesis of infection and disease, and in testing novel vaccine candidates. Ruminant animal models can be restricted by several reasons, related to space requirements, the cost of the animals, and the maintenance of the facilities. Therefore, we review the potential and limitations of the different experimental approaches currently used in paratuberculosis research, focusing on laboratory animals and cell-based models. The aim of this review is to offer a vision of the models that have been used, and what has been achieved or discovered with each one, so that the reader can choose the best model to answer their scientific questions and prove their hypotheses. Also, we bring forward new approaches that we consider worth exploring in the near future.

Keywords: paratuberculosis; in vivo; in vitro; ex vivo; infection model; macrophage; neutrophil; epithelial cell; rabbit; experimental infection

1. Introduction

Johne's disease or paratuberculosis (PTB), caused by the *Mycobacterium avium* subspecies *paratuberculosis* (MAP), is a chronic granulomatous enteritis which affects large populations of ruminants globally. The primary route of MAP transmission is the ingestion of colostrum, milk, food, or water contaminated with MAP. Following transcytosis through the gastrointestinal barrier, MAP is taken up by the resident macrophages and dendritic cells present in the subepithelial lamina propria. In this cellular niche, MAP survives and replicates, triggering granulomatous inflammation. Only a small proportion of chronically infected animals will develop a fatal, progressive form of the disease. However, subclinical PTB can negatively affect weight gain, milk production, and fertility, and can also lead to increased somatic cell counts, incidences of clinical mastitis, and susceptibility to other diseases. Prevention strategies to combat the spread of PTB among herds involve adhering to strict calving practices to ensure that young, susceptible animals do not come into contact with MAP-contaminated colostrum, milk, or feces. Achieving PTB control requires critical advances in both diagnosis and vaccination areas. It is also well accepted that progress in these areas requires a deeper knowledge of host–pathogen interactions during infection. Although PTB was first described in the late 1800s, its pathogenic mechanisms have not been clearly defined. This is probably due to the lack of economic and adequate laboratory animals that faithfully resemble ruminant PTB, and also to the absence of harmonization in the use of models among PTB research groups. It has been said that the optimization of an animal model with which to study MAP infection has been stymied by the complexity of this disease [1]. For this reason, some groups have focused on large animal models, as these are the target species for which control is desired. However, these animal models are, many times, overlooked by other research groups, in part because of the prohibitive costs of maintaining these animals in a research environment.

Extensive reviews have been previously published elsewhere describing both ruminant and laboratory animal models [2,3]. Here, we review the potential and limitations of the different experimental approaches used in PTB research, focusing on non-ruminant animals and cell-based models. Figure 1 summarizes this information and illustrates the relative contribution of each model to this field. In addition, we complete the review by bringing forward new approaches that we consider worth exploring in the near future.

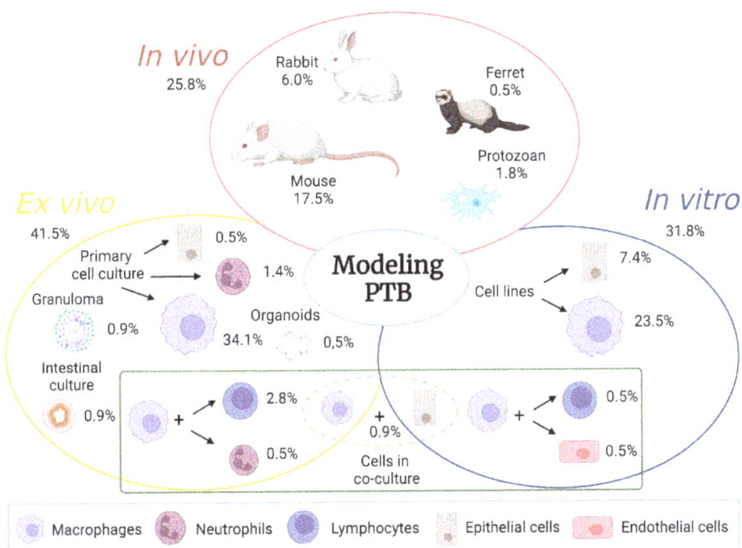

Figure 1. Modeling PTB. Relative contribution of the different models discussed in this paper. Data are expressed as the percentage of assays found for each model in PTB research. The information used to calculate these percentages was obtained by searches performed on 7 August 2023 in PubMed using the following keywords: "(mouse OR murine) AND paratuberculosis", "rabbit AND paratuberculosis",

"(protists OR amoeba) AND paratuberculosis", "ferret AND paratuberculosis", "macrophage cell line AND paratuberculosis", "macrophage cell model AND paratuberculosis", "epithelial cells AND paratuberculosis", "neutrophil AND paratuberculosis", "cells co-culture AND paratuberculosis", and "granuloma in vitro AND paratuberculosis". The inclusion criterion was considered to be a research study (not a review) on the use of an experimental model of infection to evaluate some aspects of PTB. In total, 217 assays were considered that met this criterion. It is important to note that these experiments were published in slightly fewer papers since, in some cases, the authors published results obtained from different models in the same paper. Since this review focuses on PTB or Johne's disease, those works that refer to MAP but are in the context of Crohn's disease were excluded. (Created with BioRender.com. BioRender Publication and License Rights agreement number: OB25PTVZ5Q).

2. In Vivo Animal Models in PTB Research

As previously stated, ruminant species are the first choice for study, considering their role as natural hosts of this bacterium and because the disease causes economic losses mostly in ruminant livestock, meaning that a vaccine for these species would be highly valued. However, the use of ruminant species as models, even small ones, may be restricted by several reasons related to space requirements, the cost of the animals and the maintenance of the facilities, the volume of daily excreta produced and, therefore, the need for permanent cleaning, the complexity of restraint maneuvers, among others. In addition, when the observation of clinical features is pursued, ruminant MAP infection becomes lengthy, adding to the cost and slowing down scientific progress. Laboratory animal models are normally the more realistic choice when the evaluation of new therapies, vaccines, or pathogenesis studies is the goal. Although it is impossible to reflect the complete image of PTB in laboratory animals, these non-ruminant models are advantageous because they are less expensive and, in some cases, PTB-compatible signs can be achieved in a shorter period of time. They can be used in the first steps of vaccine screening or when large animal modeling is not possible. In this review, we will focus on other alternative in vivo models, based on laboratory species. Also, we will refer to the amoebae as the simplest eukaryotic model of infection.

2.1. Mouse Model

Mice have been extensively used in PTB research for pathogenesis and vaccine prototype testing. Actually, the use of this model has been recommended for the initial steps of vaccine screening due to the similarities of granulomatous inflammation and immunological responses between mice and ruminants [2]. An in-depth review of the studies that included this animal model was performed recently [3]. Therefore, this section will only highlight the main features of this model.

Mice strains with different genetic backgrounds have been assayed, although BALB/c and C57BL/6 have been the most widely used strains [3]. MAP was administered through the intraperitoneal route (IP) in most cases (67% of the studies reviewed in [3]), followed by the oral route (20% of the studies reviewed in [3]). The oral route has been shown to be less reproducible, and it does not recreate all the classical features of intestinal PTB [3,4], only producing granulomatous lesions in the mesenteric lymph nodes [5] and disseminated infection in high-dose challenges [4]. However, the immune profile achieved with this model is similar to that observed in cattle [4]. The intraperitoneal challenge produces changes by 6–12-weeks postinfection (w.p.i.), mainly a lymphocytic inflammatory response, whereas a granulomatous inflammation is detected by 12 w.p.i. [6]. Shao et al. [7] reported differences in the CFU load in the liver from 2 to 8 w.p.i with the intraperitoneal route, with no bacterial recovery after that. They also detected differences in weight gain among the groups from 2 to 12 w.p.i. Also, a recent study compared the early stage of MAP infection among the different administration routes in the murine model, and detected increases in the weight of the spleen and the liver, and more severe histopathological changes through the IP route compared to the oral route [8].

The infection outcome in the murine model is normally evaluated by assessing the MAP burden in the tissue and histopathology of the liver, spleen, and intestines, since gross pathological lesions are normally not achieved. The drawback of this model regarding PTB is that it fails to reproduce diarrhea or severe intestinal lesions. Blood extraction in mice can be difficult, impeding the in vivo follow-up at the same time of many of the parameters that are quantifiable in blood or its components (humoral or cellular). Performing ex vivo functional cell assays can be accomplished by increasing the number of animals, dedicating each individual to a specific time-point, and by isolating bone marrow at the experiment endpoints when an immune cell analysis is required [9–11].

2.2. Rabbit Model

Rabbits are susceptible to the development of MAP infection under both experimental [12–16] and natural [17,18] conditions. Diarrhea, fecal shedding, weight loss, and intestinal colonization signs can be studied with the rabbit model.

The first studies on the experimental MAP challenge in rabbits were mostly carried out to prove infection, and to attempt to develop an appropriate model with which to study its pathogenesis and screen vaccine candidates. Therefore, different animal ages; challenge doses; and routes, including intraperitoneal, intravenous, and oral, were tested [13,15,19–21]. The oral route, mimicking the natural route of MAP infection, was the most pursued. Further studies have used this challenge route to evaluate the impact of diet [16] or a probiotic [22] on MAP infection, MAP infection on the microbiome [23], the vaccination sequence on the progression of MAP infection and the immunological response [24], and the passively transferred antibodies on MAP colonization [25]. This oral challenge model has also been complemented with ex vivo assays that have permitted the functional analysis of immune cells and the study of trained immunity exerted by vaccination [26]. When using this model, it should be considered that the reliable reproduction of the typical clinical, pathological, and microbiological signs seems to depend greatly on the strain's origin and passage, obtaining higher degrees of infection when low-passage cattle field strains are used. The key aspects of the most relevant studies involving the experimental, oral MAP infection in rabbits are summarized in Table 1.

Table 1. Experimental MAP oral challenge studies in rabbits.

Age at Challenge	Infective Strain Source	Infective Dose (Dose Number)	Study Endpoint w.p.i.	Evidence of Infection % *	Experiment Objective	Reference
4–5 w	Rabbit/ hamster	1.6–5.6 mg (7)	4–40	6–50	Pathogenesis	[20]
1–2 d	Cattle	7×10^6 CFU (5–10)	32–40	43	Pathogenesis	[13]
1–2 d	Cattle	3.6×10^8 CFU (1)	2–36	38	Pathogenesis	[14]
1–2 d	Cattle	2.6×10^8 CFU (1)	2–36	100	Pathogenesis	[14]
3 m	Cattle	5×10^8 CFU (3)	104–128	50	Pathogenesis	[15]
2 w	Cattle	1×10^8 CFU (3)	8–84	19	Pathogenesis	[15]
8 w	Cattle K10	1×10^9 CFU (3)	16–20	40–87	Diet evaluation	[16]
16 w	Cattle K10	4×10^8 CFU (3)	20	80	MAP infection and diet effect on microbiota	[23]
16 w 30 w	Cattle K10	4×10^8 CFU (6)	20	100	MAP infection and diet effect on microbiota	[23]

Table 1. Cont.

Age at Challenge	Infective Strain Source	Infective Dose (Dose Number)	Study Endpoint w.p.i.	Evidence of Infection % *	Experiment Objective	Reference
12 w	Cattle K10	4×10^8 CFU (3)	25	80	Vaccination sequence efficacy	[24]
13 w	Cattle K10	1×10^9 CFU (3)	24–25	80	Vaccination routes	[27]
13–14 w	Cattle NK-764	3×10^8 CFU (3)	19	60	Vaccination efficacy and trained immunity	[26]
15 w	Cattle NK-832	3×10^8 CFU (3)	12	100	Effect of vaccination and probiotics and trained immunity	[22]
8–11 w	Cattle	1×10^9 CFU (3)	21	100	Effect of passively transferred antibodies	[25]

w: weeks; d: days; CFU: colony-forming units; w.p.i.: weeks post-infection. * Evidence of infection was determined using post-mortem tissue culture if data were available; otherwise, MAP PCR of tissue, histology, fecal culture, or fecal PCR were considered.

The rabbit digestive tract can closely resemble that of ruminants because of the similarities between the *sacculus rotundus* in rabbits and the ileocecal valve in ruminants. Also relevant, to some extent, is the fact that the physiological body temperature of the rabbit (38.3–39.4 °C) is closer to that of the cow (37.8–39.2 °C) compared to that of the mouse (36.5–38 °C). This could somehow impact the results, since it seems that MAP infection is enhanced at 39 °C compared to 37 °C [28]. Finally, the rabbit's biological cycle allows for short-term challenge experiments permitting the development of mild PTB signs in less than 5 months. These characteristics, together with their size, handling, and ease of blood sampling, as well as the volume of blood it is possible to collect in the periodical sampling, make rabbits a convenient experimental species with which to model PTB, offering important advantages for studying key aspects of paratuberculosis that cannot be investigated with other laboratory animals because many symptoms, such as diarrhea or fecal shedding, cannot be reproduced in those models. Regarding their immunological system, two particular features place the rabbit as a model closer to ruminants than the mouse: on one hand, the presence of a primary intestine-associated lymphoid structure, which is a determinant for the generation of diversity in the B-cell repertoire (ileal Peyer's patch in ruminants, and cecum in rabbits); and, on the other hand, the fact that both rabbits and ruminants have been classified as a $\gamma\delta$ high species according to the much higher proportion of $\gamma\delta$ T lymphocytes (up to 60%) in their T-cell circulating pool than that of the $\gamma\delta$ low species, such as the mouse, in which the proportion of $\gamma\delta$ T cells is around 5% [29]. Although the role of $\gamma\delta$ T lymphocytes in PTB is not fully understood, bovine $\gamma\delta$ T lymphocytes are known to be early responders to MAP infection [30,31], and evidence for the key role of this subpopulation in the clinical phase of the disease has also been found from observations of naturally infected animals [32–34].

2.3. Ferret Model

The ferret has been used as a model for infectious diseases [35], mainly for tuberculosis (TB), for which this species has been postulated as an environmental reservoir [36,37]. In 1997, Lugton and coworkers [38] reported on the chronic, low-level natural infection of ferrets with *M. avium* subsp. *avium*. More recently, Bannantine et al. [39] published the first report using ferrets as an experimental model for MAP infection. They stated that 3–4-month-old female ferrets challenged orally with 10^8 CFU MAP lost body weight after 16 w.p.i., and developed a specific antibody response by week 13, while the IFN-γ response remained undetectable until 20 w.p.i., the final endpoint of their experiment. Taking into account these results, they proposed ferrets as a valid small animal model for testing the virulence of MAP strains. As ferrets have a very short intestinal tract, an adaptation due to their strict carnivore diet [40], Bannantine et al. [39] suggested that this anatomical feature could be an advantage for using this species as an experimental model for PTB, as the site of infection after oral inoculation would be well circumscribed. However, the usefulness of this carnivorous animal as a model for the study of PTB should be further confirmed, since it is a monogastric species that has not been shown to be susceptible in nature, there is only the one study, and the lack of species-specific immunological reagents, along with the outbred nature of the model, are major drawbacks.

2.4. Protozoan Model

Eukaryotic organisms like amoeba have been noted to interact with mycobacterial species in the environment. Based on the ability of some mycobacteria to grow within environmental amoebae [41–43], and the data suggesting the increased virulence of mycobacteria upon replication within amoebae [44,45], Phillips et al. [46] evaluated the influence of MAP infection on amoeba metabolic activity in the *Acanthamoeba castellanii* infection model, and found that this amoeba could be used as a quick initial screening tool for the selection of virulence factors with relevance to macrophage infection. The main limitation of this model is that it is a unicellular organism, and the utility of this model would seem to be limited to virulence factors research studies.

3. Ex Vivo and In Vitro Systems

Apart from in vivo models, most of the contributions to the knowledge of MAP–host interactions have been made from work on ex vivo or in vitro models, focused on the use of cells of epithelial and/or of phagocytic origin, since the target organ of this infection in ruminants is the intestine and the associated lymphoid tissue, and macrophages are the main cellular niche in which MAP can survive and replicate within its host.

Ex vivo and in vitro cellular infection assays can be fast and cost-effective alternatives to in vivo experimentation for the evaluation of MAP–host cell interactions. These assays can contribute information regarding pathogenesis and an initial notion of vaccine effects. In any case, the most relevant models found in the literature are further described in the following sections.

3.1. Ex Vivo Models

Ex vivo means "outside of a living body", implying that cells or tissue have been previously isolated from an animal and maintained in culture conditions for a study. Therefore, an in vitro assay involves an established cell system, such as a cell line, whereas an ex vivo assay is based on cells or tissue from a living organism, that is, primary cells. In the same way, the level of cellular complexity is superior in an ex vivo assay, being closer to the organism's natural conditions and, therefore, offering advantages. Although higher variability is observed due to the use of more than one individual, it is closer to reality, especially regarding PTB, a disease in which some animals are susceptible to the infection and develop clinical signs, while others are resilient and remain asymptomatic long term.

These models can be based on cells isolated from tissues and/or whole blood, or on tissue slices from different organs, conserving the original cell structures and intercellular connections. In all cases, treated or untreated, healthy or diseased animals can be used as donors, increasing the complexity of the resulting information from in vivo experiments. In PTB research, different ex vivo models have been used.

3.1.1. Intestinal Models

Ex vivo models using intact intestinal tissue have been developed, such as the culture of intestinal explants obtained with a biopsy, first described in 1959 [47], and the everted sleeve method [48]. These models were first developed from mouse samples, and then from rat, pig, and human explants as well. The use of both models has been explored in the context of PTB research: Schleig and colleagues [49] showed that explant cultures of different intestinal sections are suitable models with which to study the early events of MAP invasion, such as the cellular attachment process. In the same year, Sigurdardottir and colleagues demonstrated that MAP invasion occurs in small intestinal areas, with and without Peyer's patches, using the everted sleeve method [50]. More recently, it was described that intestinal explants can be stimulated ex vivo, inducing immune responses to a wide range of stimuli and detecting changes in expressed cellular phenotypes, as well as in the levels of biomarkers of inflammation [51]. We believe that primary cultures obtained from animals with PTB and different degrees/types of intestinal pathology versus healthy control animals could be useful for studying the differences in MAP pathogenesis.

The simplest intestinal model could be considered to be the primary culture of intestinal epithelial cells. Primary cultures of bovine epithelial cells from the small or large intestines (BIEC) have been used to evaluate the effects of the virulence factors of other microbes, toxic compounds, and the innate immune responses through Pattern Recognition Receptor (PRR) signaling. Li et al. [52] used BIEC obtained from neonatal bovine ileum to evaluate the effect of *Saccharomyces cerevisiae* components on MAP adherence. However, these cell suspensions might be contaminated with non-epithelial cells (mostly fibroblasts) and, therefore, a series of purification steps would be required to obtain relatively pure BIEC [53]. The main constraints in culturing primary cells are that they normally enter into a non-dividing state after a few passages and, also, the possible variability between donors. However, this system would be valuable for evaluating MAP intestinal invasion and early pathogenesis.

3.1.2. Immune Cell Models

Macrophages are considered key players in PTB since these immune cells act as reservoirs providing a favorable environment for MAP replication. Once phagocytosed by macrophages, MAP is able to arrest phagosome–lysosome fusion and, therefore, inhibiting the acidification of the cellular compartment and hampering pathogen destruction [54]. MAP can control monocyte and macrophage apoptosis, thereby feeding the infection cycle [55]. The other phagocytes that have been less studied, probably because they have not been considered as important cell types in MAP pathogenesis, are neutrophils.

Macrophages and Monocytes

Macrophages play a critical role in the host–pathogen interaction of PTB. Actually, they have a dual role, being effector cells that can mediate both the destruction of MAP and its survival, proliferation, and dissemination after the pathogen inhibits phagolysosome maturation. Monocyte-derived macrophages (MDMs), generated from the peripheral blood mononuclear cells (PBMCs), have been widely used to model the macrophages for ex vivo studies. As for PTB, MDMs have been extensively used to study MAP pathogenesis, screen for MAP mutants, study the immune response (phagocytosis, growth, and cytokine production) against MAP infection and, in functional assays, to aid in vaccine prototype evaluation or to complement specific resistance/susceptibility-related host trait findings.

MDMs have been infected with MAP alone, or in a co-culture with other cell types. In this section, we will discuss the studies involving MDMs cultured alone, whereas MDMs co-cultured with other primary cells or cell lines will be discussed in the co-culture section.

Regarding the methodology for monocyte isolation and enrichment, the majority of studies included protocols that isolated the PBMCs first, and selected for adherent cells on plastic culture plates for their subsequent differentiation into macrophages. A few works included protocols where MDMs were obtained after monocyte isolation from the PBMCs using an anti-CD14 antibody coupled to magnetic microbeads [56–58]. Arteche-Villasol and collaborators [58] found that this magnetic separation yielded a higher proportion of monocytes from the PBMC fraction than the plate method, although the latter achieved higher yields of MDMs than those isolated through the former in sheep and goats. It is worth mentioning that the monocyte isolation technique not only affects the yield and purity, but also impacts the resulting phenotype of the cultured cells [59]. Temperature is another important factor to consider, as it has been demonstrated that physiological temperature greatly influences MAP gene-expression profiles and the speed of macrophage invasion [28]. Many groups have further contributed methodologies that could aid in the evaluation and quantitation of bacterial intracellular growth, and that could be complementary to MAP infection assays in MDMs. Mitchell et al. [60] showed that real-time qPCR assays provide a more accurate and precise method for evaluating MAP intracellular growth dynamics in MDMs to study strain differences compared to fluorescent quantitation. Also, confocal microscopy has been suggested as a valuable tool with which to quantify MAP in MDM infection assays [61].

The first studies performed with electron microscopy showed that MAP had not degraded four weeks after infecting MDMs [62], indicating that MAP could actually survive inside macrophages. Also demonstrated was the fact that both monocytes and MDMs required the presence of serum in order to efficiently phagocytose MAP, and that MDMs showed higher phagocytosis levels than freshly adherent monocytes [63]. Later on, Khalifeh and collaborators found that IL-10 and TGF-β inhibited the destruction of intracellular MAP in the presence of IFN-γ in the MDMs from naturally infected cows, compared to the MDMs from healthy animals, suggesting important immune-regulatory roles for these cytokines during an infection with MAP [33]. Afterwards, Weiss and collaborators, using a combination of gene expression studies and functional assays, suggested that the inhibition of phagolysosomal function and apoptosis, and the expression of the inhibitors of macrophage activation, were probably important factors behind the survival of MAP in bovine macrophages [64]. The same group later studied the role of IL-10, showing that the neutralization of this cytokine enabled the macrophages to kill 57% of the MAP organisms within 96 h, along with an increase in the acidification of the phagosomes, the apoptosis of the macrophages, and the production of nitric oxide (NO), suggesting that the induction of IL-10 expression by MAP could be considered a virulence factor [65]. All these studies were important for gaining knowledge of the MAP–macrophage interaction and the role of IL-10 in PTB and, ultimately, set the conditions for future assays.

Also regarding pathogenesis elucidation, Khalifeh and coworkers evaluated the role of NO production in MAP viability within macrophages, by culturing MDMs from subclinically or clinically infected cows and healthy controls, and observed that NO production was highest in the subclinically infected cows, and that it peaked when the cells were co-stimulated with MAP and IFN-γ [66]. The authors hypothesized that in the transition from the subclinical to clinical stage, an increase in the bacterial load of the cells ceased IFN-γ upregulation, and led to an impairment of NO generation.

MDM ex vivo assays have also been used to generate knowledge for novel vaccine and therapy design, either by gaining knowledge of cell–MAP interactions at the molecular level, or by treating animals and performing ex vivo functional assays for evaluation afterwards. In many studies, it has been shown that the inhibitors of mitogen-activated protein kinases (MAPK), such as MAPKp38, increase the production of nitric oxide after a MAP infection [67], that vitamin D3 analogs promote the phagocytosis of MAP by MDMs [68], and that the antibodies directed against lipoarabinomannan (LAM) used to opsonize MAP prior to MDM contact increase macrophage apoptosis and TNF-α secretion [69]. Vaccination has been shown to enhance the microbicidal activity of macrophages against MAP ex vivo, as seen by a significant reduction in MAP viability (CFU), as well as the upregulation of iNOS and IL-10 by RTqPCR in infected MDMs from vaccinated animals, compared to non-vaccinated animals [70]. All these studies have shown that different treatments can enhance macrophage functions, at least in ex vivo conditions.

Studies focused on host genetics and MAP strain differences have also been performed using MDMs aimed at identifying markers that can be exploited for therapy or diagnosis. Actually, the possibility of finding strain differences among MDMs has been exploited to screen attenuated [57,71] and transposon mutants [72] in bovine MDMs in studies looking for novel vaccine candidates. Some studies have failed to find differences in functionality between the MDMs of the PTB-positive and control animals, but have found important strain differences [73], whereas other studies have shown that MDMs with certain SNPs (TLR2-1903 TT genotype) produce higher levels of IL-12p40 and IL-1β when stimulated with MAP, compared to cells derived from other genotypes (TLR2-1903 CT and CC), suggesting the use of these SNPs in marker-assisted breeding strategies as an additional tool in PTB control strategies [74]. The importance of host genetics has been further demonstrated in studies that have showed that the ability of MDMs to limit MAP viability is associated with specific genetic profiles [75]. Many studies have identified genetic variants in the regulatory pathways of the macrophages that may affect the susceptibility of cows that are healthy/resistant to MAP infection [76,77]. All these studies support the use of genomic and transcriptomic approaches to enable the identification of the markers associated particularly with susceptibility to MAP infection.

In summary, MDM models have been and are very useful for studying pathogenesis focused on the MAP–macrophage interplay. However, we must bear in mind that they are insufficient to provide accurate information on the events that take place during natural infection, as shown by the analysis of the MAP transcriptome from the tissue of naturally infected cows, or from in vitro-infected macrophages [78].

Granulomas

The granuloma has been considered a hallmark of mycobacterial infection, and the evaluation of the factors that affect its development and evolution is complex. A model for the in vivo dynamics of bovine paratuberculosis granulomas has been proposed by Koets et al. [79]. In an effort to simulate the earliest events in the immune response leading to tuberculous granuloma formation, a number of in vitro models have been developed (reviewed in [80]). In the field of PTB, the early interaction and coordination of macrophages upon MAP infection has been studied, starting from MDM in vitro cultures that were kept for 10 days in order to form granulomas [81,82]. In the first study, Multiplicity of Infections (MOIs) of 1:8, 1:16, and 1:33 were assayed, and found that the highest number of microgranulomas was achieved with the lowest MOI of 1:8 [81]. In the second study, the authors found that at MOIs of 1:2 and below, the macrophages displayed increased longevity compared to the uninfected cells, and formed clusters that secreted the pro-inflammatory cytokines necessary for a cell-mediated immune response [82]. While at higher MOIs, the viability of the host MDMs was negatively impacted and the intracellular MAP reproduced over the first five days of the infection.

Since maintaining granuloma structures in vitro over an extended period of time is still a limiting factor for this model [83], only the features of the early stage of granuloma formation can be studied. Although it has been proposed that granulomas are marked by the interaction of various immune cell types, which acquire different phenotypes through it, as is the case in the M1/M2 polarization of macrophages, to date, only small and simple granuloma-like structures based on MDMs have been evaluated as a model for PTB. Such a model, although still lacking the degree of realism and cellular complexity achieved in vivo, could be refined and provide, in the near future, information related to relevant aspects of PTB pathogenesis: the mechanisms involved in granuloma induction and maintenance, the activation of MAP latency genes [84], the role of different leukocyte subpopulations, and the characteristics of the granulomas that favor bactericidal functions versus bacterial persistence [85].

Neutrophils

Several studies have shown that neutrophils can play an important role in innate immune protection against mycobacterial infections [86,87]. Although neutrophils are absent in the lesions of the advanced stages of MAP infection, their presence at infection sites in the early phases has been reported [88]. Furthermore, neutrophils seem to assist in the induction of specific Th1 and Th17 cells in response to a tuberculosis vaccine [89]. In any case, their possible role in defending against MAP has been evidenced in transcriptomic studies that have described the impairment of neutrophil recruitment and activation during PTB [90–92]. These findings have recently directed the focus onto neutrophils, leading to studies that have looked into the behavior of these immune cells against MAP in ex vivo assays.

Neutrophils kill microbes through a set of mechanisms that include phagocytosis, reactive oxygen species (ROS) production, degranulation, and extracellular trap (ET) release [93]. Neutrophils isolated from healthy cattle have shown ET liberation against MAP and the effective killing of these bacteria in vitro [94]. Caprine neutrophils have also demonstrated a strong innate response against MAP, using their entire repertoire of effector functions, ET release, degranulation, chemotaxis, and phagocytosis [95]. Vaccination against PTB in rabbits increased the phagocytic activity of these immune cells, along with their ability to release ET against mycobacterial and non-mycobacterial agents [26]. In this study, the degree of protection of the different vaccine prototypes was correlated to the ex vivo antimicrobial activity of the neutrophils, indicating that this cell type may play a role in protection.

Neutrophils are tricky cell types to work with, and functional assays must be performed with freshly isolated cells, making this model less attractive than monocytes or macrophages. The number of protocols for the isolation, culture, and evaluation of functionality is continuously growing, and are largely dependent on the host species [93]. An improvement of the methodology will probably nurture further research on this phagocytic cell and its role in PTB.

3.1.3. Co-Cultures of Primary Cells Alone or Combined with Cell Lines

Co-cultures (a culture of more than one cell type) can be an option with which to study the interactions between different cell types. These kinds of models can give information about the crosstalk between the cell types evaluated and also about the host–pathogen interactions. Therefore, most ex vivo assays described in the literature have focused on macrophages, either alone or in co-culture with other cell types, including lymphocytes [96,97], PBMCs [98], neutrophils [94], or even epithelial cell lines [99,100]. The objective of each of these co-culture models varies depending on the cell type and on the origin of the cells (vaccinated, healthy, or MAP-infected animals).

Hostetter et al. [101] found that the mycobactericidal functions of MDMs were enhanced in MAP-infected cattle, and that the addition of autologous CD4+ T cells did not increase bacterial killing. However, CD4+ T cells from non-infected animals did increase bacterial killing in autologous macrophages. The same group tested the influence of $\gamma\delta$ T cells on MDMs in vitro, and found that these cells from MAP-sensitized animals were not able to produce enough IFN-γ in order to enhance the mycobacterial killing or nitrite production by the infected macrophages [102].

Co-culturing MDMs with WC1+ and WC1− $\gamma\delta$ T lymphocytes has shown that these lymphocyte subsets modulate the effector functions of MDMs, such as their MAP-killing ability [96]. Furthermore, Pooley and co-workers [97] developed a mycobacterial growth inhibition assay based on a monocyte–lymphocyte co-culture model, and used it to evaluate vaccination on sheep. They concluded that this system could quantify the ability of PBMCs to kill MAP, serving as a predictor of vaccine failure or non-response, and could be potentially useful for novel vaccine screening. Similarly, Davis and co-workers developed ex vivo platforms in cattle to study the functional activity of CD4+, CD8+, and $\gamma\delta$ T and NK cells stimulated with MAP antigen-primed antigen-presenting cells (dendritic cells (DC) present in the blood, monocyte-derived DC, and monocyte-derived macrophages) [57,103–105]. The authors highlighted the usefulness of these platforms for assessing bacterial viability, the cells' phenotype, cytotoxicity mechanisms and, consequently, for examining the potential role of the cell-mediated immune response in preventing the establishment of a persistent infection of MAP [106].

In the same way, another interesting ex vivo assay that complements an in vivo model is one based on the co-culture of MDMs with monocyte-deprived PBMCs that had been previously primed, or not, with MAP-stimulated DC. This assay showed that the macrophages from animals vaccinated with prototype vaccines, cultured alone or in a co-culture with unprimed PBMCs, showed an increased MAP-killing ability. This is in agreement with a more effective clearance of MAP infection from calf tissues in vivo [98].

The co-culture of MDMs with autologous neutrophils isolated from healthy cattle was performed to study the interaction of both cell types with MAP and *Mycobacterium bovis* alone, or in a co-culture, and showed their cooperation and synergistic effects against mycobacteria [94]. In this work, the co-cultures were performed by putting both cell types in contact in the same culture well or separately in a transwell system, sharing nutrients and metabolites but avoiding physical contact. The ET release and bacterial killing were evaluated along with cytokine secretion, offering information on the mechanisms that operate in the defense of these cell types against MAP.

Finally, using a permeable support system, Lamont and coworkers [99] were able to artificially construct an epithelial cell layer with bovine mammary epithelial cells (MAC-T cell line), along with the bovine MDMs immediately underneath them, to represent subepithelial macrophages. Thus, this system aimed at mimicking the MAP epithelial invasion a step prior to the uptake by the macrophages, and answered questions about the interaction between the bacteria and MAC-T cells, and the impact of this one on the macrophages [99,100]. This co-culture system showed that MAP leads to phagosome acidification in the epithelial cells and IL-β release that makes for an efficient epithelial transverse, ending in macrophage uptake [99]. Further studies using this model have permitted a deeper characterization of early MAP infection, identifying the metabolic, DNA repair, and virulence genes that could be considered as novel drug targets for pathogenesis studies [100].

3.1.4. Organoids

Organoids are three-dimensional organ-like structures composed of functional, stem cell-derived multicellular aggregates that can be indefinitely propagated to resemble a specific organ when cultivated with essential niche factors. An intestinal organoid culture was first developed in mice and humans following the identification of the signaling pathways involved in the maintenance and proliferation of the fast-cycling intestinal stem cells that express high levels of leucine-rich repeat-containing G protein-coupled receptor 5 (Lgr5) [107]. In these species, organoids can be obtained from adult intestinal epithelial stem cells (Lgr5+ IESC), from adult reprogrammed stromal cells or induced pluripotent stem (iPSC) cells, or from embryonic stem cells (ESC). In the case of ruminants and other farm animals, although advances made recently in the field of the development of methodologies for obtaining ESC and iPSC [108,109], to the best of our knowledge, only methods for obtaining organoids from IESC have been described. Since the organoids are generated from stem cells, they have the capacity to generate crypt-like domains, with the potential to differentiate into many of the heterogeneous epithelial cell types, contrary to the transformed or immortalized cancer-derived cell lines, and spontaneously organize into tissue-like structures that reflect the characteristics of the digestive segment of origin [110]. The methodology for obtaining and working with enteroids is beyond the scope of this review, and can be consulted in the recent literature [111].

The protocols for culturing ruminant organoids from different digestive segments have been set out in [112–120]. Since the first report in 2009, intestinal organoids have evolved as a potential alternative to in vivo models for various experimental purposes, such as imaging, molecular analysis and gene editing, but also as reductionist approaches with which to study the interaction of epithelial cells with the other relevant actors in a physiological context, such as the immune cells [121,122] or microbiota [123,124], or even to evaluate the early host–pathogen interaction in the context of many infectious diseases (protozoa [125,126], virus [127], and bacteria [128–133]).

Studies on animal organoids have gained more attention in various fields, including veterinary medicine, given recent advances in organoid technology (reviewed in [134]). In the field of PTB studies, the first steps in this direction are just beginning to be taken [119]. In that recently published work, a 3D bovine enteroid model was developed, with the apical side exposed on the exterior surface, enabling the infection of the organoids without the need for microinjection. It was concluded that this model was susceptible to infection with MAP and, therefore, useful for investigating the early stages of MAP pathogenesis. In the near future, this model could certainly be used to shed light on other aspects of pathogenesis, such as cell tropism, virulence factors, the biological differences among MAP genotypes, and exploring new signaling pathways targetable for disease prevention and, thus, facilitating the discovery of the interactions between the mycobacteria and host cells in a more physiological environment. Even so, these experimental systems still bear some limitations; namely, the lack of immune cells and other types of cells normally present in the architecture of tissue. So, the next generation of these organoids, includs co-cultures with the different cells of the immune system [122,135], could be very useful in the context of research into the immunopathogenesis of the infection and new vaccine candidate evaluation, by studying the fate of MAP infection in organoids co-cultured with immune cells, obtained from both vaccinated versus unvaccinated animals. Organoids co-cultured with macrophages could provide new evidence for the role of the M1/M2 dichotomy in the context of this infection. On the other hand, the genetic markers associated with the phenotypes of susceptibility or resistance to PTB have been described (reviewed in [136]), as well as the additional intermediate category of "tolerant hosts" [137]. Taking into account that organoids maintain an individual genetic fingerprint, the usefulness of this model for gaining knowledge of the internal differences between those phenotypes is questionable.

Among the early events of MAP infection, the bacterial interaction of the MAP enterocytes, M cells [138–140], and goblet cells [88] on the lining of the epithelium has been identified. Similarly, organoids obtained from different intestinal segments have been used to study the regional differences detected in the invasion and response to MAP infection in ileum versus jejunum [141–143]. The use of intestinal organoids could be further used to deepen the knowledge of these MAP–cell interactions.

3.2. *In Vitro Models*

Immortalized cell lines grown as monolayers are often used in research, instead of primary cells. They offer several advantages, such as eliminating donor-to-donor variation, cost effectiveness, ease of use, the provision of an unlimited supply of material, and a bypassing of the ethical concerns associated with the use of animal tissue. As pure cell populations, cell lines allow for less experimental variability and greater reproducibility of results. Since their advent in the 1950s, cell lines have revolutionized scientific research and have become a central workhorse in biomedical research. However, despite being a powerful tool, one must be careful when using cell lines. Cell lines should display and maintain functional features as close to those of the primary cells as possible. This may be particularly difficult to determine, as often the functions of the primary cells are not entirely understood. Since cell lines are genetically manipulated, this may alter their phenotype, native functions, and their responsiveness to stimuli. The serial passage of cell lines can further cause genotypic and phenotypic variation over an extended period of time, and genetic drift can also cause heterogeneity in cultures at a single point in time.

3.2.1. Macrophage Cell Lines

Ruminant macrophage cell lines have been widely used in order to evaluate the interaction of MAP with its target cell. BOMAC, derived from peritoneal macrophages transfected with SV40 plasmid DNA [144], is the most reported one. Although it has been argued that BOMAC cells are inherently dysfunctional, lack several receptors, are dissimilar to bovine primary-culture macrophages, and possess an insufficient capability to phagocytose MAP [145], different groups have reported comparable results regarding the MAP interaction of this cell line with respect to bovine MDM [70,146], and valuable data have been obtained with this model. Moreover, less assay-to-assay variation in the level of MAP infection in BOMAC, as compared to MDM cells, has been reported [147,148]. Studies of the metabolism and virulence of MAP have been conducted on BOMAC [148, 149]. The modulation by bovine antibodies directed at mycobacterial antigens or by *Saccharomyces cerevisiae* components on BOMAC cells in response to MAP infection have been analyzed by evaluating the translocation, phagocytic activity, ROS production, immune-related gene expression, cytokine secretion, and apoptosis [148–152]. The use of another ruminant macrophage cell line, MOCL-4, that was established from spontaneously proliferating adherent mononuclear cells from sheep blood, has also been evaluated as a model for MAP cellular infection [149].

Murine macrophage cell lines have long been used to elucidate host–mycobacteria interactions, and the intracellular growth characteristics of MAP [54,153]. RAW 264.7 is a murine macrophage cell line which is extremely sensitive to stimulants [154]. Despite this drawback, several studies have utilized this cell line for studying nitric oxide production in response to mycobacterial infections [155,156], and have reported differences in the immune responses and gene expression elicited from the non-pathogenic and pathogenic mycobacteria [157,158]. In addition, Everman et al. [152] used this model to evaluate the effect of antibodies to opsonize MAP. Thirunavukkarasu et al. [159] also utilized RAW 264.7 cells as a model with which to examine a series of macrophage activation parameters in response to MAP, a non-pathogenic mycobacterium, as well as mycobacterial antigens. This research demonstrated the usefulness of this model for the study of mycobacterial immunopathogenesis. Moreover, this system has been featured in the selection of attenuated vaccine candidates [160].

The phagocytic J774.A1 is another murine macrophage cell line that has been used to evaluate macrophage activation, phagosome maturation, cytokine production, the stimulation of CD4+ T cells and the intracellular survival of MAP [55,153,161,162], and to study the antimicrobial or immunomodulatory effects of the molecules on the ability of macrophages to clear MAP infection [163,164].

THP-1 is a human leukemia monocytic cell line that can be differentiated into a macrophage-like phenotype using stimuli, such as phorbol-12-myristate-13-acetate (PMA) [165]. This cell line has been evaluated by several authors with the aim of shedding light on the molecular mechanisms of MAP–host interactions [166–169]. Motamedi Boroojeni et al. [170] evaluated THP-1-derived macrophages as a model with which to evaluate the immunogenicity of constructed *Salmonella*-expressing MAP-candidate vaccine genes.

In sum, macrophage cell lines of different origins (bovine, ovine, murine, and human) have been used over time to provide valuable studies on the different aspects of the MAP–host interaction at the cellular and receptor–ligand levels, the biological differences between MAP genotypes, the importance of different bacterial virulence factors and, also, from a therapeutic perspective [164].

3.2.2. Epithelial Cell Lines

The intestinal epithelium is the largest surface that acts as a primary barrier against pathogens. A mycobacterial invasion of the intestinal epithelial cells is a complex event, requiring the participation of several bacterial and host factors. Initially, the role of the microfold epithelial cells (M cells) in MAP intestinal uptake was highlighted in [171] and, afterwards, it was learned that an invasion can also occur through other cell types, in areas with or without Peyer's patches [139], and the use of epithelial cell lines has greatly contributed to this knowledge [172,173].

Most of the studies on MAP and in vitro epithelial cell models have been conducted using intestinal epithelial lines of murine origin, or bovine epithelial cell lines from organs other than the intestines. Different groups have been working on obtaining bovine intestinal cell lines models [174,175]. More recently, Katwal et al. [176] developed and characterized an early passage, and immortalized a BIEC line obtained from the ileum of a 2-day-old calf. These cells expressed TLR 1–10, as well as cell-surface sugars relevant to the host–pathogen interaction [176,177]. The use of these epithelial models will surely improve the knowledge about the early interface of MAP intestinal invasion.

Additionally, some epithelial models could clarify the role of the interaction of MAP with different epithelial cell types in the intestinal mucosa. M-cell differentiation from BIECs has been achieved using a co-culture with murine Peyer's patch lymphocytes [178] or treatment with the supernatant from bovine PBMC cultured with IL-2 [174]. In the field of mycobacterial infections, the model proposed by Kerneis et al. [178] has been used concomitantly with other mouse models to prove that the translocation by M cell is a vital entry mechanism that contributes to the pathogenesis of *Mycobacterium tuberculosis* [179]. This model would also be valuable for studying early MAP pathogenesis. Unexpectedly, despite all these reports, there is no bovine intestinal epithelial line available in culture collections. Nevertheless, other non-intestinal cell lines have been used as models for early events in a MAP invasion. Bannantine et al. [180] first used Madin–Darby bovine kidney epithelial cells (MDBK) as a model for bovine intestinal mucosa. In addition, Patel et al. [181] described that MAP was capable of infecting the confluent monolayers of MDBK cells. Since epithelial cells are not phagocytic in nature, a MAP invasion of MDBK cells indicates that the bacteria might trigger their own uptake, probably by inducing cytoskeleton reorganization [181]. This model was used to evaluate the relevance of MAP lipids and proteins on pathogenesis [182], and the role of specific antibodies in MAP invasion, translocation, and cell-mediated killing [152]. Likewise, Everman et al. [183] developed an interesting cell culture model that simulated the passage of bacteria from their uptake by the intestinal epithelium (modeled using a primary infection of MDBK monolayer infection), their spread to the tissue phagocytes (mimicked by the infection of a RAW 264.7 cell line with MAP obtained from MDBK lysates), and their ultimate return to the intestinal epithelium during the later stages of infection (modeled using a secondary infection of MDBK with MAP recovered from the intracellular compartment of the RAW 264.7). This model offered an alternative with which to study both the host and bacterial mechanisms used during an invasion, and the infection process of the ruminant intestine, and discovered unidentified changes and interactions that occur during the disease. For example, this research group performed a gene expression analysis and detected different profiles of the immune signals, as well as the different bacterial phenotypes between the primary- and secondary-infected MDBK.

MAP has been reported to infect mammary tissue [184,185]. Various studies have demonstrated that MAP can infect the MAC-T cell line from both the apical and basolateral surfaces, with comparable efficiency and survival inside these cells, causing an impact on their gene expression [181,186]. Moreover, Patel et al. [181] described how the prior incubation of MAP in MAC-T cells enhanced the efficiency of an invasion of MDBK cells, hypothesizing that MAP present in milk could have an invasive phenotype, enhancing the infection of the suckling calves. This model was also used to evaluate the co-infection processes of MAP and other pathogens, such as *Escherichia coli*, *Staphylococcus aureus*, and *Streptococcus agalactiae*, and demonstrated the rapid baso-apical translocation of MAP in epithelial cells [187], and the fast internalization of *S. aureus* in MAC-T cells previously infected by MAP [188]. The knockout variants of TLR 4 or IL-10Rα of this model have recently been used to demonstrate the role of those receptors in the regulation of the innate response to MAP [189,190].

3.2.3. Co-Culture of Cell Lines

The co-culture of different cell lines constitutes the simplest manner by which to include some elements of the microenvironment into the model, as an over-simplification of the complex in vivo crosstalk of the different cell types in a tissue context. Even so, such in vitro cell culture systems still offer more controllable, versatile, and reproducible setups compared to in vivo and ex vivo systems. To our knowledge, only two research studies using cell lines in co-cultures related to PTB have been published. These studies have allowed for an increased understanding of the modulation of the CD4+ T cell stimulatory capacity of MAP-infected macrophages [161], and the ability of MAP to stimulate the synthesis of integrins in macrophages, permitting their efficient attachment and translocation throughout the endothelial layer [191].

Thus far, we have reviewed and discussed the different experimental approaches used in PTB research. Table 2 summarizes the analysis carried out in this work.

Table 2. Summary of the main experimental models for the study of PTB considered in this review.

Models		Reported Application * P	D	V	T	References	Main Limitations	Scientific Potential
In vivo non-ruminant animal models	Mouse	√		√	√	[4–11]	Lack of development of clinical signs.	Large amount of immunological reagents. Well-defined model. Initial vaccine screening platform.
	Rabbit	√	√	√	√	[13–16,20,22–27]	Scarcity of immunological reagents.	Clinical outcome closely resembles ruminant PTB.
	Ferret	√	√			[39]	Scarcity of immunological reagents. Carnivore diet. Outbred nature of the model.	Strain virulence factor research.
	Protozoa	√				[41,43,46]	Unicellular organism.	Strain virulence factor research.
Ex vivo models	Intestinal	√			√	[49,50,52]	Difficult to manage. No standardized protocols available.	Early-stage pathogenesis.
	Immune cells: macrophages/ monocytes	√		√	√	[33,54–76,78]	High variability between donors and lower reproducibility.	Strain virulence factor research. Host genetics MAP–host interaction at receptor–ligand level.
	Granulomas	√			√	[81,82]	Limited growth potential. Not all granuloma cell types included.	Early-stage pathogenesis. Initial vaccine screening platform.
	Immune cells: neutrophils	√		√		[26,93–95]	Role in PTB not fully understood. Short life-span cells. Inability to cryopreserve or expand them in vitro.	Early-stage host–pathogen interaction.
	Co-cultures of primary cells	√		√		[57,94,96–106]	Limited growth potential. High variability between donors.	Dissection of cross-talk between key cell types upon MAP infection. Initial vaccine screening platform.
	Organoids	√				[119]	No standardized protocols available.	Early-stage pathogenesis.
In vitro models	Macrophage cell lines	√		√	√	[54,55,69,144–153,159,161,162]	Genetically transformed cells. Lower or different response in relation to primary cells.	Strain virulence factor research. MAP–host interaction at receptor–ligand level.
	Epithelial cell lines	√		√	√	[139,172,173,180–183,186–190]	Only used to study initial interactions as they are not the target cells.	MAP intestinal invasion (cellular and molecular level) and early pathogenesis and therapy.
	Co-culture of cell lines	√				[161,191]	Simple and controllable setup.	Early-stage pathogenesis.

* P: pathogenesis; D: diagnosis; V: vaccination; and T: therapy.

4. Modeling PTB: Future Perspectives

The pressure to avoid animal experimentation and search for alternatives is growing. Added to this, the in vitro alternatives that are based on cell lines present limitations that have been mentioned throughout this work. In this context, other approaches are possible. The following are models that we think are worth considering in the future for PTB research.

4.1. Nematode Model

The nematode *Caenorhabditis elegans* can be infected by a wide variety of bacterial species, including mycobacteria, and has, therefore, been used in recent years as a model for studying some aspects of the host–pathogen interaction. Even when the absence of a clear macrophage cell lineage is posed in *C. elegans* [192], evidence from studies of different intracellular pathogens, such as *Legionella pneumophila* [193], indicates that *C. elegans*, as a host, allows for the analysis of specific virulence mechanisms which are of relevance for the pathogen–macrophage interplay. Although no study has been performed to date using this model for MAP studies, several authors have reported positive experiences using this model to evaluate other mycobacteria of pathological interest, such as *Mycobacterium avium* Subsp. *hominissuis* (MAH) [194] and *Mycobacterium marinum* [195]. Interestingly, host defense molecules, such as ROS and lysozymes, produced in response to bacteria, are delivered to the lumen of the worm intestine [196] rather than to an endosomal compartment, as would occur within macrophages. It should also be noted that the intestine of this worms is covered with an epithelium quite similar to that of the mammalian intestine, sharing a comparable morphology, structure, and function. In those experiments that evaluated *C. elegans* as a model for the pathogenesis of mycobacteria, the worms were fed with them and the infection occurred at the intestinal level.

Galbadage et al. [197] demonstrated that *C. elegans* has particular value for studying the role of MAPK pathways in mycobacterial pathogenesis. Bermudez et al. [194] tested the pathogenicity of different adherence–deficient clones of MAH in *C. elegans*, and found that all of them had an altered ability to colonize this host. In the light of these results, *C. elegans* appears as a promising, genetically tractable pathogenicity model for both the host and the pathogen. It would be interesting to test the capacity of MAP to infect this nematode, causing a pathology, as a first step toward considering this model for pathogenesis studies, as well as for the preliminary selection of attenuated vaccine candidates for PTB.

4.2. Organs-on-a-Chip

Other advanced culture systems, such as organoid-derived 2D monolayer [198] and organ-on-a-chip systems, have emerged to overcome the weakness of the current organoid systems, such as their static nature, lack of media flow, and application of mechanical strain. These advanced culture platforms offer a great potential to widen the scope of research through the dissociation of 3D organoids into single cells, and their integration into advanced culture systems [199,200]. Organs-on-a-chip are self-contained and modular in vitro models with easily controllable features, involving microfluidic approaches coupled with cell culture.

Regarding PTB, both gut-on-a-chip and immune system-on-a-chip can be attractive models. Gut-on-a-chip is fairly advanced in human medicine, and has been used to study inflammatory bowel disease [201], a gastrointestinal inflammatory disorder that shows similarity to PTB. Recent work has described the study of host–pathogen interactions in the context of intestine-on-a-chip models [202]. Studies based on human primary intestinal cells have shown that these systems have a closer transcriptomic profile and functionality to the intestine in vivo, when compared to other in vitro models. So, intestine-on-a-chip models appear as a promising tool for further research in the PTB field [134,203]. As for immune cells or immune organs-on-a-chip, these have been recently reviewed [204] and, regarding the potential application of these models in PTB, it seems that the lymph-node-on-a-chip and the inflammation-on-a-chip could be options that should be explored.

4.3. Precision-Cut Intestinal Slices (PCIS)

Ruminant animal model experiments require huge budgets and the appropriate facilities, as mentioned before. Although laboratory animal models can support PTB research, sometimes the differences in results between these models and the target species can hamper translation. Added to this, there is an increasing pressure to avoid animal experimentation in any format and to search for alternatives. The in vitro alternatives based on cell lines present limitations that have been pointed out, such as lacking in vivo characteristics like fluidic flow or periodic peristalsis of limited crosstalk between the host and microorganism, and between the different cell types. In this context, there is a clear need for alternative models with higher complexity that can better resemble the disease in the natural host. As for PTB, the source of the intestine could be animals going to slaughter, taking advantage of tissue that would otherwise be residue. Precision-cut tissue slices (PCTS) should be explored, as they have been described as an invaluable model providing the cellular and structural complexity of the host tissue [205].

A type of PCTS, precision-cut intestinal slices (PCIS) have been described as a novel ex vivo model with which to study the transport, metabolism, and toxicology of drugs [206]. This technique was first reported in 2005 by de Kanter et al. [207], who used a rat small intestine and colon. The intestinal slices are obtained by embedding the excised intestinal tissue in agarose and cutting very thin sheets (200–400 μm) while it is submerged in an oxygenated buffer that guarantees cell viability, using a specialized microtome. PCIS can be incubated and remain viable for up to 24 h, permitting the study of solute transport and metabolism.

Although PCIS has been used for decades to study pharmacological and toxicological effects, the use of this technique to assess host–pathogen interactions has not been fully developed, and its literature shows far fewer studies than those using precision-cut slices of tissue from other origins, such as the lungs. Actually, precision-cut lung slices (PCLS) have been used to evaluate antimycobacterial agents against *Mycobacterium abscessus* in mice [208], and to study the early lung response to infection by *M. bovis* and *Mycobacterium tuberculosis* in cattle [209]. As of now, PCIS has been used to study viral infections, such as avian influenza [210], coronavirus [211], and hepatitis C [212], but bacterial infections have not been assessed. This is probably due to technical issues that must be resolved, and that could include the optimization of the culture medium and the development of preservation techniques that require more research, as pointed out by Li and collaborators [206].

In the near future, PCTS will be an invaluable model with which to assess host–pathogen interactions, responses to vaccines in a "whole organ" mode [205] and, hopefully, PCIS can be used to study PTB pathogenesis and evaluate vaccine candidates and/or therapies, as well.

5. Conclusions

PTB control has been difficult due to the long incubation period of the disease and failures both in diagnosis and immunoprophylaxis. This review has been intended to give a view of the wide content of experimental models used to study PTB. Having standardized experimental infection models will aid in a deeper understanding of this pathology and benefit the development of control tools. So far, laboratory animal and cell-based models have contributed to an increased knowledge of MAP pathogenesis, host–MAP interaction, and the associated genetic traits. Furthermore, these models have contributed to initial phases of therapy evaluation, as well as to the testing of novel vaccine candidates. In the years to come, advances in omics approaches and their combination with the models described herein, or even with novel tissue engineering techniques, will certainly contribute to the development of new tools in the context of PTB research.

Author Contributions: Conceptualization, A.J., B.F., S.L.M. and N.E.; methodology, A.J., B.F., S.L.M. and N.E.; writing—original draft preparation, A.J., B.F., S.L.M. and N.E.; writing—review and editing, A.J., B.F., S.L.M. and N.E.; funding acquisition, N.E., S.L.M. and B.F. All authors have read and agreed to the published version of the manuscript.

Funding: This research was funded by the Agencia Estatal de Investigación, Ministerio de Ciencia e Innovación of Spain, with additional funding from FEDER of the European Union (grant number PID2021-125807OB-C22), the Agencia Nacional de Promoción Científica y Técnica of Argentina (BID-PICT-2018-03595), and the Universidad de Buenos Aires of Argentina (UBACyT 20020190200158BA).

Institutional Review Board Statement: Not Applicable.

Informed Consent Statement: Not Applicable.

Data Availability Statement: No new data were created. The results of the research engine searches run to obtain the material that has been reviewed in this work can be shared upon request to the authors.

Conflicts of Interest: The authors declare no conflict of interest. The funders had no role in the design of the study; in the collection, analyses, or interpretation of data; in the writing of the manuscript; or in the decision to publish the results.

References

1. Stabel, J.R.; Bannantine, J.P.; Hostetter, J.M. Comparison of sheep, goats, and calves as infection models for *Mycobacterium avium* subsp. *paratuberculosis*. *Vet. Immunol. Immunopathol.* **2020**, *225*, 110060. [CrossRef] [PubMed]
2. Hines, M.E., II; Stabel, J.R.; Sweeney, R.W.; Griffin, F.; Talaat, A.M.; Bakker, D.; Benedictus, G.; Davis, W.C.; de Lisle, G.W.; Gardner, I.A.; et al. Experimental challenge models for Johne's disease: A review and proposed international guidelines. *Vet. Microbiol.* **2007**, *122*, 197–222. [CrossRef] [PubMed]
3. Talaat, A.M.; Wu, C.; Hines, M.E., 2nd. Chapter 16: Experimental animal models of paratuberculosis. In *Paratuberculosis: Organism, Disease and Control*, 2nd ed.; Behr, M.A., Stevenson, K., Kapur, V., Eds.; CABI: Boston, MA, USA, 2020; pp. 213–247.
4. Cooney, M.A.; Stelle, J.L.; Steinberg, H.; Talaat, A.M. A murine oral model for *Mycobacterium avium* subsp. *paratuberculosis* infection and immunomodulation with *Lactobacillus casei* ATCC 334. *Front. Cell. Infect. Microbiol.* **2014**, *4*, 11. [CrossRef] [PubMed]
5. Veazey, R.S.; Horohov, D.W.; Krahenbuhl, J.L.; Taylor, H.W.; Oliver, J.L., III; Snider, T.G., III. Comparison of the resistance of C57BL/6 and C3H/He mice to infection with *Mycobacterium paratuberculosis*. *Vet. Microbiol.* **1995**, *47*, 79–87. [CrossRef] [PubMed]
6. Shin, S.J.; Wu, C.; Steinberg, H.; Talaat, A.M. Identification of novel virulence determinants in *Mycobacterium paratuberculosis* by screening a library of insertional mutants. *Infect. Immun.* **2006**, *74*, 3825–3833. [CrossRef] [PubMed]
7. Shao, M.; Cui, N.; Tang, Y.; Chen, F.; Cui, Y.; Dang, G.; Liu, S. A candidate subunit vaccine induces protective immunity against *Mycobacterium avium* subspecies *paratuberculosis* in mice. *NPJ Vaccines* **2023**, *8*, 72. [CrossRef] [PubMed]
8. Lee, J.H.; Park, H.T.; Shim, S.; Kim, S.; Woo, S.H.; Kim, D.Y.; Yoo, H.S. Immunopathological mechanisms in the early stage of *Mycobacterium avium* subsp. *paratuberculosis* infection via different administration routes in a murine model. *PLoS ONE* **2023**, *18*, e0281880. [CrossRef] [PubMed]
9. Hussain, T.; Zhao, D.; Ali Shah, S.Z.; Wang, J.; Yue, R.; Liao, Y.; Sabir, N.; Yang, L.; Zhou, X. Responses of Murine Macrophages Infected by *Mycobacterium avium* subspecies *paratuberculosis* by Targeting Interleukin-10 and TGF-β-Activated Protein Kinase 1 Binding Protein 2. *Front. Immunol.* **2018**, *8*, 915. [CrossRef]
10. Hussain, T.; Zhao, D.; Shah, S.Z.A.; Sabir, N.; Wang, J.; Liao, Y.; Song, Y.; Dong, H.; Hussain, M.M.; Ni, J.; et al. Nilotinib: A Tyrosine Kinase Inhibitor Mediates Resistance to Intracellular Mycobacterium Via Regulating Autophagy. *Cells* **2019**, *8*, 506. [CrossRef]
11. Abdissa, K.; Ruangkiattikul, N.; Ahrend, W.; Nerlich, A.; Beineke, A.; Laarmann, K.; Janze, N.; Lobermeyer, U.; Suwandi, A.; Falk, C.; et al. Relevance of inducible nitric oxide synthase for immune control of *Mycobacterium avium* subspecies *paratuberculosis* infection in mice. *Virulence* **2020**, *11*, 465–481. [CrossRef]
12. Harding, H.P. Experimental infection with *Mycobacterium johneei*. *J. Comp. Pathol. Therap.* **1957**, *67*, 37–52, IN9–IN11. [CrossRef]
13. Mokresh, A.H.; Czuprynski, C.J.; Butler, D.G. A rabbit model for study of *Mycobacterium paratuberculosis* infection. *Infect. Immun.* **1989**, *57*, 3798–3807. [CrossRef] [PubMed]
14. Mokresh, A.H.; Butler, D.G. Granulomatous enteritis following oral inoculation of newborn rabbits with *Mycobacterium paratuberculosis* of bovine origin. *Can. J. Vet. Res.* **1990**, *54*, 313–319. [PubMed]
15. Vaughan, J.A.; Lenghaus, C.; Stewart, D.J.; Tizard, M.L.; Michalski, W.P. Development of a Johne's disease infection model in laboratory rabbits following oral administration of *Mycobacterium avium* subspecies *paratuberculosis*. *Vet. Microbiol.* **2005**, *105*, 207–213. [CrossRef] [PubMed]
16. Arrazuria, R.; Molina, E.; Mateo-Adad, M.; Arostegui, I.; Garrido, J.; Juste, R.A.; Elguezabal, N. Effect of various dietary regimens infection on oral challenge with *Mycobacterium avium* subsp. *paratuberculosis* in a rabbit model. *Res. Vet. Sci.* **2015**, *101*, 80–83. [CrossRef] [PubMed]

17. Beard, P.M.; Rhind, S.M.; Buxton, D.; Daniels, M.J.; Henderson, D.; Pirie, A.; Rudge, K.; Greig, A.; Hutchings, M.R.; Stevenson, K.; et al. Natural paratuberculosis infection in rabbits in Scotland. *J. Comp. Pathol.* **2001**, *124*, 290–299. [CrossRef] [PubMed]
18. Greig, A.; Stevenson, K.; Henderson, D.; Perez, V.; Hughes, V.; Pavlik, I.; Hines, M.E.; McKendrick, I.; Sharp, J.M. Epidemiological study of paratuberculosis in wild rabbits in Scotland. *J. Clin. Microbiol.* **1999**, *37*, 1746–1751. [CrossRef]
19. Francis, J. Infection of laboratory animals with *Mycobacterium johnei*. *J. Comp. Pathol.* **1943**, *53*, 140–150. [CrossRef]
20. Hirch, A. Infection of hamsters and rabbits with *Mycobacterium johnei*. *J. Comp. Pathol.* **1956**, *66*, 260–269. [CrossRef]
21. Collins, P.; Matthews, P.R.; McDiarmid, A.; Brown, A. The pathogenicity of *Mycobacterium avium* and related mycobacteria for experimental animals. *J. Med. Microbiol.* **1983**, *16*, 27–35. [CrossRef]
22. Oyanguren, M.; Molina, E.; Mugica, M.; Badiola, A.; Ladero-Auñon, I.; Elguezabal, N. Probiotic bacteria modulate immune responses to paratuberculosis vaccination. In Proceedings of the 15th International Association for Paratuberculosis Colloquium, Dublin, Ireland, 12–16 June 2022.
23. Arrazuria, R.; Elguezabal, N.; Juste, R.A.; Derakhshani, H.; Khafipour, E. *Mycobacterium avium* Subspecies *paratuberculosis* Infection Modifies Gut Microbiota under Different Dietary Conditions in a Rabbit Model. *Front Microbiol.* **2016**, *7*, 446. [CrossRef] [PubMed]
24. Arrazuria, R.; Molina, E.; Garrido, J.M.; Pérez, V.; Juste, R.A.; Elguezabal, N. Vaccination sequence effects on immunological response and tissue bacterial burden in paratuberculosis infection in a rabbit model. *Vet Res.* **2016**, *47*, 77. [CrossRef] [PubMed]
25. Mundo, S.L. Anticuerpos en Paratuberculosis bovina: ¿Qué sabemos? In Proceedings of the XV Annual Meeting of the Argentine Association of Veterinary Immunology (AAIV), Santa Fe City, Argentine, November 2023.
26. Ladero-Auñon, I.; Molina, E.; Oyanguren, M.; Barriales, D.; Fuertes, M.; Sevilla, I.A.; Luo, L.; Arrazuria, R.; De Buck, J.; Anguita, J.; et al. Oral vaccination stimulates neutrophil functionality and exerts protection in a *Mycobacterium avium* subsp. *paratuberculosis* infection model. *NPJ Vaccines* **2021**, *6*, 102. [CrossRef] [PubMed]
27. Arrazuria, R.; Ladero, I.; Molina, E.; Fuertes, M.; Juste, R.; Fernández, M.; Pérez, V.; Garrido, J.; Elguezabal, N. Alternative Vaccination Routes against Paratuberculosis Modulate Local Immune Response and Interference with Tuberculosis Diagnosis in Laboratory Animal Models. *Vet. Sci.* **2020**, *7*, 7. [CrossRef] [PubMed]
28. Lamont, E.A.; Sreevatsan, S. Paradigm redux–*Mycobacterium avium* subspecies *paratuberculosis*-macrophage interactions show clear variations between bovine and human physiological body temperatures. *Microb. Pathog.* **2010**, *48*, 143–149. [CrossRef] [PubMed]
29. Su, C.; Jakobsen, I.; Gu, X.; Nei, M. Diversity and evolution of T-cell receptor variable region genes in mammals and birds. *Immunogenetics* **1999**, *50*, 301–308. [CrossRef]
30. Charavaryamath, C.; Gonzalez-Cano, P.; Fries, P.; Gomis, S.; Doig, K.; Scruten, E.; Potter, A.; Napper, S.; Griebel, P.J. Host responses to persistent *Mycobacterium avium* subspecies *paratuberculosis* infection in surgically isolated bovine ileal segments. *Clin. Vaccine Immunol.* **2013**, *20*, 156–165. [CrossRef]
31. Krueger, L.A.; Beitz, D.C.; Humphrey, S.B.; Stabel, J.R. Gamma delta T cells are early responders to *Mycobacterium avium* ssp. *paratuberculosis* in colostrum-replete Holstein calves. *J. Dairy Sci.* **2016**, *99*, 9040–9050. [CrossRef]
32. Khalifeh, M.S.; Stabel, J.R. Effects of gamma interferon, interleukin-10, and transforming growth factor beta on the survival of *Mycobacterium avium* subsp. *paratuberculosis* in monocyte-derived macrophages from naturally infected cattle. *Infect. Immun.* **2004**, *72*, 1974–1982. [CrossRef]
33. Albarrak, S.M.; Waters, W.R.; Stabel, J.R.; Hostetter, J.M. Evaluating the cytokine profile of the WC1+ $\gamma\delta$ T cell subset in the ileum of cattle with the subclinical and clinical forms of MAP infection. *Vet. Immunol. Immunopathol.* **2018**, *201*, 26–31. [CrossRef]
34. Ludwig, L.; Egan, R.; Baquero, M.; Mansz, A.; Plattner, B.L. WC1+ and WC1neg $\gamma\delta$ T lymphocytes in intestinal mucosa of healthy and *Mycobacterium avium* subspecies *paratuberculosis*-infected calves. *Vet. Immunol. Immunopathol.* **2019**, *216*, 109919. [CrossRef] [PubMed]
35. Albrecht, R.A.; Liu, W.C.; Sant, A.J.; Tompkins, S.M.; Pekosz, A.; Meliopoulos, V.; Cherry, S.; Thomas, P.G.; Schultz-Cherry, S. Moving Forward: Recent Developments for the Ferret Biomedical Research Model. *mBio* **2018**, *9*, e01113–e01118. [CrossRef] [PubMed]
36. Cross, M.L.; Labes, R.E.; Mackintosh, C.G. Oral infection of ferrets with virulent *Mycobacterium bovis* or *Mycobacterium avium*: Susceptibility, pathogenesis and immune response. *J. Comp. Pathol.* **2000**, *123*, 15–21. [CrossRef] [PubMed]
37. McCallan, L.; Corbett, D.; Andersen, P.L.; Aagaard, C.; McMurray, D.; Barry, C.; Thompson, S.; Strain, S.; McNair, J. A New Experimental Infection Model in Ferrets Based on Aerosolised *Mycobacterium bovis*. *Vet. Med. Int.* **2011**, *2011*, 981410. [CrossRef] [PubMed]
38. Lugton, I.W.; Wobeser, G.; Morris, R.S.; Caley, P. Epidemiology of *Mycobacterium bovis* infection in feral ferrets (*Mustela furo*) in New Zealand: I. Pathology and diagnosis. *N. Z. Vet. J.* **1997**, *45*, 140–150. [CrossRef] [PubMed]
39. Bannantine, J.P.; Gupta, T.; Zinniel, D.K.; Hikal, A.; Quinn, F.D.; Barletta, R.G. Use of a Ferret Model to Test Efficacy and Immunogenicity of Live Attenuated *Mycobacterium avium* Subspecies *paratuberculosis* Vaccines. *Methods Mol. Biol.* **2022**, *2411*, 95–104. [CrossRef] [PubMed]
40. Mayer, J.; Marinim, R.P.; Fox, J.G. Biology and Diseases of Ferrets. *Lab. Anim. Med.* **2015**, *2015*, 577–622. [CrossRef]
41. Mura, M.; Bull, T.J.; Evans, H.; Sidi-Boumedine, K.; McMinn, L.; Rhodes, G.; Pickup, R.; Hermon-Taylor, J. Replication and long-term persistence of bovine and human strains of *Mycobacterium avium* subsp. *paratuberculosis* within Acanthamoeba polyphaga. *Appl. Environ. Microbiol.* **2006**, *72*, 854–859. [CrossRef]
42. Drancourt, M. Looking at amoebae as a source of mycobacteria. *Microb. Pathog.* **2014**, *77*, 119–124. [CrossRef]

43. Samba-Louaka, A.; Robino, E.; Cochard, T.; Branger, M.; Delafont, V.; Aucher, W.; Wambeke, W.; Bannantine, J.P.; Biet, F.; Héchard, Y. Environmental *Mycobacterium avium* subsp. *paratuberculosis* Hosted by Free-Living Amoebae. *Front. Cell. Infect. Microbiol.* **2018**, *8*, 28. [CrossRef]
44. Tenant, R.; Bermudez, L.E. *Mycobacterium avium* genes upregulated upon infection of *Acanthamoeba castellanii* demonstrate a common response to the intracellular environment. *Curr. Microbiol.* **2006**, *52*, 128–133. [CrossRef] [PubMed]
45. Dubois, V.; Pawlik, A.; Bories, A.; Le Moigne, V.; Sismeiro, O.; Legendre, R.; Varet, H.; Rodríguez-Ordóñez, M.D.P.; Gaillard, J.L.; Coppée, J.Y.; et al. *Mycobacterium abscessus* virulence traits unraveled by transcriptomic profiling in amoeba and macrophages. *PLoS Pathog.* **2019**, *15*, e1008069. [CrossRef] [PubMed]
46. Phillips, I.L.; Everman, J.L.; Bermudez, L.E.; Danelishvili, L. *Acanthamoeba castellanii* as a Screening Tool for *Mycobacterium avium* Subspecies *paratuberculosis* Virulence Factors with Relevance in Macrophage Infection. *Microorganisms* **2020**, *8*, 1571. [CrossRef]
47. Randall, K.J.; Turton, J.; Foster, J.R. Explant culture of gastrointestinal tissue: A review of methods and applications. *Cell Biol. Toxicol.* **2011**, *27*, 267–284. [CrossRef]
48. Karasov, W.H.; Diamond, J.M. A simple method for measuring intestinal nutrient uptake in vitro. *J. Comp. Physiol. B.* **1983**, *152*, 105–116. [CrossRef]
49. Schleig, P.M.; Buergelt, C.D.; Davis, J.K.; Williams, E.; Monif, G.R.; Davidson, M.K. Attachment of *Mycobacterium avium* subspecies *paratuberculosis* to bovine intestinal organ cultures: Method development and strain differences. *Vet. Microbiol.* **2005**, *108*, 271–279. [CrossRef] [PubMed]
50. Sigurdardóttir, O.G.; Bakke-McKellep, A.M.; Djønne, B.; Evensen, O. *Mycobacterium avium* subsp. *paratuberculosis* enters the small intestinal mucosa of goat kids in areas with and without Peyer's patches as demonstrated with the everted sleeve method. *Comp. Immunol. Microbiol. Infect. Dis.* **2005**, *28*, 223–230. [CrossRef]
51. Brand, R.M.; Biswas, N.; Siegel, A.; Myerski, A.; Engstrom, J.; Jeffrey Metter, E.; Brand, R.E.; Cranston, R.D.; McGowan, I. Immunological responsiveness of intestinal tissue explants and mucosal mononuclear cells to ex vivo stimulation. *J. Immunol. Methods.* **2018**, *463*, 39–46. [CrossRef]
52. Li, Z.; You, Q.; Ossa, F.; Mead, P.; Quinton, M.; Karrow, N.A. Assessment of yeast *Saccharomyces cerevisiae* component binding to *Mycobacterium avium* subspecies *paratuberculosis* using bovine epithelial cells. *BMC Vet. Res.* **2016**, *12*, 42. [CrossRef]
53. Villena, J.; Aso, H.; Rutten, V.P.M.G.; Takahashi, H.; van Eden, W.; Kitazawa, H. Immunobiotics for the Bovine Host: Their Interaction with Intestinal Epithelial Cells and Their Effect on Antiviral Immunity. *Front Immunol.* **2018**, *9*, 326. [CrossRef]
54. Hostetter, J.; Steadham, E.; Haynes, J.; Bailey, T.; Cheville, N. Phagosomal maturation and intracellular survival of *Mycobacterium avium* subspecies *paratuberculosis* in J774 cells. *Comp. Immunol. Microbiol. Infect. Dis.* **2003**, *26*, 269–283. [CrossRef] [PubMed]
55. Kabara, E.; Coussens, P.M. Infection of Primary Bovine Macrophages with *Mycobacterium avium* Subspecies *paratuberculosis* Suppresses Host Cell Apoptosis. *Front. Microbiol.* **2012**, *3*, 215. [CrossRef] [PubMed]
56. Langelaar, M.F.; Weber, C.N.; Overdijk, M.B.; Müller, K.E.; Koets, A.P.; Rutten, V.P. Cytokine gene expression profiles of bovine dendritic cells after interaction with *Mycobacterium avium* ssp. *paratuberculosis* (M.a.p.), *Escherichia coli* (*E. coli*) or recombinant M.a.p. heat shock protein 70. *Vet. Immunol. Immunopathol.* **2005**, *107*, 153–161. [CrossRef] [PubMed]
57. Abdellrazeq, G.S.; Fry, L.M.; Elnaggar, M.M.; Bannantine, J.P.; Schneider, D.A.; Chamberlin, W.M.; Mahmoud, A.H.A.; Park, K.T.; Hulubei, V.; Davis, W.C. Simultaneous cognate epitope recognition by bovine CD4 and CD8 T cells is essential for primary expansion of antigen-specific cytotoxic T-cells following ex vivo stimulation with a candidate *Mycobacterium avium* subsp. *paratuberculosis* peptide vaccine. *Vaccine* **2020**, *38*, 2016–2025. [CrossRef] [PubMed]
58. Arteche-Villasol, N.; Benavides, J.; Espinosa, J.; Vallejo, R.; Royo, M.; Ferreras, M.D.C.; Pérez, V.; Gutiérrez-Expósito, D. Optimized in vitro isolation of different subpopulation of immune cells from peripheral blood and comparative techniques for generation of monocyte-derived macrophages in small ruminants. *Vet. Immunol. Immunopathol.* **2020**, *230*, 110131. [CrossRef] [PubMed]
59. Nielsen, M.C.; Andersen, M.N.; Møller, H.J. Monocyte isolation techniques significantly impact the phenotype of both isolated monocytes and derived macrophages in vitro. *Immunology* **2020**, *159*, 63–74. [CrossRef] [PubMed]
60. Mitchell, R.M.; Gollnick, N.S.; Sreevatsan, S.; Russell, D.G.; Schukken, Y.H. Quantification of *Mycobacterium avium* subsp. *paratuberculosis* (MAP) survival in monocyte-derived macrophages. *Vet. Immunol. Immunopathol.* **2011**, *139*, 73–78. [CrossRef]
61. Mathie, H.A.; Jensen, K.; Stevens, J.M.; Glass, E.J.; Hope, J.C. Quantifying *Mycobacterium avium* subspecies *paratuberculosis* infection of bovine monocyte derived macrophages by confocal microscopy. *J. Microbiol. Methods.* **2020**, *168*, 105779. [CrossRef]
62. Bendixen, P.H.; Bloch, B.; Jorgensen, J.B. Lack of intracellular degradation of *Mycobacterium paratuberculosis* by bovine macrophages infected in vitro and in vivo: Light microscopic and electron microscopic observations. *Am. J. Vet. Res.* **1981**, *42*, 109–113.
63. Zurbrick, B.G.; Czuprynski, C.J. Ingestion and intracellular growth of *Mycobacterium paratuberculosis* within bovine blood monocytes and monocyte-derived macrophages. *Infect. Immun.* **1987**, *55*, 1588–1593. [CrossRef]
64. Weiss, D.J.; Evanson, O.A.; Deng, M.; Abrahamsen, M.S. Sequential patterns of gene expression by bovine monocyte-derived macrophages associated with ingestion of mycobacterial organisms. *Microb. Pathog.* **2004**, *37*, 215–224. [CrossRef] [PubMed]
65. Weiss, D.J.; Evanson, O.A.; de Souza, C.; Abrahamsen, M.S. A critical role of interleukin-10 in the response of bovine macrophages to infection by *Mycobacterium avium* subsp. *paratuberculosis*. *Am. J. Vet. Res.* **2005**, *66*, 721–726. [CrossRef] [PubMed]
66. Khalifeh, M.S.; Al-Majali, A.M.; Stabel, J.R. Role of nitric oxide production in dairy cows naturally infected with *Mycobacterium avium* subsp. *paratuberculosis*. *Vet. Immunol. Immunopathol.* **2009**, *131*, 97–104. [CrossRef] [PubMed]

67. Souza, C.D. Blocking the mitogen activated protein kinase-p38 pathway is associated with increase expression of nitric oxide synthase and higher production of nitric oxide by bovine macrophages infected with *Mycobacterium avium* subsp. *paratuberculosis*. *Vet. Immunol. Immunopathol.* **2015**, *164*, 1–9. [CrossRef] [PubMed]
68. Wherry, T.L.T.; Dassanayake, R.P.; Casas, E.; Mooyottu, S.; Bannantine, J.P.; Stabel, J.R. Exogenous Vitamin D3 Modulates Response of Bovine Macrophages to *Mycobacterium avium* subsp. *paratuberculosis* Infection and Is Dependent Upon Stage of Johne's Disease. *Front. Cell. Infect. Microbiol.* **2022**, *11*, 773938. [CrossRef] [PubMed]
69. Jolly, A.; Lompardía, S.; Hajos, S.E.; Mundo, S.L. Evidence of a pro-apoptotic effect of specific antibodies in a bovine macrophage model of infection with *Mycobacterium avium* subsp. *paratuberculosis*. *Vet. Immunol. Immunopathol.* **2016**, *169*, 47–53. [CrossRef] [PubMed]
70. Arteche-Villasol, N.; Gutiérrez-Expósito, D.; Vallejo, R.; Espinosa, J.; Elguezabal, N.; Ladero-Auñon, I.; Royo, M.; Del Carmen Ferreras, M.; Benavides, J.; Pérez, V. Early response of monocyte-derived macrophages from vaccinated and non-vaccinated goats against in vitro infection with *Mycobacterium avium* subsp. *paratuberculosis*. *Vet. Res.* **2021**, *52*, 69. [CrossRef]
71. Lamont, E.A.; Talaat, A.M.; Coussens, P.M.; Bannantine, J.P.; Grohn, Y.T.; Katani, R.; Li, L.L.; Kapur, V.; Sreevatsan, S. Screening of *Mycobacterium avium* subsp. *paratuberculosis* mutants for attenuation in a bovine monocyte-derived macrophage model. *Front. Cell. Infect. Microbiol.* **2014**, *4*, 87. [CrossRef]
72. Rathnaiah, G.; Lamont, E.A.; Harris, N.B.; Fenton, R.J.; Zinniel, D.K.; Liu, X.; Sotos, J.; Feng, Z.; Livneh-Kol, A.; Shpigel, N.Y.; et al. Generation and screening of a comprehensive *Mycobacterium avium* subsp. *paratuberculosis* transposon mutant bank. *Front. Cell. Infect. Microbiol.* **2014**, *4*, 144. [CrossRef]
73. Gollnick, N.S.; Mitchell, R.M.; Baumgart, M.; Janagama, H.K.; Sreevatsan, S.; Schukken, Y.H. Survival of *Mycobacterium avium* subsp. *paratuberculosis* in bovine monocyte-derived macrophages is not affected by host infection status but depends on the infecting bacterial genotype. *Vet. Immunol. Immunopathol.* **2007**, *120*, 93–105. [CrossRef]
74. Koets, A.; Santema, W.; Mertens, H.; Oostenrijk, D.; Keestra, M.; Overdijk, M.; Labouriau, R.; Franken, P.; Frijters, A.; Nielen, M.; et al. Susceptibility to paratuberculosis infection in cattle is associated with single nucleotide polymorphisms in Toll-like receptor 2 which modulate immune responses against *Mycobacterium avium* subspecies *paratuberculosis*. *Prev. Vet. Med.* **2010**, *93*, 305–315. [CrossRef] [PubMed]
75. Badia-Bringué, G.; Canive, M.; Alonso-Hearn, M. Control of *Mycobacterium avium* subsp. *paratuberculosis* load within infected bovine monocyte-derived macrophages is associated with host genetics. *Front. Immunol.* **2023**, *14*, 1042638. [CrossRef] [PubMed]
76. Ariel, O.; Brouard, J.S.; Marete, A.; Miglior, F.; Ibeagha-Awemu, E.; Bissonnette, N. Genome-wide association analysis identified both RNA-seq and DNA variants associated to paratuberculosis in Canadian Holstein cattle in vitro experimentally infected macrophages. *BMC Genom.* **2021**, *22*, 162. [CrossRef] [PubMed]
77. Marete, A.; Ariel, O.; Ibeagha-Awemu, E.; Bissonnette, N. Identification of Long Non-coding RNA Isolated From Naturally Infected Macrophages and Associated With Bovine Johne's Disease in Canadian Holstein Using a Combination of Neural Networks and Logistic Regression. *Front. Vet. Sci.* **2021**, *8*, 639053. [CrossRef] [PubMed]
78. Janagama, H.K.; Lamont, E.A.; George, S.; Bannantine, J.P.; Xu, W.W.; Tu, Z.J.; Wells, S.J.; Schefers, J.; Sreevatsan, S. Primary transcriptomes of *Mycobacterium avium* subsp. *paratuberculosis* reveal proprietary pathways in tissue and macrophages. *BMC Genom.* **2010**, *11*, 561. [CrossRef] [PubMed]
79. Koets, A.P.; Eda, S.; Sreevatsan, S. The within host dynamics of *Mycobacterium avium* ssp. *paratuberculosis* infection in cattle: Where time and place matter. *Vet. Res.* **2015**, *46*, 61. [CrossRef] [PubMed]
80. Elkington, P.; Lerm, M.; Kapoor, N.; Mahon, R.; Pienaar, E.; Huh, D.; Kaushal, D.; Schlesinger, L.S. In Vitro Granuloma Models of Tuberculosis: Potential and Challenges. *J. Infect. Dis.* **2019**, *219*, 1858–1866. [CrossRef] [PubMed]
81. Abendaño, N.; Tyukalova, L.; Barandika, J.F.; Balseiro, A.; Sevilla, I.A.; Garrido, J.M.; Juste, R.A.; Alonso-Hearn, M. *Mycobacterium avium* subsp. *paratuberculosis* isolates induce in vitro granuloma formation and show successful survival phenotype, common anti-inflammatory and antiapoptotic responses within ovine macrophages regardless of genotype or host of origin. *PLoS ONE* **2014**, *9*, e104238. [CrossRef]
82. Rice, H.J.; McDaniel, M.M.; Holland, A.; Eda, S. Modelling Bovine Granuloma Formation In Vitro upon Infection with *Mycobacterium avium* Subspecies *paratuberculosis*. *Vet. Sci.* **2019**, *6*, 80. [CrossRef]
83. Ganesan, N.; Ronsmans, S.; Vanoirbeek, J.; Hoet, P.H.M. Assessment of Experimental Techniques That Facilitate Human Granuloma Formation in an In Vitro System: A Systematic Review. *Cells* **2022**, *11*, 864. [CrossRef]
84. Lamont, E.A.; Bannantine, J.P.; Armién, A.; Ariyakumar, D.S.; Sreevatsan, S. Identification and characterization of a spore-like morphotype in chronically starved *Mycobacterium avium* subsp. *paratuberculosis* cultures. *PLoS ONE* **2012**, *7*, e30648. [CrossRef] [PubMed]
85. Lugo-Villarino, G.; Hudrisier, D.; Benard, A.; Neyrolles, O. Emerging trends in the formation and function of tuberculosis granulomas. *Front. Immunol.* **2013**, *3*, 405. [CrossRef] [PubMed]
86. Kroon, E.E.; Coussens, A.K.; Kinnear, C.; Orlova, M.; Möller, M.; Seeger, A.; Wilkinson, R.J.; Hoal, E.G.; Schurr, E. Neutrophils: Innate effectors of TB resistance? *Front. Immunol.* **2018**, *9*, 2637. [CrossRef] [PubMed]
87. Hilda, J.N.; Das, S.; Tripathy, S.P.; Hanna, L.E. Role of neutrophils in tuberculosis: A bird's eye view. *Innate Immun.* **2020**, *26*, 240–247. [CrossRef] [PubMed]

88. Khare, S.; Nunes, J.S.; Figueiredo, J.F.; Lawhon, S.D.; Rossetti, C.A.; Gull, T.; Rice-Ficht, A.C.; Adams, L.G. Early phase morphological lesions and transcriptional responses of bovine ileum infected with *Mycobacterium avium* subsp. *paratuberculosis*. *Vet. Pathol.* **2009**, *46*, 717–728. [CrossRef]
89. Trentini, M.M.; de Oliveira, F.M.; Kipnis, A.; Junqueira-Kipnis, A.P. The role of neutrophils in the induction of specific Th1 and Th17 during vaccination against tuberculosis. *Front. Microbiol.* **2016**, *7*, 898. [CrossRef] [PubMed]
90. David, J.; Barkema, H.W.; Guan, L.L.; De Buck, J. Gene-expression profiling of calves 6 and 9 months after inoculation with *Mycobacterium avium* subspecies *paratuberculosis*. *Vet. Res.* **2014**, *45*, 96. [CrossRef]
91. Gossner, A.; Watkins, C.; Chianini, F.; Hopkins, J. Pathways and genes associated with immune dysfunction in sheep Paratuberculosis. *Sci. Rep.* **2017**, *7*, 1–12. [CrossRef]
92. Alonso-Hearn, M.; Canive, M.; Blanco-Vazquez, C.; Torremocha, R.; Balseiro, A.; Amado, J.; Varela-Martinez, E.; Ramos, R.; Jugo, B.M.; Casais, R. RNA-Seq analysis of ileocecal valve and peripheral blood from Holstein cattle infected with *Mycobacterium avium* subsp. *paratuberculosis* revealed dysregulation of the CXCL8/IL8 signaling pathway. *Sci. Rep.* **2019**, *9*, 14845. [CrossRef]
93. Ladero-Auñon, I.; Elguezabal, N. Neutrophil functional assays. In *Encyclopedia of Infection and Immunity*, 1st ed.; Razei, N., Ed.; Elsevier: Amsterdam, The Netherlands, 2022; pp. 141–152.
94. Ladero-Auñon, I.; Molina, E.; Holder, A.; Kolakowski, J.; Harris, H.; Urkitza, A.; Anguita, J.; Werling, D.; Elguezabal, N. Bovine Neutrophils Release Extracellular Traps and Cooperate With Macrophages in *Mycobacterium avium* subsp. *paratuberculosis* clearance In Vitro. *Front. Immunol.* **2021**, *12*, 645304. [CrossRef]
95. Criado, M.; Pérez, V.; Arteche-Villasol, N.; Elguezabal, N.; Molina, E.; Benavides, J.; Gutiérrez-Expósito, D. Evaluation of the innate immune response of caprine neutrophils against *Mycobacterium avium* subspecies *paratuberculosis* in vitro. *Vet. Res.* **2023**, *54*, 61. [CrossRef] [PubMed]
96. Baquero, M.M.; Plattner, B.L. Bovine peripheral blood WC1+ and WC1neg γδ T lymphocytes modulate monocyte-derived macrophage effector functions during in vitro *Mycobacterium avium* subspecies *paratuberculosis* infection. *Cell. Immunol.* **2017**, *315*, 34–44. [CrossRef] [PubMed]
97. Pooley, H.B.; Plain, K.M.; Purdie, A.C.; Begg, D.J.; Whittington, R.J.; de Silva, K. Integrated vaccine screening system: Using cellular functional capacity in vitro to assess genuine vaccine protectiveness in ruminants. *Pathog. Dis.* **2018**, *76*, fty029. [CrossRef] [PubMed]
98. Eshraghisamani, R.; Arrazuria, R.; Luo, L.; De Buck, J. Evaluation of *Mycobacterium avium* subsp. *paratuberculosis* isocitrate lyase (IcL) and ABC transporter (BacA) knockout mutants as vaccine candidates. *Front. Cell. Infect. Microbiol.* **2023**, *30*, 1149419. [CrossRef] [PubMed]
99. Lamont, E.A.; O'Grady, S.M.; Davis, W.C.; Eckstein, T.; Sreevatsan, S. Infection with *Mycobacterium avium* subsp. *paratuberculosis* results in rapid interleukin-1β release and macrophage transepithelial migration. *Infect. Immun.* **2012**, *80*, 3225–3235. [CrossRef] [PubMed]
100. Lamont, E.A.; Xu WW, X.; Sreevatsan, S. Host-*Mycobacterium avium* subsp. *paratuberculosis* interactome reveals a novel iron assimilation mechanism linked to nitric oxide stress during early infection. *BMC Genom.* **2013**, *14*, 694. [CrossRef] [PubMed]
101. Hostetter, J.; Zhang, W.; Simutis, F. *Mycobacterium avium* subspecies *paratuberculosis* infection of cattle does not diminish peripheral blood-derived macrophage mycobactericidal activity. *Immunol. Lett.* **2006**, *107*, 76–79. [CrossRef]
102. Simutis, F.J.; Jones, D.E.; Hostetter, J.M. Failure of antigen-stimulated gammadelta T cells and CD4+ T cells from sensitized cattle to upregulate nitric oxide and mycobactericidal activity of autologous *Mycobacterium avium* subsp. *paratuberculosis*-infected macrophages. *Vet. Immunol. Immunopathol.* **2007**, *116*, 1–12. [CrossRef]
103. Park, K.T.; Allen, A.J.; Bannantine, J.P.; Seo, K.S.; Hamilton, M.J.; Abdellrazeq, G.S.; Rihan, H.M.; Grimm, A.; Davis, W.C. Evaluation of two mutants of *Mycobacterium avium* subsp. *paratuberculosis* as candidates for a live attenuated vaccine for Johne's disease. *Vaccine* **2011**, *29*, 4709–4719. [CrossRef]
104. Park, K.T.; ElNaggar, M.M.; Abdellrazeq, G.S.; Bannantine, J.P.; Mack, V.; Fry, L.M.; Davis, W.C. Phenotype and Function of CD209+ Bovine Blood Dendritic Cells, Monocyte-Derived-Dendritic Cells and Monocyte-Derived Macrophages. *PLoS ONE* **2016**, *11*, e0165247. [CrossRef]
105. Abdellrazeq, G.S.; Elnaggar, M.M.; Bannantine, J.P.; Park, K.T.; Souza, C.D.; Backer, B.; Hulubei, V.; Fry, L.M.; Khaliel, S.A.; Torky, H.A.; et al. A *Mycobacterium avium* subsp. *paratuberculosis* relA deletion mutant and a 35 kDa major membrane protein elicit development of cytotoxic T lymphocytes with ability to kill intracellular bacteria. *Vet. Res.* **2018**, *49*, 53. [CrossRef] [PubMed]
106. Davis, W.C.; Mahmoud, A.H.; Abdellrazeq, G.S.; Elnaggar, M.M.; Dahl, J.L.; Hulubei, V.; Fry, L.M. Ex vivo Platforms to Study the Primary and Recall Immune Responses to Intracellular Mycobacterial Pathogens and Peptide-Based Vaccines. *Front. Vet. Sci.* **2022**, *9*, 878347. [CrossRef] [PubMed]
107. Sato, T.; Vries, R.G.; Snippert, H.J.; van de Wetering, M.; Barker, N.; Stange, D.E.; van Es, J.H.; Abo, A.; Kujala, P.; Peters, P.J.; et al. Single Lgr5 stem cells build crypt-villus structures in vitro without a mesenchymal niche. *Nature* **2009**, *459*, 262–265. [CrossRef] [PubMed]
108. Gao, X.; Nowak-Imialek, M.; Chen, X.; Chen, D.; Herrmann, D.; Ruan, D.; Chen, A.C.H.; Eckersley-Maslin, M.A.; Ahmad, S.; Lee, Y.L.; et al. Establishment of porcine and human expanded potential stem cells. *Nat. Cell Biol.* **2019**, *21*, 687–699. [CrossRef] [PubMed]
109. Zhao, L.; Gao, X.; Zheng, Y.; Wang, Z.; Zhao, G.; Ren, J.; Zhang, J.; Wu, J.; Wu, B.; Chen, Y.; et al. Establishment of bovine expanded potential stem cells. *Proc. Natl. Acad. Sci. USA* **2021**, *118*, e2018505118. [CrossRef] [PubMed]

110. Beumer, J.; Clevers, H. Cell fate specification and differentiation in the adult mammalian intestine. *Nat. Rev. Mol. Cell Biol.* **2020**, *22*, 39–53. [CrossRef] [PubMed]
111. Beaumont, M.; Blanc, F.; Cherbuy, C.; Egidy, G.; Giuffra, E.; Lacroix-Lamandé, S.; Wiedemann, A. Intestinal organoids in farm animals. *Vet. Res.* **2021**, *52*, 33. [CrossRef] [PubMed]
112. Powell, R.H.; Behnke, M.S. WRN conditioned media is sufficient for in vitro propagation of intestinal organoids from large farm and small companion animals. *Biol. Open.* **2017**, *6*, 698–705. [CrossRef]
113. Hamilton, C.A.; Young, R.; Jayaraman, S.; Sehgal, A.; Paxton, E.; Thomson, S.; Katzer, F.; Hope, J.; Innes, E.; Morrison, L.J.; et al. Development of in vitro enteroids derived from bovine small intestinal crypts. *Vet. Res.* **2018**, *49*, 54. [CrossRef]
114. Alfajaro, M.M.; Kim, J.Y.; Barbé, L.; Cho, E.H.; Park, J.G.; Soliman, M.; Baek, Y.B.; Kang, M.I.; Kim, S.H.; Kim, G.J.; et al. Dual Recognition of Sialic Acid and αGal Epitopes by the VP8* Domains of the Bovine Rotavirus G6P[5] WC3 and of Its Mono-reassortant G4P[5] RotaTeq Vaccine Strains. *J. Virol.* **2019**, *93*, e00941-19. [CrossRef]
115. Derricott, H.; Luu, L.; Fong, W.Y.; Hartley, C.S.; Johnston, L.J.; Armstrong, S.D.; Randle, N.; Duckworth, C.A.; Campbell, B.J.; Wastling, J.M.; et al. Developing a 3D intestinal epithelium model for livestock species. *Cell. Tissue Res.* **2019**, *375*, 409–424. [CrossRef] [PubMed]
116. Töpfer, E.; Pasotti, A.; Telopoulou, A.; Italiani, P.; Boraschi, D.; Ewart, M.A.; Wilde, C. Bovine colon organoids: From 3D bioprinting to cryopreserved multi-well screening platforms. *Toxicol. Vitr.* **2019**, *61*, 104606. [CrossRef] [PubMed]
117. Lee, B.R.; Yang, H.; Lee, S.I.; Haq, I.; Ock, S.A.; Wi, H.; Lee, H.C.; Lee, P.; Yoo, J.G. Robust Three-Dimensional (3D) Expansion of Bovine Intestinal Organoids: An In Vitro Model as a Potential Alternative to an In Vivo System. *Animals* **2021**, *11*, 2115. [CrossRef] [PubMed]
118. Smith, D.; Price, D.R.G.; Burrells, A.; Faber, M.N.; Hildersley, K.A.; Chintoan-Uta, C.; Chapuis, A.F.; Stevens, M.; Stevenson, K.; Burgess, S.T.G.; et al. The Development of Ovine Gastric and Intestinal Organoids for Studying Ruminant Host-Pathogen Interactions. *Front. Cell. Infect. Microbiol.* **2021**, *11*, 733811. [CrossRef] [PubMed]
119. Blake, R.; Jensen, K.; Mabbott, N.; Hope, J.; Stevens, J. The Development of 3D Bovine Intestinal Organoid Derived Models to Investigate *Mycobacterium avium* ssp. *paratuberculosis* Pathogenesis. *Front. Vet. Sci.* **2022**, *9*, 921160. [CrossRef] [PubMed]
120. Zhang, J.; Li, J.; Yan, P.; He, L.; Zhang, X.; Wang, X.; Shi, Y.; Deng, L.; Zhang, Z.; Zhao, B. In-depth analysis of the relationship between bovine intestinal organoids and enteroids based on morphology and transcriptome. *J. Tissue Eng. Regen. Med.* **2022**, *16*, 1032–1046. [CrossRef] [PubMed]
121. Noel, G.; Baetz, N.W.; Staab, J.F.; Donowitz, M.; Kovbasnjuk, O.; Pasetti, M.F.; Zachos, N.C. A primary human macrophage-enteroid co-culture model to investigate mucosal gut physiology and host-pathogen interactions. *Sci. Rep.* **2017**, *7*, 45270. [CrossRef]
122. Bar-Ephraim, Y.E.; Kretzschmar, K.; Clevers, H. Organoids in immunological research. *Nat. Rev. Immunol.* **2020**, *20*, 279–293. [CrossRef]
123. Son, Y.S.; Ki, S.J.; Thanavel, R.; Kim, J.J.; Lee, M.O.; Kim, J.; Jung, C.R.; Han, T.S.; Cho, H.S.; Ryu, C.M.; et al. Maturation of Human Intestinal Organoids in Vitro Facilitates Colonization by Commensal Lactobacilli by Reinforcing the Mucus Layer. *FASEB J.* **2020**, *34*, 9899–9910. [CrossRef]
124. Puschhof, J.; Pleguezuelos-Manzano, C.; Martinez-Silgado, A.; Akkerman, N.; Saftien, A.; Boot, C.; de Waal, A.; Beumer, J.; Dutta, D.; Heo, I.; et al. Intestinal organoid cocultures with microbes. *Nat. Protoc.* **2021**, *16*, 4633–4649. [CrossRef]
125. Heo, I.; Dutta, D.; Schaefer, D.A.; Iakobachvili, N.; Artegiani, B.; Sachs, N.; Boonekamp, K.E.; Bowden, G.; Hendrickx, A.P.A.; Willems, R.J.L.; et al. Modelling *Cryptosporidium* infection in human small intestinal and lung organoids. *Nat. Microbiol.* **2018**, *3*, 814–823. [CrossRef]
126. Guzmán-Delgado, N.E.; Carranza-Torres, I.E.; García-Davis, S.; Rivera, G.; Morán-Martínez, J.; Betancourt-Martínez, N.D.; Groothuis, G.M.M.; de Graaf, I.A.M.; Carranza-Rosales, P. Development of a Novel Ex-vivo 3D Model to Screen Amoebicidal Activity on Infected Tissue. *Sci. Rep.* **2019**, *9*, 8396. [CrossRef] [PubMed]
127. Crawford, S.E.; Ramani, S.; Blutt, S.E.; Estes, M.K. Organoids to Dissect Gastrointestinal Virus-Host Interactions: What Have We Learned? *Viruses* **2021**, *13*, 999. [CrossRef] [PubMed]
128. Kotloff, K.L.; Nataro, J.P.; Blackwelder, W.C.; Nasrin, D.; Farag, T.H.; Panchalingam, S.; Wu, Y.; Sow, S.O.; Sur, D.; Breiman, R.F.; et al. Burden and aetiology of diarrhoeal disease in infants and young children in developing countries (the Global Enteric Multicenter Study, GEMS): A prospective, case-control study. *Lancet* **2013**, *382*, 209–222. [CrossRef]
129. Hill, D.R.; Huang, S.; Nagy, M.S.; Yadagiri, V.K.; Fields, C.; Mukherjee, D.; Bons, B.; Dedhia, P.H.; Chin, A.M.; Tsai, Y.H.; et al. Bacterial colonization stimulates a complex physiological response in the immature human intestinal epithelium. *Elife* **2017**, *6*, e29132. [CrossRef] [PubMed]
130. Roodsant, T.; Navis, M.; Aknouch, I.; Renes, I.B.; van Elburg, R.M.; Pajkrt, D.; Wolthers, K.C.; Schultsz, C.; van der Ark, K.C.H.; Sridhar, A.; et al. A Human 2D Primary Organoid-Derived Epithelial Monolayer Model to Study Host-Pathogen Interaction in the Small Intestine. *Front. Cell. Infect. Microbiol.* **2020**, *10*, 272. [CrossRef] [PubMed]
131. Abuaita, B.H.; Lawrence, A.L.E.; Berger, R.P.; Hill, D.R.; Huang, S.; Yadagiri, V.K.; Bons, B.; Fields, C.; Wobus, C.E.; Spence, J.R.; et al. Comparative transcriptional profiling of the early host response to infection by typhoidal and non-typhoidal Salmonella serovars in human intestinal organoids. *PLoS Pathog.* **2021**, *17*, e100998. [CrossRef]
132. Co, J.Y.; Margalef-Català, M.; Monack, D.M.; Amieva, M.R. Controlling the polarity of human gastrointestinal organoids to investigate epithelial biology and infectious diseases. *Nat Protoc.* **2021**, *16*, 5171–5192. [CrossRef]

133. Aguirre Garcia, M.; Hillion, K.; Cappelier, J.M.; Neunlist, M.; Mahe, M.M.; Haddad, N. Intestinal Organoids: New Tools to Comprehend the Virulence of Bacterial Foodborne Pathogens. *Foods* **2022**, *11*, 108. [CrossRef]
134. Kawasaki, M.; Goyama, T.; Tachibana, Y.; Nagao, I.; Ambrosini, Y.M. Farm and Companion Animal Organoid Models in Translational Research: A Powerful Tool to Bridge the Gap Between Mice and Humans. *Front. Med. Technol.* **2022**, *4*, 895379. [CrossRef]
135. Hentschel, V.; Seufferlein, T.; Armacki, M. Intestinal organoids in coculture: Redefining the boundaries of gut mucosa ex vivo modeling. Am. J. Physiol. Gastrointest. *Liver Physiol.* **2021**, *321*, G693–G704. [CrossRef]
136. Kravitz, A.; Pelzer, K.; Sriranganathan, N. The Paratuberculosis Paradigm Examined: A Review of Host Genetic Resistance and Innate Immune Fitness in *Mycobacterium avium* subsp. *paratuberculosis* Infection. *Front. Vet. Sci.* **2021**, *8*, 721706. [CrossRef]
137. Alonso-Hearn, M.; Badia-Bringué, G.; Canive, M. Genome-wide association studies for the identification of cattle susceptible and resilient to paratuberculosis. *Front. Vet. Sci.* **2022**, *9*, 935133. [CrossRef] [PubMed]
138. Sigurdardóttir, O.G.; Valheim, M.; Press, C.M. Establishment of *Mycobacterium avium* subsp. *paratuberculosis* infection in the intestine of ruminants. *Adv. Drug Deliv. Rev.* **2004**, *56*, 819–834. [CrossRef]
139. Bermudez, L.E.; Petrofsky, M.; Sommer, S.; Barletta, R.G. Peyer's patch-deficient mice demonstrate that *Mycobacterium avium* subsp. *paratuberculosis* translocates across the mucosal barrier via both M cells and enterocytes but has inefficient dissemination. *Infect. Immun.* **2010**, *78*, 3570–3577. [CrossRef]
140. Ponnusamy, D.; Periasamy, S.; Tripathi, B.N.; Pal, A. *Mycobacterium avium* subsp. *paratuberculosis* invades through M cells and enterocytes across ileal and jejunal mucosa of lambs. *Res. Vet. Sci.* **2013**, *94*, 306–312. [CrossRef] [PubMed]
141. Facciuolo, A.; Gonzalez-Cano, P.; Napper, S.; Griebel, P.J.; Mutharia, L.M. Marked Differences in Mucosal Immune Responses Induced in Ileal versus Jejunal Peyer's Patches to *Mycobacterium avium* subsp. *paratuberculosis* Secreted Proteins following Targeted Enteric Infection in Young Calves. *PLoS ONE* **2016**, *11*, e0158747. [CrossRef] [PubMed]
142. Facciuolo, A.; Lee, A.H.; Gonzalez Cano, P.; Townsend, H.G.G.; Falsafi, R.; Gerdts, V.; Potter, A.; Napper, S.; Hancock, R.E.W.; Mutharia, L.M.; et al. Regional Dichotomy in Enteric Mucosal Immune Responses to a Persistent *Mycobacterium avium* ssp. *paratuberculosis* Infection. *Front. Immunol.* **2020**, *11*, 1020. [CrossRef]
143. Ibeagha-Awemu, E.M.; Bissonnette, N.; Do, D.N.; Dudemaine, P.L.; Wang, M.; Facciuolo, A.; Griebel, P. Regionally Distinct Immune and Metabolic Transcriptional Responses in the Bovine Small Intestine and Draining Lymph Nodes During a Subclinical *Mycobacterium avium* subsp. *paratuberculosis* Infection. *Front. Immunol.* **2021**, *12*, 760931. [CrossRef] [PubMed]
144. Stabel, J.R.; Stabel, T.J. Immortalization and characterization of bovine peritoneal macrophages transfected with SV40 plasmid DNA. *Vet. Immunol. Immunopathol.* **1995**, *45*, 211–220. [CrossRef] [PubMed]
145. Woo, S.R.; Sotos, J.; Hart, A.P.; Barletta, R.G.; Czuprynski, C.J. Bovine monocytes and a macrophage cell line differ in their ability to phagocytose and support the intracellular survival of *Mycobacterium avium* subsp. *paratuberculosis*. *Vet. Immunol. Immunopathol.* **2006**, *110*, 109–120. [CrossRef] [PubMed]
146. Alonso-Hearn, M.; Magombedze, G.; Abendaño, N.; Landin, M.; Juste, R.A. Deciphering the virulence of *Mycobacterium avium* subsp. *paratuberculosis* isolates in animal macrophages using mathematical models. *J. Theor. Biol.* **2019**, *468*, 82–91. [CrossRef] [PubMed]
147. Tooker, B.C.; Coussens, P.M. Phagocytosis of *M. paratuberculosis* fails to activate expression of NADH dehydrogenase and nucleolin-related protein in bovine macrophages. *Immunol Lett.* **2004**, *93*, 137–142. [CrossRef] [PubMed]
148. Abendaño, N.; Sevilla, I.A.; Prieto, J.M.; Garrido, J.M.; Juste, R.A.; Alonso-Hearn, M. *Mycobacterium avium* subspecies *paratuberculosis* isolates from sheep and goats show reduced persistence in bovine macrophages than cattle, bison, deer and wild boar strains regardless of genotype. *Vet. Microbiol.* **2013**, *163*, 325–334. [CrossRef] [PubMed]
149. Alonso-Hearn, M.; Abendaño, N.; Ruvira, M.A.; Aznar, R.; Landin, M.; Juste, R.A. *Mycobacterium avium* subsp. *paratuberculosis* (Map) Fatty Acids Profile Is Strain-Dependent and Changes Upon Host Macrophages Infection. *Front. Cell. Infect. Microbiol.* **2017**, *7*, 89. [CrossRef]
150. Mundo, S.L.; Fontanals, A.M.; García, M.; Durrieu, M.; Alvarez, E.; Gentilini, E.R.; Hajos, S.E. Bovine IgG1 antibodies against *Mycobacterium avium* subsp. *paratuberculosis* protein p34-cx improve association of bacteria and macrophages. *Vet. Res.* **2008**, *39*, 6. [CrossRef] [PubMed]
151. Jolly, A.; Colavecchia, S.B.; Fernández, B.; Fernández, E.; Mundo, S.L. Antibodies Induced by Lipoarabinomannan in Bovines: Characterization and Effects on the Interaction between *Mycobacterium avium* subsp. *paratuberculosis* and Macrophages In Vitro. *Vet. Med. Int.* **2011**, *2011*, 258479. [CrossRef] [PubMed]
152. Everman, J.L.; Bermudez, L.E. Antibodies against invasive phenotype-specific antigens increase *Mycobacterium avium* subspecies *paratuberculosis* translocation across a polarized epithelial cell model and enhance killing by bovine macrophages. *Front. Cell. Infect. Microbiol.* **2015**, *5*, 58. [CrossRef]
153. Kuehnel, M.P.; Goethe, R.; Habermann, A.; Mueller, E.; Rohde, M.; Griffiths, G.; Valentin-Weigand, P. Characterization of the intracellular survival of *Mycobacterium avium* ssp. *paratuberculosis*: Phagosomal pH and fusogenicity in J774 macrophages compared with other mycobacteria. *Cell. Microbiol.* **2001**, *3*, 551–566. [CrossRef]
154. Raschke, W.C.; Baird, S.; Ralph, P.; Nakoinz, I. Functional macrophage cell lines transformed by Abelson leukemia virus. *Cell* **1978**, *15*, 261–267. [CrossRef]
155. Park, E.; Levis, W.R.; Quinn, M.R.; Park, S.Y.; Schuller-Levis, G.B. Regulation of nitric oxide induced by mycobacterial lipoarabinomannan in murine macrophages: Effects of interferon-beta and taurine-chloramine. *Int. J. Lepr. Other Mycobact. Dis.* **2000**, *68*, 444–451. [PubMed]

156. Means, T.K.; Jones, B.W.; Schromm, A.B.; Shurtleff, B.A.; Smith, J.A.; Keane, J.; Golenbock, D.T.; Vogel, S.N.; Fenton, M.J. Differential effects of a Toll-like receptor antagonist on *Mycobacterium tuberculosis*-induced macrophage responses. *J. Immunol.* **2001**, *166*, 4074–4082. [CrossRef] [PubMed]
157. Helguera-Repetto, A.C.; Chacon-Salinas, R.; Cerna-Cortes, J.F.; Rivera-Gutierrez, S.; Ortiz-Navarrete, V.; Estrada-Garcia, I.; Gonzalez-y-Merchand, J.A. Differential macrophage response to slow- and fast-growing pathogenic mycobacteria. *Biomed. Res. Int.* **2014**, *2014*, 916521. [CrossRef] [PubMed]
158. Taka, S.; Gazouli, M.; Sotirakoglou, K.; Liandris, E.; Andreadou, M.; Triantaphyllopoulos, K.; Ikonomopoulos, J. Functional analysis of 3′UTR polymorphisms in the caprine SLC11A1 gene and its association with the *Mycobacterium avium* subsp. *paratuberculosis* infection. *Vet. Immunol. Immunopathol.* **2015**, *167*, 75–79. [CrossRef] [PubMed]
159. Thirunavukkarasu, S.; Plain, K.M.; Purdie, A.C.; Whittington, R.J.; de Silva, K. IFN-γ fails to overcome inhibition of selected macrophage activation events in response to pathogenic mycobacteria. *PLoS ONE* **2017**, *12*, e0176400. [CrossRef] [PubMed]
160. Barletta, R.G.; Bannantine, J.P.; Stabel, J.R.; Muthukrishnan, E.; Anderson, D.K.; Dutta, E.; Manthena, V.; Hanafy, M.; Zinniel, D.K. *Mycobacterium avium* subsp. *paratuberculosis* Candidate Vaccine Strains Are Pro-apoptotic in RAW 264.7 Murine Macrophages. *Vaccines* **2023**, *11*, 1085. [CrossRef] [PubMed]
161. Zur Lage, S.; Goethe, R.; Darji, A.; Valentin-Weigand, P.; Weiss, S. Activation of macrophages and interference with CD4+ T-cell stimulation by *Mycobacterium avium* subspecies *paratuberculosis* and *Mycobacterium avium* subspecies *avium*. *Immunology* **2003**, *108*, 62–69. [CrossRef]
162. Ghosh, P.; Wu, C.W.; Talaat, A.M. Key role for the alternative sigma factor, SigH, in the intracellular life of *Mycobacterium avium* subsp. *paratuberculosis* during macrophage stress. *Infect. Immun.* **2013**, *81*, 2242–2257. [CrossRef]
163. Hostetter, J.M.; Steadham, E.M.; Haynes, J.S.; Bailey, T.B.; Cheville, N.F. Cytokine effects on maturation of the phagosomes containing *Mycobacterium avium* subspecies *paratuberculosis* in J774 cells. *FEMS Immunol. Med. Microbiol.* **2002**, *34*, 127–134. [CrossRef]
164. Cirone, K.M.; Lahiri, P.; Holani, R.; Tan, Y.L.; Arrazuria, R.; De Buck, J.; Barkema, H.W.; Cobo, E.R. Synthetic cathelicidin LL-37 reduces *Mycobacterium avium* subsp. *paratuberculosis* internalization and pro-inflammatory cytokines in macrophages. *Cell Tissue Res.* **2020**, *379*, 207–217. [CrossRef]
165. Chanput, W.; Mes, J.J.; Wichers, H.J. THP-1 cell line: An in vitro cell model for immune modulation approach. *Int. Immunopharmacol.* **2014**, *23*, 37–45. [CrossRef] [PubMed]
166. Shin, M.; Shin, S.W.; Jung, M.; Park, H.; Park, H.; Yoo, H.S. Host gene expression for *Mycobacterium avium* subsp. *paratuberculosis* infection in human THP-1 macrophages. *Pathog. Dis.* **2015**, *73*, ftv031. [CrossRef] [PubMed]
167. Mostoufi-Afshar, S.; Tabatabaei, M.; Ghahramani Seno, M.M. *Mycobacterium avium* subsp. *paratuberculosis* induces differential cytosine methylation at miR-21 transcription start site region. *Iran. J. Vet. Res. IVJR* **2018**, *19*, 262–269.
168. Park, H.T.; Lee, S.M.; Ko, S.; Kim, S.; Park, H.E.; Shin, M.K.; Kim, D.; Yoo, H.S. Delineating transcriptional crosstalk between *Mycobacterium avium* subsp. *paratuberculosis* and human THP-1 cells at the early stage of infection via dual RNA-seq analysis. *Vet Res.* **2022**, *53*, 71. [CrossRef] [PubMed]
169. Borrmann, E.; Möbius, P.; Diller, R.; Köhler, H. Divergent cytokine responses of macrophages to *Mycobacterium avium* subsp. *paratuberculosis* strains of Types II and III in a standardized in vitro model. *Vet. Microbiol.* **2011**, *152*, 101–111. [CrossRef] [PubMed]
170. Motamedi Boroojeni, A.; Derakhshandeh, A.; Haghkhah, M.; Jalali, S.A.H.; Eraghi, V.A. A Novel Approach to Deliver a *Mycobacterium avium* subsp. *paratuberculosis* Antigen in Eukaryotic Cells. *Mol. Biotechnol.* **2019**, *61*, 506–512. [CrossRef]
171. Momotani, E.; Whipple, D.L.; Thiermann, A.B.; Cheville, N.F. Role of M cells and macrophages in the entrance of *Mycobacterium paratuberculosis* into domes of ileal Peyer's patches in calves. *Vet. Pathol.* **1988**, *25*, 131–137. [CrossRef]
172. Alonso-Hearn, M.; Patel, D.; Danelishvili, L.; Meunier-Goddik, L.; Bermudez, L.E. The *Mycobacterium avium* subsp. paratuberculosis MAP3464 gene encodes an oxidoreductase involved in invasion of bovine epithelial cells through the *activation* of host cell Cdc42. *Infect. Immun.* **2008**, *76*, 170–178. [CrossRef]
173. Pott, J.; Basler, T.; Duerr, C.U.; Rohde, M.; Goethe, R.; Hornef, M.W. Internalization-dependent recognition of *Mycobacterium avium* ssp. *paratuberculosis* by intestinal epithelial cells. *Cell. Microbiol.* **2009**, *11*, 1802–1815. [CrossRef]
174. Miyazawa, K.; Hondo, T.; Kanaya, T.; Tanaka, S.; Takakura, I.; Itani, W.; Rose, M.T.; Kitazawa, H.; Yamaguchi, T.; Aso, H. Characterization of newly established bovine intestinal epithelial cell line. *Histochem. Cell Biol.* **2010**, *133*, 125–134. [CrossRef]
175. Zhan, K.; Lin, M.; Liu, M.M.; Sui, Y.N.; Zhao, G.Q. Establishment of primary bovine intestinal epithelial cell culture and clone method. *Vitr. Cell. Dev. Biol. Anim.* **2017**, *53*, 54–57. [CrossRef]
176. Katwal, P.; Thomas, M.; Uprety, T.; Hildreth, M.B.; Kaushik, R.S. Development and biochemical and immunological characterization of early passage and immortalized bovine intestinal epithelial cell lines from the ileum of a young calf. *Cytotechnology* **2019**, *71*, 127–148. [CrossRef] [PubMed]
177. Katwal, P.; Uprety, T.; Okda, F.; Antony, L.; Thomas, M.; Chase, C.; Diel, D.G.; Nelson, E.; Young, A.; Li, F.; et al. Characterization of bovine ileal epithelial cell line for lectin binding, susceptibility to enteric pathogens, and TLR mediated immune responses. *Comp. Immunol. Microbiol. Infect. Dis.* **2021**, *74*, 101581. [CrossRef] [PubMed]
178. Kernéis, S.; Bogdanova, A.; Kraehenbuhl, J.P.; Pringault, E. Conversion by Peyer's patch lymphocytes of human enterocytes into M cells that transport bacteria. *Science* **1997**, *277*, 949–952. [CrossRef] [PubMed]
179. Nair, V.R.; Franco, L.H.; Zacharia, V.M.; Khan, H.S.; Stamm, C.E.; You, W.; Marciano, D.K.; Yagita, H.; Levine, B.; Shiloh, M.U. Microfold Cells Actively Translocate *Mycobacterium tuberculosis* to Initiate Infection. *Cell Rep.* **2016**, *6*, 1253–1258. [CrossRef]

180. Bannantine, J.P.; Huntley, J.F.J.; Miltner, E.; Stabel, J.R.; Bermudez, L.E. The Mycobacterium avium subsp. paratuberculosis 35 kDa protein plays a role in invasion of bovine epithelial cells. *Microbiology* **2003**, *149*, 2061–2069. [CrossRef] [PubMed]
181. Patel, D.; Danelishvili, L.; Yamazaki, Y.; Alonso, M.; Paustian, M.L.; Bannantine, J.P.; Meunier-Goddik, L.; Bermudez, L.E. The ability of *Mycobacterium avium* subsp. *paratuberculosis* to enter bovine epithelial cells is influenced by preexposure to a hyperosmolar environment and intracellular passage in bovine mammary epithelial cells. *Infect. Immun.* **2006**, *74*, 2849–2855. [CrossRef]
182. Everman, J.L.; Danelishvili, L.; Flores, L.G.; Bermudez, L.E. MAP1203 Promotes *Mycobacterium avium* Subspecies *paratuberculosis* Binding and Invasion to Bovine Epithelial Cells. *Front. Cell. Infect. Microbiol.* **2018**, *8*, 217. [CrossRef]
183. Everman, J.L.; Eckstein, T.M.; Roussey, J.; Coussens, P.; Bannantine, J.P.; Bermudez, L.E. Characterization of the inflammatory phenotype of *Mycobacterium avium* subspecies *paratuberculosis* using a novel cell culture passage model. *Microbiology* **2015**, *161*, 1420–1434. [CrossRef]
184. Sweeney, R.W.; Whitlock, R.H.; Rosenberger, A.E. *Mycobacterium paratuberculosis* isolated from fetuses of infected cows not manifesting signs of the disease. *Am. J. Vet. Res.* **1992**, *53*, 477–480.
185. Thompson, B.R.; Clark, R.G.; Mackintosh, C.G. Intra-uterine transmission of *Mycobacterium avium* subsp *paratuberculosis* in subclinically affected red deer (*Cervus elaphus*). *N. Z. Vet. J.* **2007**, *55*, 308–313. [CrossRef] [PubMed]
186. Kugadas, A.; Lamont, E.A.; Bannantine, J.P.; Shoyama, F.M.; Brenner, E.; Janagama, H.K.; Sreevatsan, S. A *Mycobacterium avium* subsp. *paratuberculosis* Predicted Serine Protease Is Associated with Acid Stress and Intraphagosomal Survival. *Front. Cell. Infect. Microbiol.* **2016**, *6*, 85. [CrossRef] [PubMed]
187. Schwarz, D.G.G.; Shoyama, F.M.; Oliveira, L.L.; Sreevatsan, S.; Moreira, M.A.S. Rapid baso-apical translocation of *Mycobacterium avium* ssp. *paratuberculosis* in mammary epithelial cells in the presence of *Escherichia coli*. *J. Dairy Sci.* **2018**, *101*, 6287–6295. [CrossRef] [PubMed]
188. Pena, J.L.; Gonçalves Schwarz, D.G.; Willian de Lima Brasil, A.; Licursi de Oliveira, L.; Albuquerque Caldeira, J.L.; Scatamburlo Moreira, M.A. Differences in the coinfective process of *Staphylococcus aureus* and *Streptococcus agalactiae* in bovine mammary epithelial cells infected by *Mycobacterium avium* subsp. *paratuberculosis*. *Microb Pathog.* **2020**, *149*, 104476. [CrossRef] [PubMed]
189. Shandilya, U.K.; Wu, X.; McAllister, C.; Mutharia, L.; Karrow, N.A. Role of Toll-Like Receptor 4 in *Mycobacterium avium* subsp. *paratuberculosis* Infection of Bovine Mammary Epithelial (MAC-T) Cells In Vitro. *Microbiol. Spectr.* **2023**, *11*, e0439322. [CrossRef] [PubMed]
190. Shandilya, U.K.; Wu, X.; McAllister, C.; Mutharia, L.; Karrow, N.A. Impact of *Mycobacterium avium* subsp. *paratuberculosis* infection on bovine IL10RA knockout mammary epithelial (MAC-T) cells. *Vitr. Cell Dev. Biol. Anim.* **2023**, *59*, 214–223. [CrossRef] [PubMed]
191. Phillips, I.L.; Danelishvili, L.; Bermudez, L.E. Macrophage Proteome Analysis at Different Stages of *Mycobacterium avium* Subspecies *paratuberculosis* Infection Reveals a Mechanism of Pathogen Dissemination. *Proteomes* **2021**, *9*, 20. [CrossRef]
192. Engelmann, I.; Pujol, N. Innate immunity in *C. elegans*. *Adv. Exp. Med. Biol.* **2010**, *708*, 105–121. [CrossRef]
193. Komura, T.; Yasui, C.; Miyamoto, H.; Nishikawa, Y. *Caenorhabditis elegans* as an alternative model host for legionella pneumophila, and protective effects of *Bifidobacterium infantis*. *Appl. Environ. Microbiol.* **2010**, *76*, 4105–4108. [CrossRef]
194. Bermudez, L.E.; Rose, S.J.; Everman, J.L.; Ziaie, N.R. Establishment of a Host-to-Host Transmission Model for *Mycobacterium avium* subsp. *hominissuis* Using *Caenorhabditis elegans* and Identification of Colonization-Associated Genes. *Front. Cell. Infect. Microbiol.* **2018**, *8*, 123. [CrossRef]
195. Viri, V.; Arveiler, M.; Lehnert, T.; Gijs, M.A.M. An In Vivo Microfluidic Study of Bacterial Load Dynamics and Absorption in the *C. elegans* Intestine. *Micromachines* **2021**, *12*, 832. [CrossRef] [PubMed]
196. Mallo, G.V.; Kurz, C.L.; Couillault, C.; Pujol, N.; Granjeaud, S.; Kohara, Y.; Ewbank, J.J. Inducible antibacterial defense system in *C. elegans*. *Curr. Biol.* **2002**, *12*, 1209–1214. [CrossRef] [PubMed]
197. Galbadage, T.; Shepherd, T.F.; Cirillo, S.L.; Gumienny, T.L.; Cirillo, J.D. The *Caenorhabditis elegans* p,8 MAPK Gene plays a key role in protection from mycobacteria. *Microbiologyopen* **2016**, *5*, 436–452. [CrossRef] [PubMed]
198. Sutton, K.M.; Orr, B.; Hope, J.; Jensen, S.R.; Vervelde, L. Establishment of bovine 3D enteroid-derived 2D monolayers. *Vet. Res.* **2022**, *53*, 15. [CrossRef] [PubMed]
199. Candarlioglu, P.L.; Dal Negro, G.; Hughes, D.; Balkwill, F.; Harris, K.; Screen, H.; Morgan, H.; David, R.; Beken, S.; Guenat, O.; et al. Organ-on-a-chip: Current gaps and future directions. *Biochem. Soc. Trans.* **2022**, *50*, 665–673. [CrossRef] [PubMed]
200. Csukovich, G.; Pratscher, B.; Burgener, I.A. The World of Organoids: Gastrointestinal Disease Modelling in the Age of 3R and One Health with Specific Relevance to Dogs and Cats. *Animals* **2022**, *2*, 2461. [CrossRef] [PubMed]
201. Xian, C.; Zhang, J.; Zhao, S.; Li, X.G. Gut-on-a-chip for disease models. *J. Tissue Engineer.* **2023**, *14*, 1–14. [CrossRef]
202. Baddal, B.; Marrazzo, P. Refining Host-Pathogen Interactions: Organ-on-Chip Side of the Coin. *Pathogens.* **2021**, *10*, 203. [CrossRef]
203. Pimenta, J.; Ribeiro, R.; Almeida, R.; Costa, P.F.; da Silva, M.A.; Pereira, B. Organ-on-Chip Approaches for Intestinal 3D In Vitro Modeling. *Cell. Mol. Gastroenterol. Hepatol.* **2022**, *13*, 351–367. [CrossRef]
204. Morsink, M.A.J.; Willemen, N.G.A.; Leijten, J.; Bansal, R.; Shin, S.R. Immune organs and immune cells on a chip: An overview of biomedical applications. *Micromachines* **2020**, *11*, 849. [CrossRef]
205. Majorova, D.; Atkins, E.; Martineau, H.; Vokral, I.; Oosterhuis, D.; Olinga, P.; Wren, B.; Cuccui, J.; Dirk Werling, D. Use of Precision-Cut Tissue Slices as a Translational Model to Study Host-Pathogen Interaction. *Front. Vet. Sci.* **2021**, *8*, 686088. [CrossRef] [PubMed]
206. Li, M.; de Graaf, I.A.M.; Groothuis, G.M.M. Precision-cut intestinal slices: Alternative model for drug transport, metabolism, and toxicology research. *Expert. Opin. Drug. Metab. Toxicol.* **2016**, *12*, 175–190. [CrossRef] [PubMed]

207. de Kanter, R.; Tuin, A.; van de Kerkhof, E.; Martignoni, M.; Draaisma, A.L.; de Jager, M.H.; de Graaf, I.A.; Meijer, D.K.; Groothuis, G.M. A new technique for preparing precision-cut slices from small intestine and colon for drug biotransformation studies. *J. Pharmacol. Toxicol. Methods.* **2005**, *51*, 65–72. [CrossRef] [PubMed]
208. Molina-Torres, C.A.; Flores-Castillo, O.N.; Carranza-Torres, I.E.; Guzmán-Delgado, N.E.; Viveros-Valdez, E.; Vera-Cabrera, L.; Ocampo-Candiani, J.; Verde-Star, J.; Castro-Garza, J.; Carranza-Rosales, P. Ex vivo infection of murine precision-cut lung tissue slices with *Mycobacterium abscessus*: A model to study antimycobacterial agents. *Ann. Clin. Microbiol. Antimicrob.* **2020**, *19*, 52. [CrossRef] [PubMed]
209. Remot, A.; Carreras, F.; Coupé, A.; Doz-Deblauwe, É.; Boschiroli, M.L.; Browne, J.A.; Marquant, Q.; Descamps, D.; Archer, F.; Aseffa, A.; et al. Mycobacterial Infection of Precision-Cut Lung Slices Reveals Type 1 Interferon Pathway Is Locally Induced by *Mycobacterium bovis* but Not *M. tuberculosis* in a Cattle Breed. *Front. Vet. Sci.* **2021**, *8*, 696525. [CrossRef] [PubMed]
210. Punyadarsaniya, D.; Winter, C.; Mork, A.K.; Amiri, M.; Naim, H.Y.; Rautenschlein, S.; Herrler, G. Precision-cut intestinal slices as a culture system to analyze the infection of differentiated intestinal epithelial cells by avian influenza viruses. *J. Virol. Methods.* **2015**, *212*, 71–75. [CrossRef] [PubMed]
211. Krimmling, T.; Beineke, A.; Schwegmann-Weßels, C. Infection of porcine precision cut intestinal slices by transmissible gastroenteritis coronavirus demonstrates the importance of the spike protein for enterotropism of different virus strains. *Vet. Microbiol.* **2017**, *205*, 1–5. [CrossRef] [PubMed]
212. Martinec, O.; Huliciak, M.; Staud, F.; Cecka, F.; Vokral, I.; Cerveny, L. Anti-HIV and Anti-Hepatitis C Virus Drugs Inhibit P-Glycoprotein Efflux Activity in Caco-2 Cells and Precision-Cut Rat and Human Intestinal Slices. *Antimicrob. Agents. Chemother.* **2019**, *63*, e00910–e00919. [CrossRef]

MDPI AG
Grosspeteranlage 5
4052 Basel
Switzerland
Tel.: +41 61 683 77 34

Animals Editorial Office
E-mail: animals@mdpi.com
www.mdpi.com/journal/animals

www.ingramcontent.com/pod-product-compliance
Lightning Source LLC
Chambersburg PA
CBHW040816250926
51850CB00007BA/1185